W9-CSP-208

CRC Handbook of Marine Science

Robert J. Moore
Editor-in-Chief
Director, Marine Science Institute
University of Texas

Compounds from Marine Organisms
Volume II

Authors
Joseph T. Baker
Vreni Murphy
Roche Research Institute of Marine Pharmacology
Sydney, Australia

With a special contribution from Synnøve Liaaen-Jensen
Professor, Organic Chemistry Laboratories
Norwegian Institute of Technology
Trondheim, Norway

CRC Press, Inc.
Boca Raton, Florida

CHEM

69170460

Library of Congress Cataloging in Publication Data

Baker, Joseph T
Compounds from marine organisms.

Bibliography: v. 1, p.
Includes indexes.
1. Marine pharmacology- -Handbooks, manuals, etc.
2. Chemistry, Organic- -Handbooks, manuals, etc.
I. Murphy, Vreni, joint author. II. Title.
III. Series.
GC24.C17 vol. 1 [QH345] 551.46s [615'.3] 76-10180
ISBN 0-87819-391-X

This book represents information obtained from authentic and highly regarded sources. Reprinted material is quoted with permission, and sources are indicated. A wide variety of references are listed. Every reasonable effort has been made to give reliable data and information, but the author and the publisher cannot assume responsibility for the validity of all materials or for the consequences of their use.

All rights reserved. This book, or any parts thereof, may not be reproduced in any form without written consent from the publisher.

Direct all inquiries to CRC Press, Inc., 2000 N.W. 24th Street, Boca Raton, Florida 33431.

© 1981 by CRC Press, Inc.

7/30/81 (gdB)

International Standard Book Number 0-87819-391-x (Volume I)
International Standard Book Number 0-8493-0214-5 (Volume II)

Library of Congress Card Number 76-10180
Printed in the United States

RS160
.7
B3
V.2
CHEM

THE EDITOR-IN-CHIEF

J. Robert Moore, Ph.D., is Director of the Marine Science Institute and Professor of Marine Studies at the University of Texas in Austin.

Dr. Moore holds degrees from the University of Houston (B.S.), Harvard University (M.S.), and the University of Wales in Cardiff (Ph.D.).

Before joining the faculty of the University of Texas, Dr. Moore was Director of the Institute of Marine Science at the University of Alaska, and before that, Director of the Marine Research Laboratory and Marine Minerals Program at the University of Wisconsin. Still earlier, he was with Texaco Research. He has been involved in marine minerals exploration research for the past twenty-five years, including projects in the Bering Sea, the Irish Sea, the Central Pacific, the Great Lakes, Venezuela, the Gulf of Mexico, and the coastal waters of the U.S. Atlantic coast, and Alaska.

Since its founding, Dr. Moore has been Editor-in-Chief of the journal, *Marine Mining,* and he is an active consultant to American and international firms engaged in marine minerals exploration and mining. Since 1966, he has been an active member of several national and international panels and committees dealing with marine mineral resources, including special assignments for the United Nations.

PREFACE

Scientific literature dealing with natural products from marine fauna and flora continues to expand. This second volume in the series *Compounds from Marine Organisms* is integrated with Volume I, and where relevant, references are made in Volume II to information in Volume I. The aims established in Volume I have been maintained, viz. (1) to draw attention to the marine environment as a source of novel organic substances, in many cases displaying structural types differing from those available from nonmarine sources; (2) to present an accurate record of organic compounds derived from marine organisms; and (3) to indicate where biological, microbiological, or pharmacological activity has been reported.

To allow rapid referral to the 397 compounds tabulated in order of increasing molecular complexity, the formula, structure, name, source, activity, original references, and additionally, reported availability of each compound by synthesis or from nonmarine sources is indicated. In separate sections important classes of compounds are grouped and discussed.

Compounds which appeared in Volume I, and which are mentioned in references listed in Volume II, are listed in Table 28. Their formulas have not been re-drawn.

In the text, references to compounds reported in Volume I are referred to as, e.g., I/268, I/465, etc. All other compound numbers refer to new entries in Volume II.

Original references have been used wherever possible, and this second volume incorporates compounds reported to the end of 1974. Our preparation for Volume III is scheduled for completion in April 1981 and will contain compounds reported to the end of 1978.

In Volume II a specialist contribution on marine carotenoids is provided by Professor Synnøve Liaaen-Jensen. This concept of specialist contributions will be expanded in Volume III.

The comprehensive reference system adopted in Volume I is extended in Volume II to include a Taxonomic Index. An explanation of the method of use of this specific index is incorporated at the commencement of the Index. References are given to compounds which appear in both Tables 27 and 28.

We gratefully acknowledge the assistance of many colleagues in the various facets of preparation of this volume. In particular, we thank Dr. R. Urban of F. Hoffmann-La Roche, Basel, for advice on the systematic nomenclature of some compounds; Mr. I. G. Skinner, formerly of Roche Research Institute of Marine Pharmacology, for careful checking of the Taxonomic Index; Dr. P. T. Murphy for proofreading; Mrs. P. Comans for library material; Mrs. S. Osborn for assistance in preparing the indexes, proofreading, and typing; and Mrs. B. Brown and Mrs. D. Labone for typing.

J. T. Baker
V. Murphy

THE AUTHORS

Joseph Thomas Baker, M.Sc., Ph.D., F.R.A.C.I., is Director, Roche Research Institute of Marine Pharmacology, Sydney, Australia, and Director of Research for Roche in Australia. He was formerly Associate Professor of Chemistry, James Cook University of North Queensland, Australia.

Dr. Baker is a member of the Great Barrier Reef Marine Park Authority, the Great Barrier Reef Consultative Committee, President of The Australian Museum Trust, Chairman of the Lizard Island Research Station, Executive Chairman of the Visiting Committee, School of Chemistry, University of New South Wales, and Vice President of The Australian Museum Sciences Association.

His publications are in the area of marine coloring matters and insect scent gland constituents.

Vreni Murphy is Research Assistant, Roche Research Institute of Marine Pharmacology, Sydney, Australia. She is a graduate of Langenthal Technical College, Switzerland. Before joining Roche she was a research assistant with the Neurological University Clinic, Basel, Switzerland, and in the Department of Chemistry, James Cook University of North Queensland, Australia.

To Valerie and Peter

TABLE OF CONTENTS

Discussion 1
Non-Isoprenoid Hydrocarbons and Their Derivatives 1
Aromatic Compounds 7
Isoprene-Derived Hydrocarbons and Their Derivatives 19
Monoterpenes 19
Sesquiterpenes 20
Diterpenes 28
Carotenoids 33
Sterols 40
Compounds Containing Nitrogen 62
Simple Amines 62
Imidazole Derivatives 62
Choline Derivatives 64
Catecholamines 64
Amino Acids 65
Porphyrins and Chlorophylls 67
Tabulation of Compounds (Table 27) 69
Compound Name Index for Table 27 167
Tabulation of Compounds from Volume I Mentioned in References Listed in Volume II (Table 28) 175
Compound Name Index for Table 28 187
References 191
Author Index 201
Taxonomic Index 207

DISCUSSION

Non-Isoprenoid Hydrocarbons and Their Derivatives

Youngblood and Blumer[283] investigated the benthic marine algae in the region of Cape Cod in the U.S., with one aim being to assess any difference in natural hydrocarbon content compared with hydrocarbons that have been found in fossil fuels, and therefore to consider if it would be possible to determine a background spectrum of hydrocarbons that could be expected in the marine environment from natural sources. The hydrocarbons are minor but universal components of all plants and animals of the seawater and of the sediments. The hydrocarbon concentration and also the alkene to alkane ratio as well as the content of polyolefin were found to be highest in the young plants or in the most rapidly growing tissue of the older plants. Youngblood and Blumer have reported on the occurrence of the twelve saturated open-chain hydrocarbons from C_{13} to C_{24} and of the unsaturated hydrocarbons $C_{19}H_{30}$ (189), $C_{19}H_{32}$ (190), and $C_{21}H_{34}$ (244).

Cellular chemotaxis plays an important role in the reproduction of brown algae. It is believed that the gynogametes (eggs) release volatile substances that act as attractants for the androgametes (sperm).[134] This phenomenon has been long recognized in the case of the brown alga *Fucus serratus,* and Müller and Jaenicke[218,135] have been able to identify the active principle released by the eggs as fucoserratene (1,3-*trans*-5-*cis*-octatriene) (62).

(62) fucoserratene

Jaenicke et al.,[134] in studying the brown alga *Cutleria multifida,* isolated two hydrocarbons, of which one was found to be a male-attracting substance and the other was found to be biologically inactive in this particular alga. The active compound (108) multifidene [*trans*-4-vinyl-5-(*cis*-1′-butenyl)cyclopentene] is isomeric with the inactive compound aucantene (107) [*trans*-4-vinyl-5-(*trans*-1′-propenyl)-cyclohexene].

(108) multifidene

(107) aucantene

Multifidene (108) had been previously isolated by Müller[219] and the structure confirmed in the collaborative study with Moore and co-workers.[134] In Volume I[11] we previously reported on a number of C_{11} hydrocarbons isolated from *Dictyopteris* sp., some of which were projected as being important in the reproduction of the algae. The substances *cis,trans*-1,3,5-undecatriene (111) and *trans,trans,trans*-2,4,6-undecatriene (112) have also been isolated as minor components of the essential oils of the Hawaiian *Dictyopteris.*[214] Dictyopterene A and dictyopterene C′ were represented as compounds (173) and (174), respectively, in Volume I of this series, and the synthesis of the racemic forms of each of these compounds has been reported.[219] Moore and Yost[213] have isolated and identified the two dihydrotropones (109) and (110) from *Dic-*

tyopteris australis and *D. plagiogramma,* the dihydrotropones being metabolites of dictyopterene C′ (I/174).

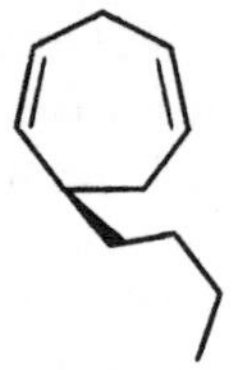

(I/174) dictyopterene C′

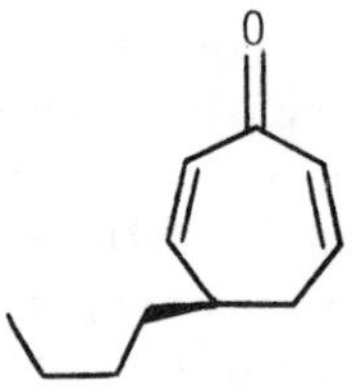

(109)

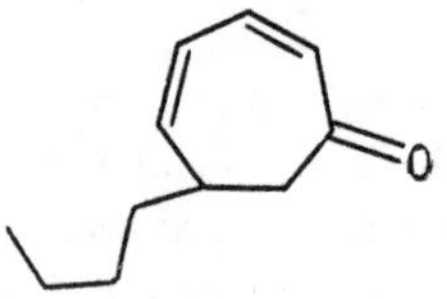

(110)

Red algae have yielded a number of interesting halogenated metabolites, e.g., the red alga *Asparagopsis taxiformis* is believed to contain a wide spectrum of haloforms, and the simple compound bromoform (1)[85] has been detected. *Asparagopsis taxiformis,* the favorite edible seaweed of most native Hawaiians, has always been known for its strong flavor and for the so-called "iodine" odor which it develops on standing. It is interesting to note that this alga has yielded a number of simple halogenated acetones (3), (4), (5), (6), (7), (8), (9), (10),[82,85] all of which have some antimicrobial activity. Another red alga, *Bonnemaisonia hamifera,* has yielded five halogenated C_7 ketones (42), (43), (44), (45), (48).[232]

We have previously commented[11] on a number of interesting cyclic compounds derived from pentadeca-3,9,12-trien-1-yn-6,7-diol. The open-chain compounds (145), (146), (147) and (148) have been reported by Irie and co-workers[177] from *Laurencia nipponica.*

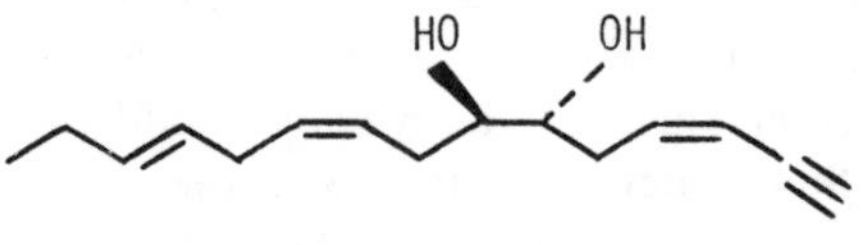

(145) 6*R*,7*R* *cis*-laurediol

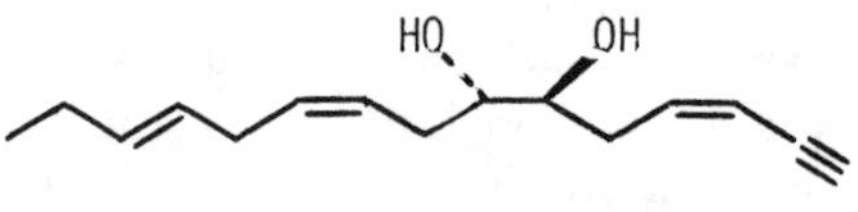

(146) 6*S*,7*S* *cis*-laurediol

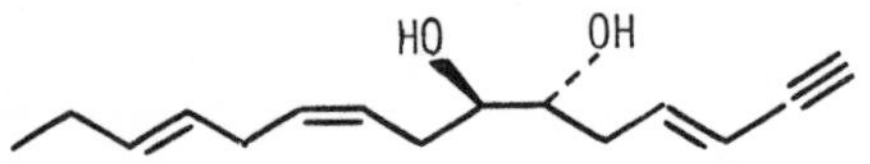

(147) 6*R*,7*R* *trans*-laurediol

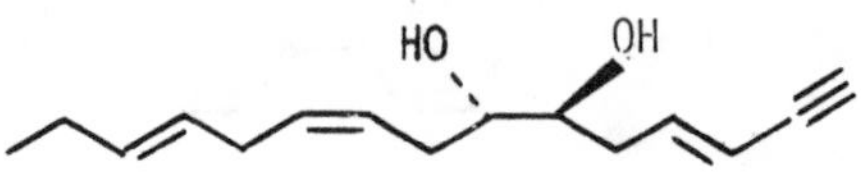

(148) 6*S*,7*S* *trans*-laurediol

Further examples of cyclic ethers derived from pentadeca-3,9,12-trien-1-yn-6,7-diol are given in (136), (137) and (138),[177] all being from the red alga *Laurencia nipponica.*

Dactylyne (135) has been isolated from the sea hare *Aplysia dactylomela,* and the structure has been confirmed as containing a six-membered ether ring[185] by use of single-crystal X-ray diffraction methods, correcting the previous structure allocation[244] which contained a four-membered ether ring.

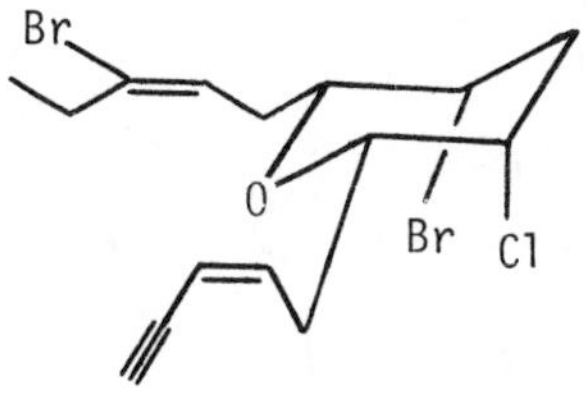

Br Br Cl O

(135) dactylyne

Dactylyne, although having been isolated from the sea hare, is assumed to be of algal origin. It has been demonstrated that halogenated compounds from sea hares are either present in the algae on which the animals feed, or represent simple modifications of such algal constituents.

Fenical[80] isolated the very stable halogenated peroxide (131) from a collection of a *Laurencia* sp. in the Gulf of California. This compound, rhodophytin, represents the first example of a vinyl peroxide moiety, and Fenical suggests that peroxidation of a laurediol, followed by bromonium ion-induced ring closure, would generate the cyclic peroxide structure deduced for (131).*

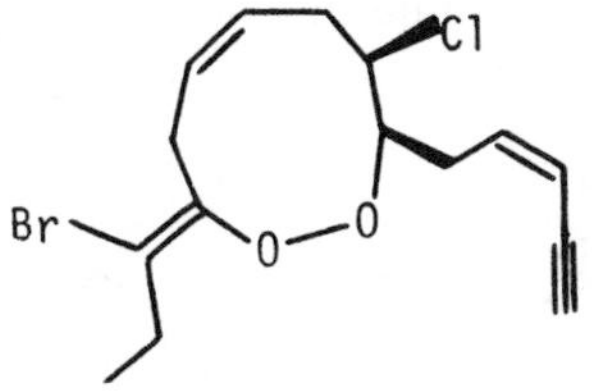

Cl Br O O

(131) rhodophytin

Marine prostaglandins have continued to attract attention, particularly in the soft coral *Plexaura homomalla,* but also in other marine species.

Corey and Washburn[52] considered the question of whether prostaglandin synthesis occurs in *P. homomalla* or in the symbiotic algae. They isolated approximately 50 to 100 mg of clean algal cells from 15 g of the soft coral and, following culture in sterile *Gonyaulax* medium, extracted the acid components and esterified with diazomethane, Arachidonic acid represented only 0.7% of the fatty acid constituents, this leading the investigators to conclude that, if the algae were the source of the arachidonic acid used by *P. homomalla* the substance could not be stored by the algae, but must be transmitted directly to the coral.

Corey and Washburn[52] also demonstrated that the algal cells do not contain prostaglandin synthetase. Further studies are necessary to resolve the questions posed by this study as to the relative roles of algae and soft coral in prostaglandin synthesis.

Weinheimer[271] has reviewed the discovery of 15-*epi*-PGA_2 (15*R*-PGA_2, I/292) in *P. homomalla.*

* This structure has since been revised; see Volume III.

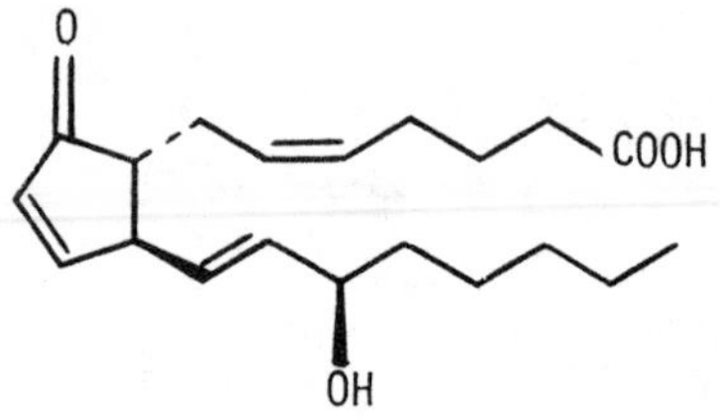

(I/292) 15R-PGA$_2$

Several pharmaceutical companies showed interest in the possible utilization of *P. homomalla* as a natural source of prostaglandins (e.g., Reference 10) but the commercial potential is not clearly demonstrated at this time.

Nomura et al.[224] conducted a survey of the prostaglandins in fish testes and semen as their first step in a plan to elucidate the occurrence and biological significance of prostaglandins in relation to essential fatty acids and to reproduction in marine animals.

Extracts were initially assayed for smooth-muscle stimulating activity and from the active fractions PGE$_2$ (I/296) was found in the testes of the flounder *Paralichthys olivaceus*, PGF$_{1\alpha}$ (219) in the semen of the chum salmon *Oncorhynchus keta*, and PGE$_2$ and PGF$_{2\alpha}$ (218) in the testes of the bluefin tuna *Thunnus thynnus*.

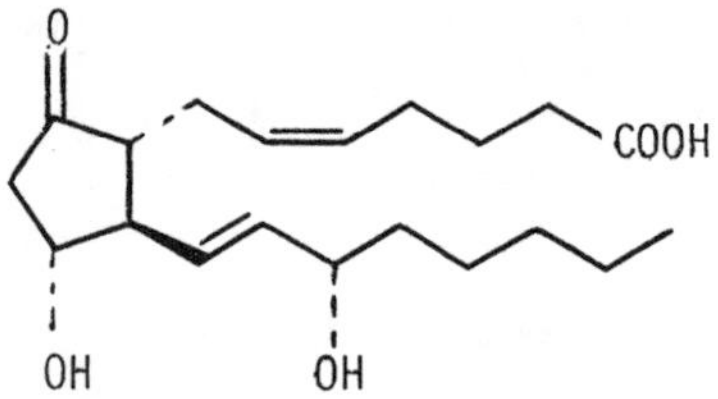

(I/296) PGE$_2$

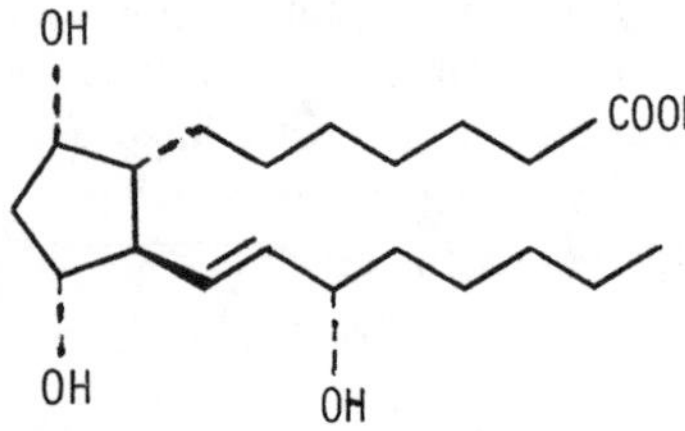

(219) PGF$_{1\alpha}$

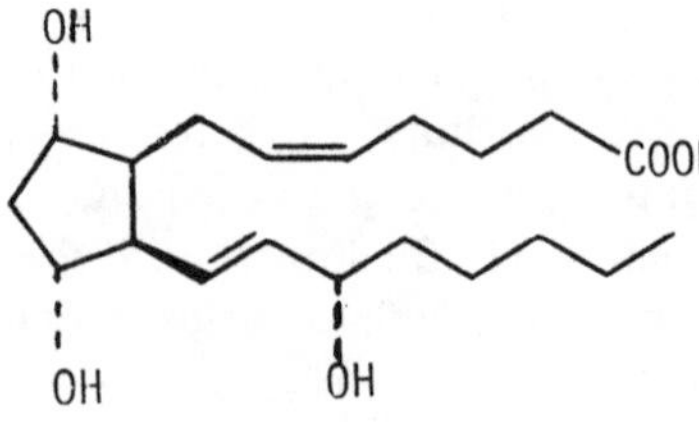

(218) PGF$_{2\alpha}$

Acrylic acid (11), which is often referred to as a toxic substance, has been reported as occurring in a mucilaginous colonial alga of the genus *Phaeocystis* at a level of 0.38% wet weight, or 7.4% dry weight.[245] This level was apparently nontoxic to *Euphausia superba* which grazes upon the alga, and which is, in turn, that part of the diet of the pygoscelid penguin which is understood to inhibit the penguin's gastrointestinal microflora. Sieburth[245] details the antimicrobial activity of the acrylic acid which accounts for the activity of the alga and of the stomach content of the breeding euphausids. Reference is also made to the occurrence of acrylic acid (11) in the epiphytic alga *Polysiphonia lanosa,* where the acrylic acid may be derived by cleavage of dimethyl-β-propiothetine (I/55).

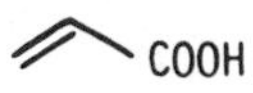

(11) acrylic acid

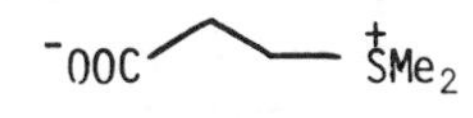

(I/55) dimethyl-β-propiothetine

Attempts to determine whether dimethyl-β-propiothetine is present in the *Phaeocystis* sp. were inconclusive.

The biosynthesis of the acrylic acid, its function in the alga, and its effect on the physiological ecology of the marine habitat are indicated as areas of further study.

The spermatozoa of the giant octopus of the North Pacific Ocean, *Octopus dofleini martini,* are capable of long survival, both in vitro and in vivo, retaining their mobility for several days. Mann et al.[191] investigated the metabolism of sperm suspensions incubated anaerobically in vitro and identified D(−)-lactic acid (15) as a major metabolite. Evidence was also produced to support the postulate that D(−)-lactic acid is also produced in vivo. These octopus spermatozoa also possess a highly active stereospecific, NAD-dependent D(−)-lactate dehydrogenase which clearly differs from the L(+)-lactate dehydrogenase commonly found in animal tissues and in body fluids.[191]

Two previously unreported C_{26} acids have been identified in a study of lipid metabolism in the marine sponge *Microciona prolifera.*[136] These have been characterized as *cis,cis*-5,9-hexacosadienoic acid (296) and *cis,cis,cis*-5,9,19-hexacosatrienoic acid (295).

Because the acids are not reported as components of any planktonic species ingested by the sponge, it is speculated that the acids are biosynthesized in the sponge with a parallel origin from palmitic and palmitoleic acids by elongation and desaturation. The total lipids in which the C_{26} acids occur are found in high concentration and may have considerable influence on membrane structure.

Based on a similar total concentration of "26:2" and "26:3" acids, but with a wide variation in the levels of the two individual acids, in samples of *Microciona prolifera* collected at two different latitudes, the authors suggest that there could be interchangeability between the two C_{26} acids, perhaps to maintain constant membrane flexibility with variations in environmental temperature.

Except for the examples provided by sponges, C_{26} polyunsaturated fatty acids are rarely found in nature. Initially there was some consideration that C_{26} and C_{28} unsaturated fatty acids might well be characteristic of sponges, but it is doubtful if sufficient analyses of different species have been conducted to justify such a chemotaxonomic criterion.

Marsh and Ciereszko[194] have analyzed for α-glyceryl ethers in a number of coelenterates and in two symbiotic algae occurring in sea anemones. The α-glyceryl ether content of the unsaponifiable lipids ranged from 7% in the sea anemone *Metridium senile* to 38% in the alcyonacean *Nephthea* sp. Glyceryl ethers are of interest because of their suggested beneficial effects in several areas, such as wound healing, radiation sickness, and hemolysis, but their therapeutic potential requires further demonstration of efficacy under experimental conditions.

In the work of Marsh and Ciereszko, the major α-glyceryl ethers contained 16 or 18 carbon atoms in the alkyl chain, and trace amounts of ethers containing 17, 19, and 20 carbon atoms in the alkyl chain were also detected in some cases. Alkyl alcohols containing 14 to 20 carbon atoms also occurred in most of the species studied, with

C_{16} and C_{18} alcohols predominating. Unsaturation was reported only in the case of the ethers and alcohols containing 18 carbons in the R-group.

Methoxy-substituted glyceryl ethers have been reported[113] as constituents of Greenland shark liver oils, co-occurring as the minor component with α-glyceryl ethers in a ratio of approximately 1:24 and distributed as approximately 60% 1-*O*-(2-methoxy-4-hexadecenyl)glycerol (221), 15% 1-*O*-(2-methoxyhexadecyl)glycerol (225) and 20% 1-*O*-(2-methoxy-4-octadecenyl)glycerol (257). Subsequently[114] Hallgren et al. isolated 1-*O*-(2-methoxyalkyl)glycerols from the lipids of herring fillets, Baltic herring fillets, mackerel fillets, marine crayfish, shrimps, sea mussels, and cod liver oil. As with the unsubstituted glyceryl ethers from shark liver oil, the major components contained 16 carbon atoms in the alkyl chain and, together with the compounds containing 18 carbon atoms in the alkyl chain, represented over 90% of the total methoxy glyceryl ether fraction.

A glyceryl ether with a phytanyl alkyl group (264) was isolated from the liver oil of cod caught in the Baltic Sea.[114]

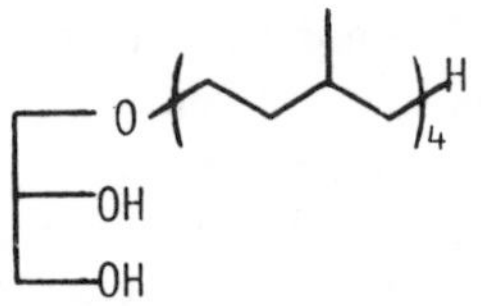

(264) 1-*O*-phytanylglycerol

Methoxy-substituted glyceryl ethers have shown antibiotic activity and inhibited the dissemination and growth of several experimental tumors in mice.[25,26] These methoxy-substituted glyceryl ethers occur in relatively high concentration in shark liver oil, as compared with the minute amounts found in most mammalian tissue.

Compounds isolated were of the type

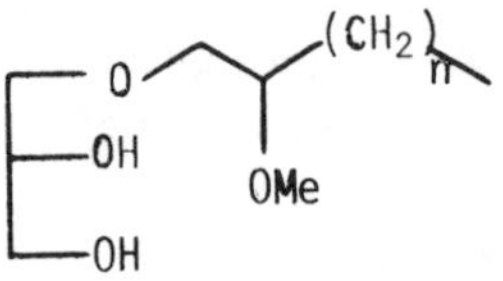

with n = 11 (188), 12 (196), 13 (225), 14 (248), 15 (260), 16 (265), 17 (271), 18 (281), and 19 (297).

1-*O*-(2-Hydroxyalkyl)glycerols were also reported from Greenland shark liver oil,[115] the major component being the unsaturated 1-*O*-(2-hydroxy-4-hexadecenyl)glycerol (191). Minor hydroxy-substituted glyceryl ethers isolated were (176), (180), (195), and (246).

In sea urchins, sperm agglutination and especially the acrosome reaction, which are considered essential at the time of fertilization, are due to the interaction of the spermatozoa with the glycoprotein coating of the egg. This glycoprotein contains sialic acid and fucose sulfate as the major carbohydrate residues.[132] In the Japanese sea urchins *Hemicentrotus pulcherrimus* and *Pseudocentrotus depressus,* the structure of the fucose sulfate is proposed as L-fucose-4-sulfate (33).

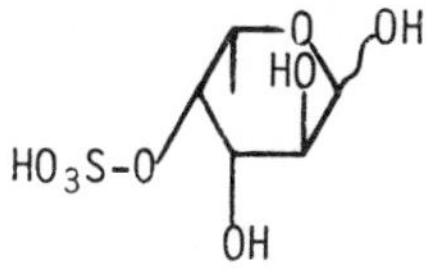

(33) L-fucose-4-sulfate

The authors suggest that, on the basis of i.r. analysis, several Mediterranean sea urchins contain the same fucose sulfate, whereas the Japanese sea urchin *Anthocidaris crassispina* may contain a fucose sulfate differing in the position of the ester sulfate.[132]

Aromatic Compounds

Glombitza and co-workers, in extending their studies on the red alga *Polysiphonia lanosa,* have demonstrated the presence of 3-bromo-4,5-dihydroxybenzaldehyde (I/92), 3-bromo-4,5-dihydroxybenzylalcohol (41), 2,3-dibromo-4,5-dihydroxybenzaldehyde (I/91), and 2,3-dibromo-4,5-dihydroxybenzylalcohol (I/94).[252] The compound (41), which was not reported in Volume I of this series, has antibiotic properties and is found in other red algae, e.g., *Halopithys incurvus, Polysiphonia nigrescens,* and *P. urceolata.*[97]

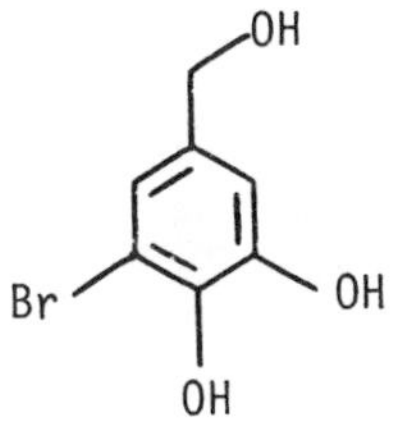

(41) 3-bromo-4,5-dihydroxybenzylalcohol

Pedersen et al.[229] conducted a survey for simple bromophenols in 23 species of red algae, using tlc and glc/ms for detection and analysis. Their work confirmed that members of the Rhodomelaceae are particularly rich in bromine-containing phenols. Bromophenols were also present in the families Ceramiaceae, Delesseriaceae, Bonnemaisoniaceae, Rhodophyllaceae, and Corallinaceae. No bromophenols were found in the family Dasyaceae, nor in the orders Porphyridiales, Bangiales, Goniotrichales, Nemalionales, or Rhodymeniales.

Trailliella intricata and *Falkenbergia rufolanosa* were regarded as interesting algae, containing several bromophenols not previously reported. These algae are now under further investigation.[229]

The possibility of annual fluctuations in the content of bromophenols was not investigated, nor was any relationship between age of the algae and bromophenol content. Investigations of the influence of bromophenols on algal growth under axenic conditions are being pursued.

Proposed structures for the mono- and dibromophenols detected in this survey are

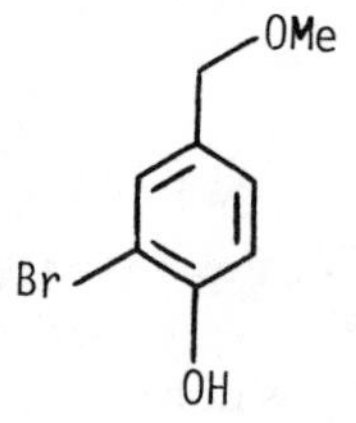

OMe
Br
OH

Artifact?

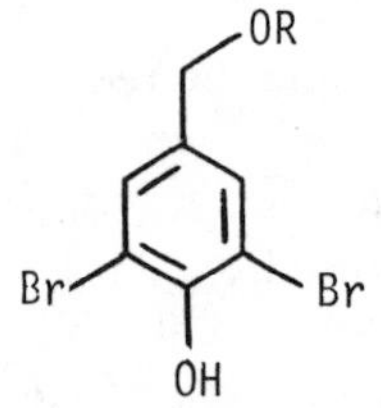

OR
Br Br
OH

R = H (I/93)
R = Me Artifact?
R = Et Artifact?

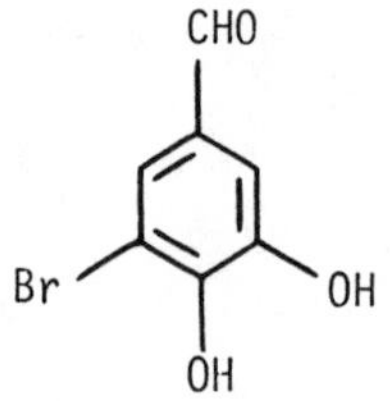

CHO
Br OH
OH

(I/92)

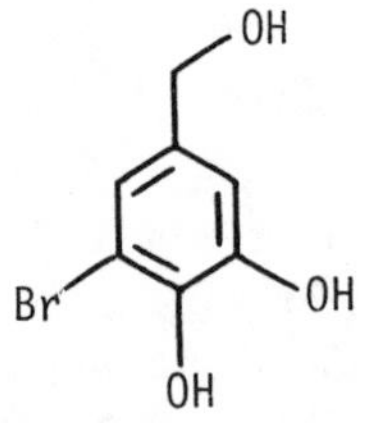

OH
Br OH
OH

(41)

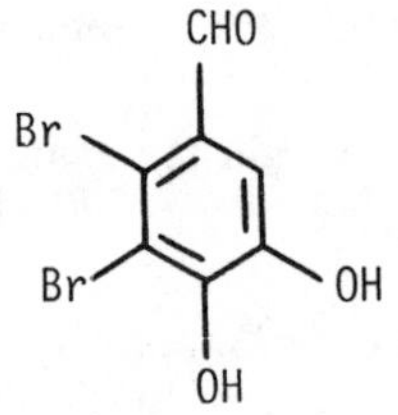

CHO
Br
Br OH
OH

(I/91)

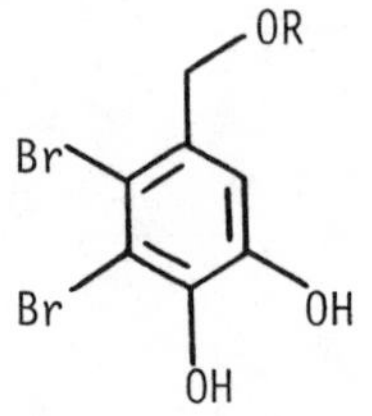

OR
Br
Br OH
OH

R = H (I/94)
R = Me Artifact?
R = Et Artifact?

Substances containing the designation "Artifact?" may have been produced in the extraction procedure and their positive identification as natural constituents of the algae requires further study. The method of extraction would also appear to decompose any constituents naturally present as the sulfates of either phenols or alcohols.

Brominated phenols were also reported from the red alga *Halopithys incurvus*.[43] This species is a member of the family Rhodomelaceae and yielded 3,5-dibromo-4-hydroxyphenylacetic acid (54) and 3,5-dibromo-4-hydroxyphenylpyruvic acid (66).

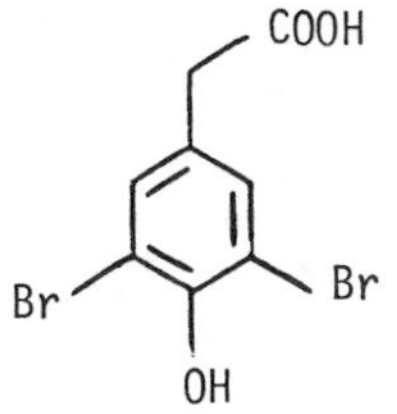

COOH
Br Br
OH

(54)

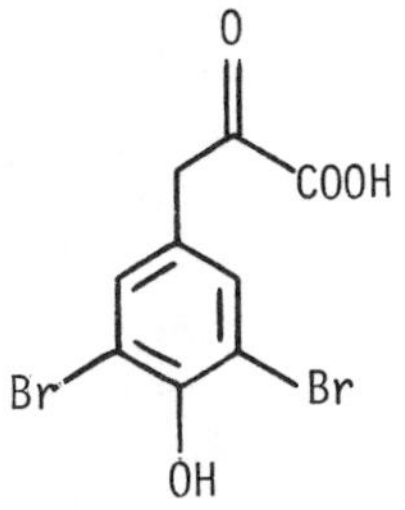

O
COOH
Br Br
OH

(66)

Seasonal variations in the concentration of the two components were noted, with more of both (54) and (66) being present in October than in March in the Atlantic Ocean (Rade de Brest). This aspect, together with a study of the function of (54) and (66) in the algae, will be the subject of continuing investigation.[43]

It appears that different extraction procedures, adopted by different workers, may selectively exclude certain constituents. Systematic studies on biosynthesis and function of the brominated phenols will require a knowledge of all substances naturally present in the algae.

A sponge, *Verongia lacunosa,* has yielded a novel derivative of a bromophenol containing two 2-oxazolidone rings (120).[29] The substance, isolated as part of the Lederle screen for marine antibiotics, has no significant antimicrobial activity and has been designated by the code number LL-PAA216. It is considered as possibly derived biogenetically from dibromotyrosine.

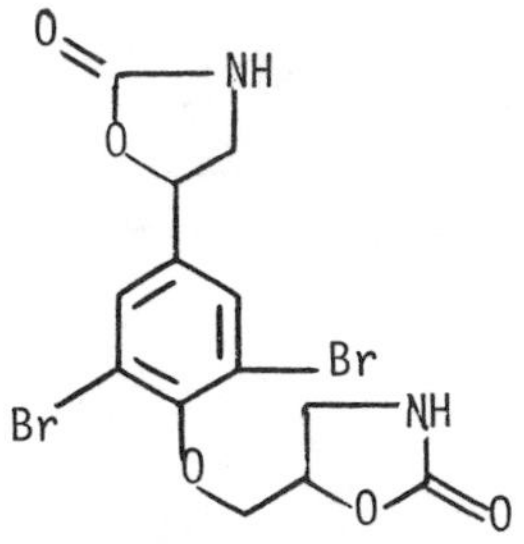

O NH
O
Br
Br NH
O
O O

(120) LL-PAA216

The marine segmented worm *Thelepus setosus* has yielded five brominated metabolites,[119] the simplest being the known alcohol (I/93), and the corresponding previously unreported aldehyde (36).

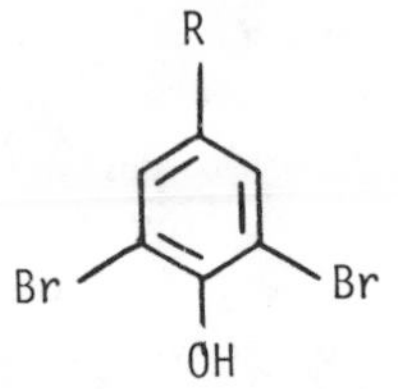

R = CH_2OH (I/93)

R = CHO (36)

From the same chromatographic fraction as the aldehyde (36), bis(3,5-dibromo-4-hydroxyphenyl)methane (119) was reported by tlc, its structure being confirmed by comparison with synthetic material.

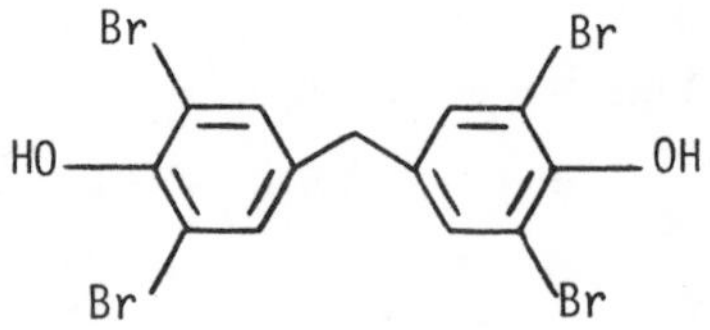

(119)

The two remaining products isolated were 2,4′-dihydroxy-5-hydroxymethyl-3,3′,5′-tribromodiphenylmethane (124) and the spiro compound named as thelepin (123), the structure of which bears many similarities to that of the antimycotic, griseofulvin. No indication of the biological activity of (123) was given.

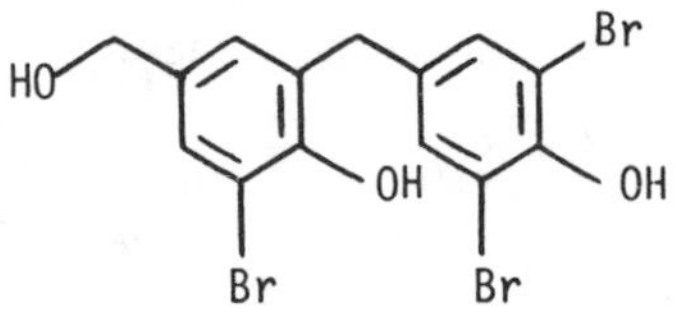

(124)

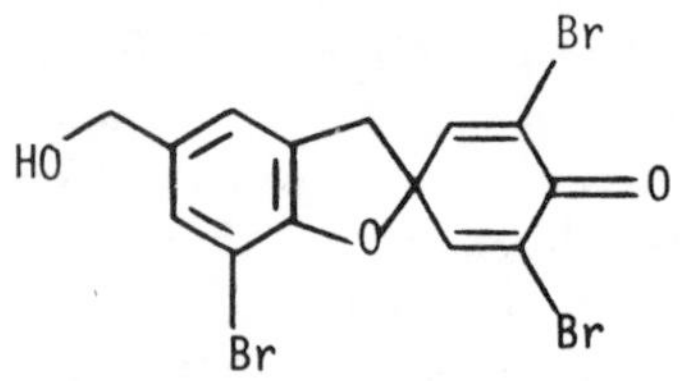

(123) thelepin

Of the marine-derived phenols *p*-hydroxyphenylacetic acid (56) and phenylacetic acid (55) have been isolated from the brown alga *Undaria pinnatifida,* and were found to be active plant growth regulators in the elongation of *Avena* coleoptile sections, with maximum activity of the *p*-hydroxyphenylacetic acid (56) being at a concentration of 50 ppm.[1]

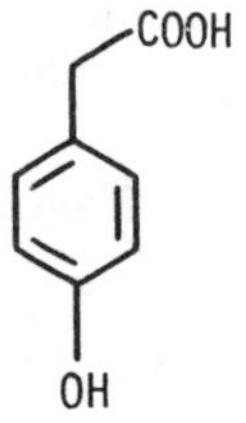

(56) *p*-hydroxyphenylacetic acid

Glombitza and co-workers, working on brown algae, have isolated phloroglucinol (30) in 17 of 26 species investigated.[94]

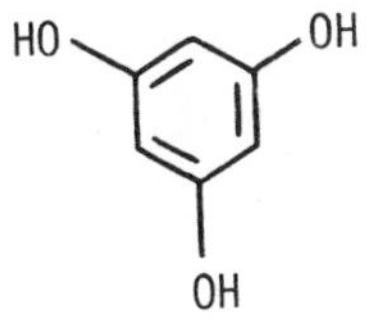

(30) phloroglucinol

From a species of brown alga *Bifurcaria bifurcata*, [96] bifuhalol (116) (3,4,5,2′,4′,6′-hexahydroxydiphenyl ether) has been characterized and suggested as a low molecular weight precursor of the phaeophyte tannins. Another brown alga, *Halidrys siliquosa*, yielded trifuhalol (181),[95] and some higher molecular weight substances, two of which are believed to be the homologous tetraphenyltriether and the hexaphenylpentaether, respectively.

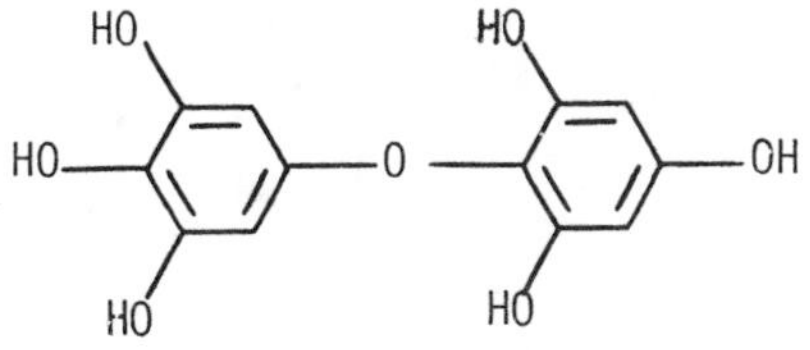

(116) bifuhalol

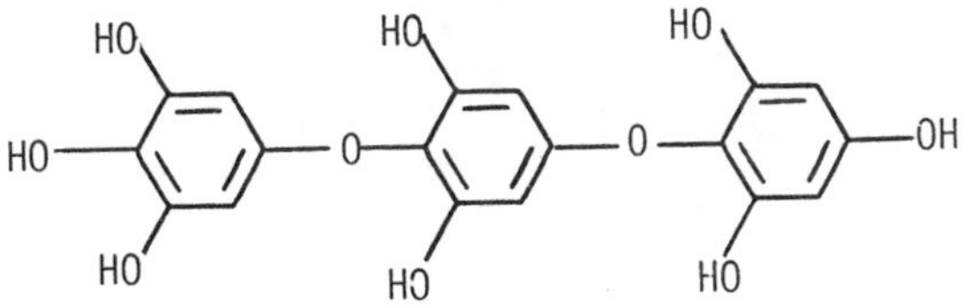

(181) trifuhalol

Table 1
PHENOLS

Ring system	Cpd. no.	C_2	C_3	C_4	C_5	C_6	Source	References
OH	132		H	Br	Me	H	Red alga	162
	364	H	R	H	H	H	Sea hare	159, 160
	363	H	R	Br	H	H	Sea hare	159, 160
	183	H			H	H	Dogfish Mollusc Sea urchin Starfish	31 32 202, 273
	182	H			H	H	Dogfish Mollusc	32, 202 273

R =

Table 2
1,2-BENZOQUINOLS

Ring system	Cpd. no.	C_3	C_4	C_5	C_6	Source	References
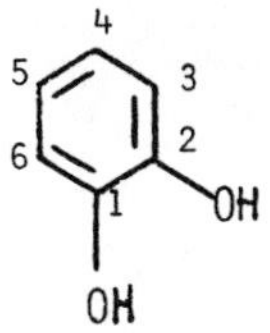 4, 5, 3, 6, 2, 1, OH, OH	60	H	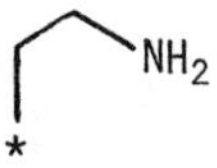 NH_2, *	H	H	Crab Lobster Molluscs Sea urchin Starfish	58 163 202
	61	H	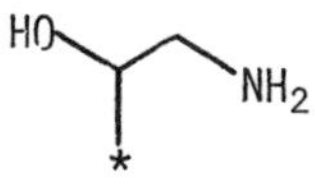 HO, NH_2, *	H	H	Clam Octopuses Sea urchin Squid Starfish	58 163 202
	41	Br	H	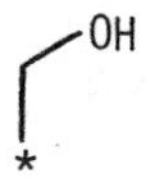 OH, *	H	Red algae	97, 252
	80	Br	Br	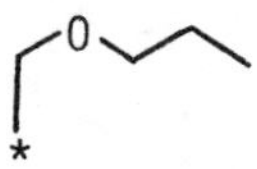 O, *	H	Red algae	97
	37	Br	Br	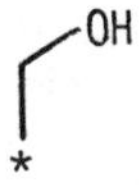 OH, *	Br	Red algae	97

Table 3
RESORCINOLS

Ring system	Cpd. no.	C_2	C_4	C_5	C_6	Source	References
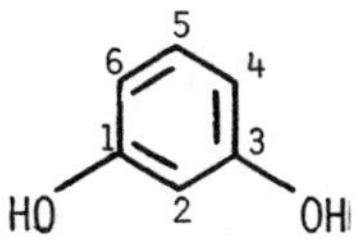 5, 6, 4, 1, 3, 2, HO, OH	57	H	H	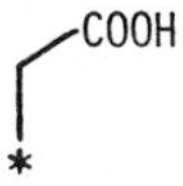 COOH, *	H	Octopus Squid	163
	261	H	H	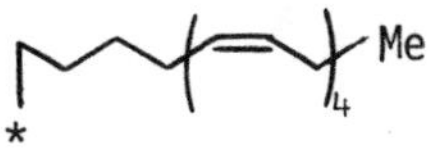 Me, 4, *		Brown alga	162

Table 4
1,4-BENZOQUINOLS

Ring system	Cpd. no.	C_2	C_3	C_5	C_6	Source	References
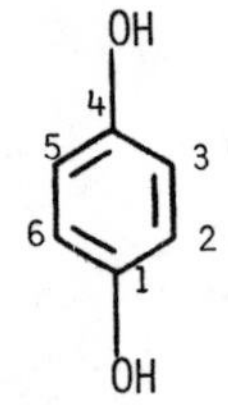 OH, 4, 5, 3, 6, 2, 1, OH	163	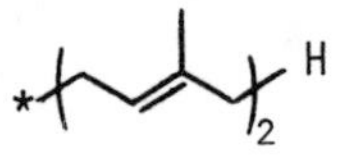 H	H	H	H	Tunicates	83 162
	361	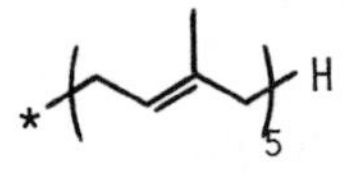 H	H	H	H	Sponge	162

Table 5
MODIFIED 1,4-BENZOQUINOLS

Cpd. no.	Formula	Source	References
298	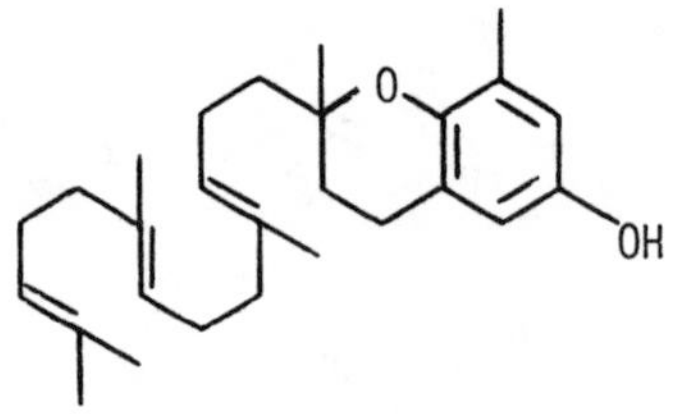 O, OH	Brown alga	176
318	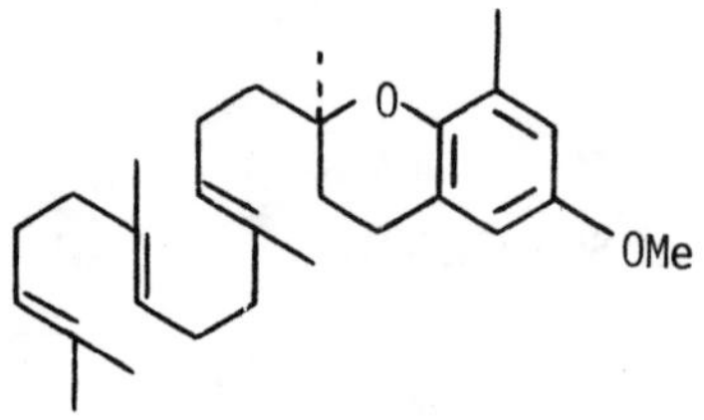 O, OMe	Brown alga	162
299	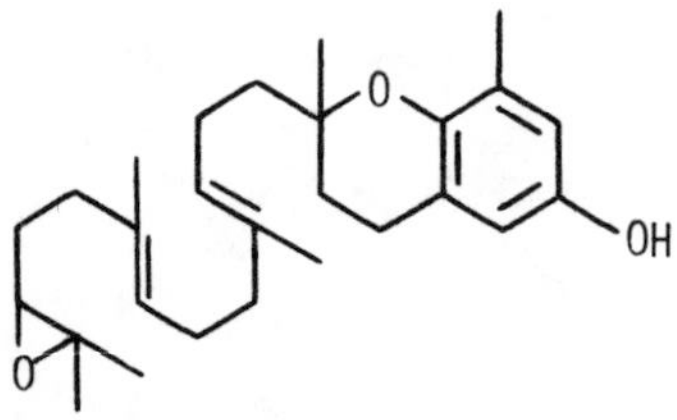 O, OH, O	Brown alga	176
337	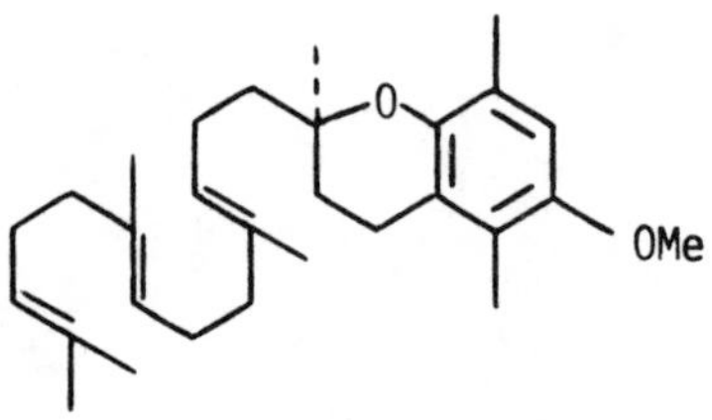 O, OMe	Brown alga	162

Table 5 (continued)
MODIFIED 1,4-BENZOQUINOLS

Cpd. no.	Formula	Source	References
319	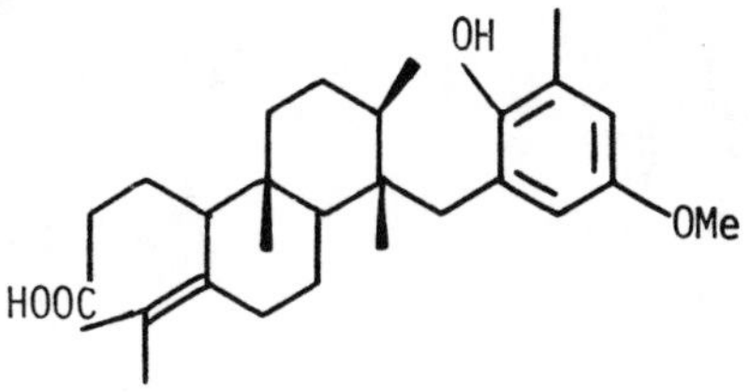	Brown alga	100
310	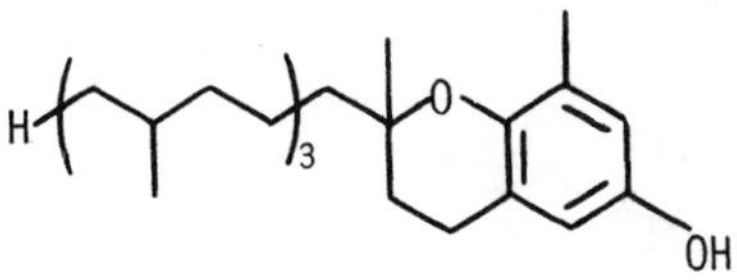	Algae	139 202
333	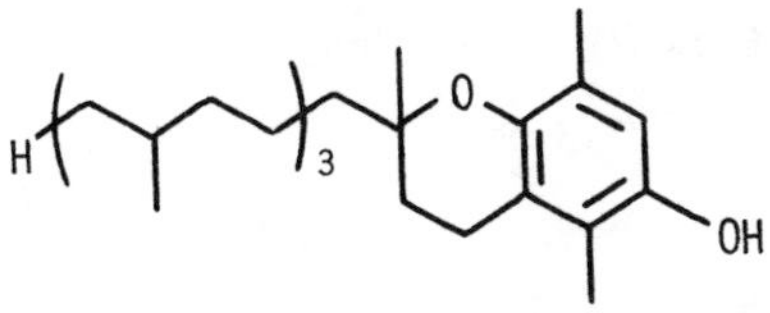	Algae	139 202
334	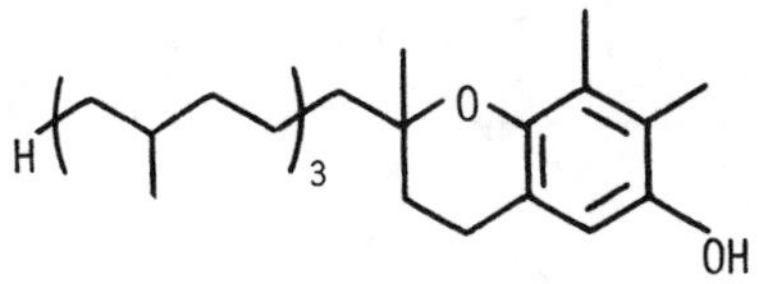	Algae	139 202
351	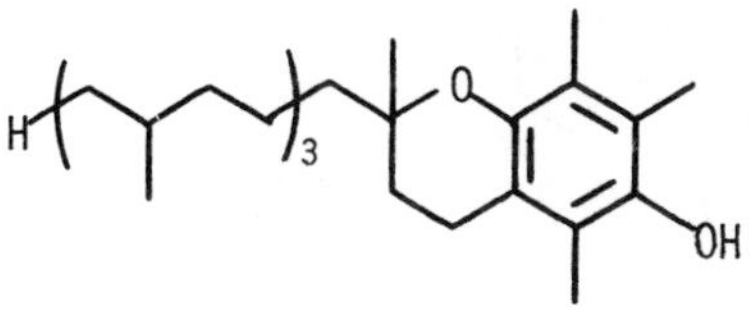	Algae	139

Table 6
TRIHYDRIC PHENOLS

Cpd. no.	Formula	Source	References
29	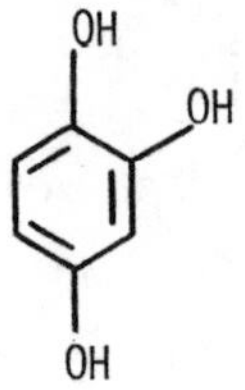 OH, OH, OH	Sponge	46 202
30	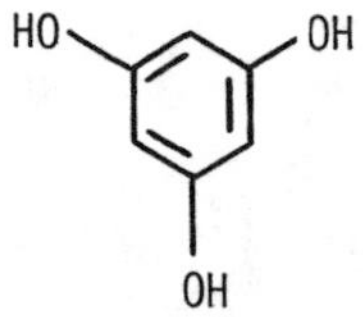 HO, OH, OH	Brown algae	94 202

Table 7
COMPLEX PHENOLS

Cpd. no.	Formula	Source	References
124	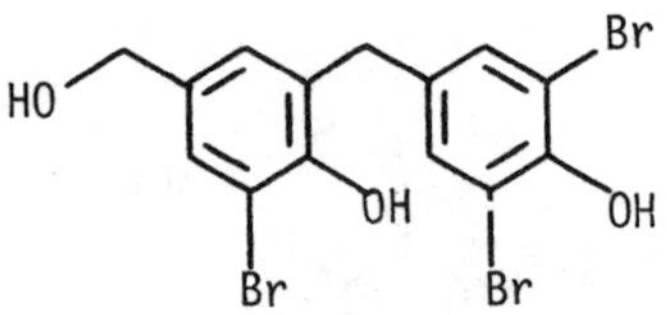 HO, Br, OH, OH, Br, Br	Sea worm	119
119	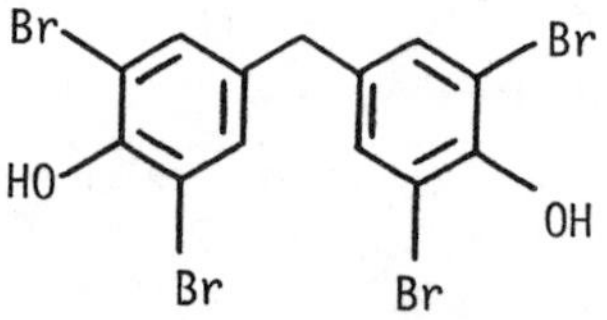 Br, Br, HO, OH, Br, Br	Sea worm	119
116	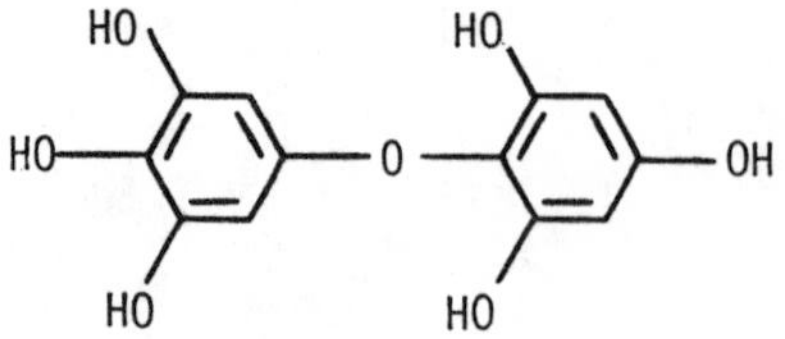 HO, HO, HO, O, OH, HO, HO	Brown alga	96
181	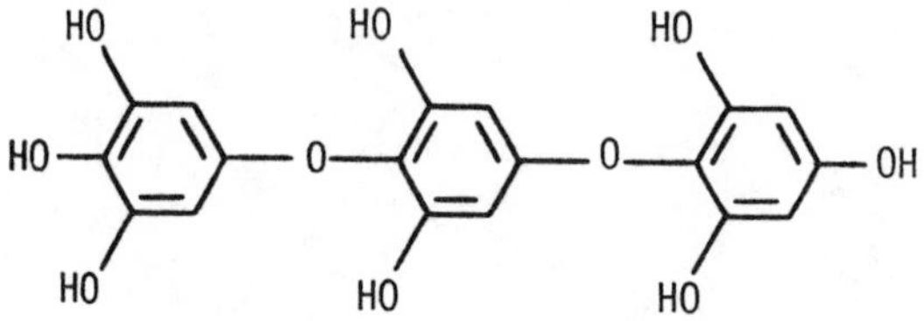 HO, HO, HO, HO, O, O, OH, HO, HO, HO	Brown alga	95

Table 8
1,4-BENZOQUINONES

Ring system	Cpd. no.	Substituent R at C_2*	Source	References
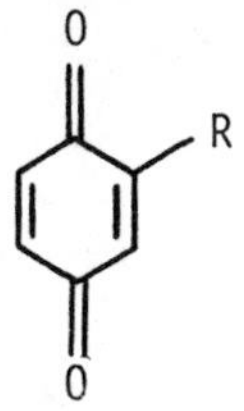 O, R, O	227	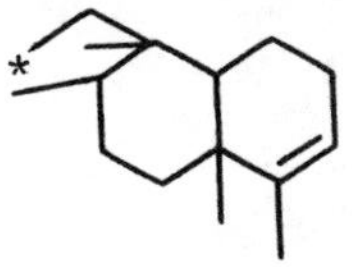 *	Sponge	207
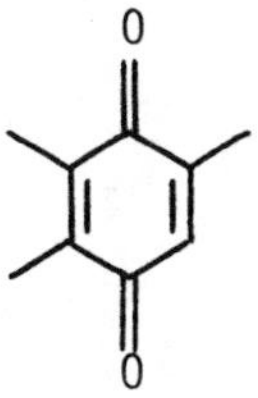 O, R, O	395	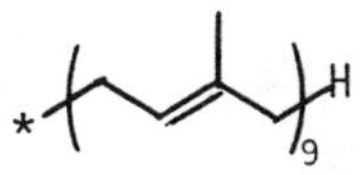 * H 9	Algae	27 73 215

Table 9
COMPLEX 1,4-BENZOQUINONES

Cpd. no.	Formula	Source	References
250	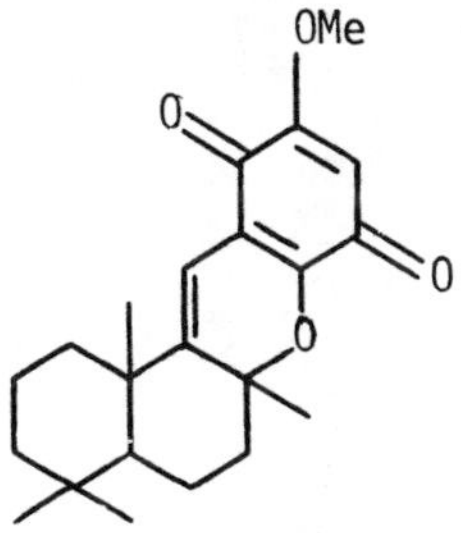 OMe, O, O, O	Sponge	162
251	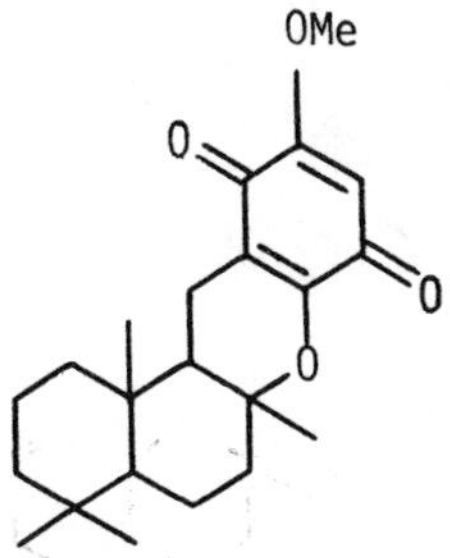 OMe, O, O, O	Sponge	162

Table 9 (continued)
COMPLEX 1,4-BENZOQUINONES

Cpd. no.	Formula	Source	References
252	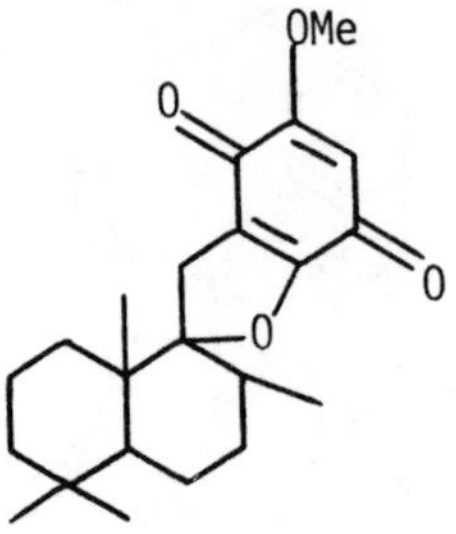	Sponge	162
253	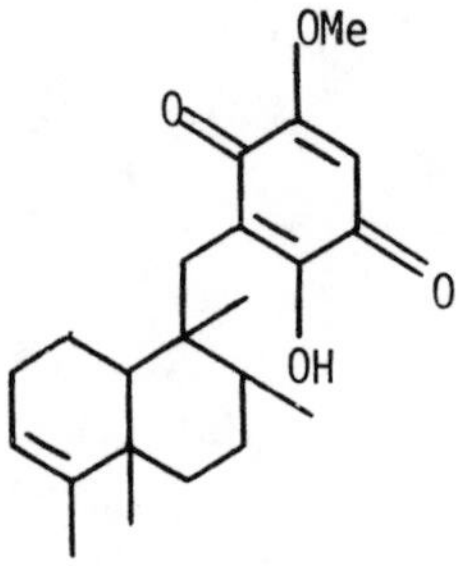	Sponge	162
254	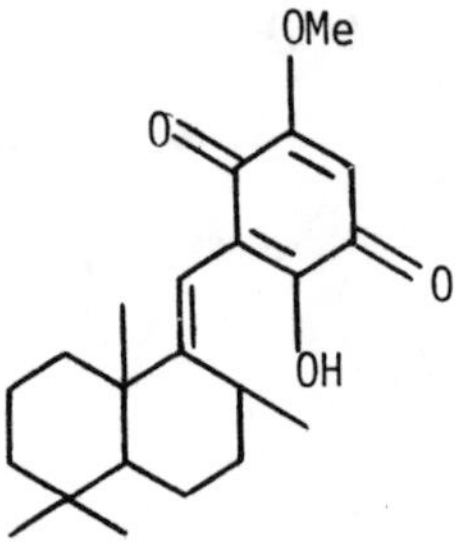	Sponge	162

Table 10
1,4-NAPHTHAQUINONES

Ring system	Cpd. no.	C_2	C_3	C_6	C_7	Source	References
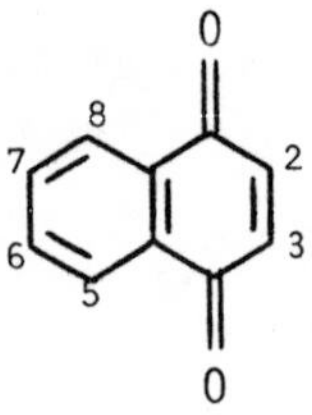	362	Me	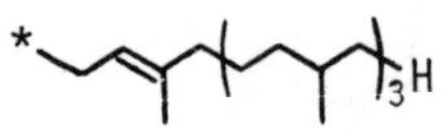	H	H	Algae	73 202 215

Isoprene-Derived Hydrocarbons and Their Derivatives

Monoterpenes

In Volume I,[11] we foreshadowed work further to that reported by Faulkner and co-workers on polyhalogenated monoterpenes. Ichikawa et al.[126] investigated the constituents of the fragrant subtidal red alga *Desmia hornemanni* Mertens [*Chondrococcus hornemanni* (Mertens) Schmitz] and the structures of 11 halogenated monoterpenes were established. In addition to the halogenated compounds, myrcene (97) was also present, as well as the fatty acid ester methyl palmitate (175).

The series of monoterpenes isolated is summarized below:

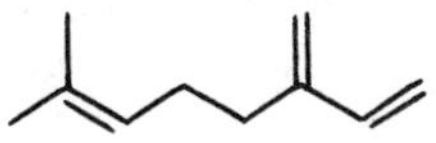

myrcene (97) 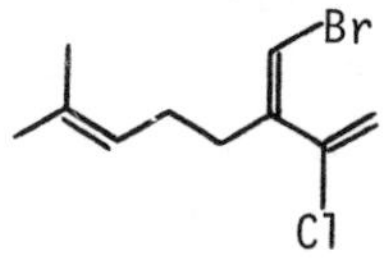 (90)

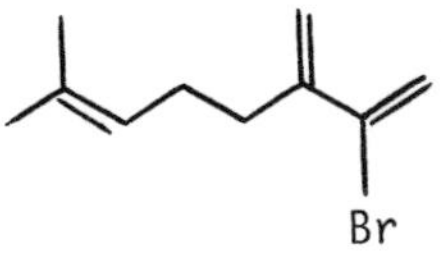

(92) 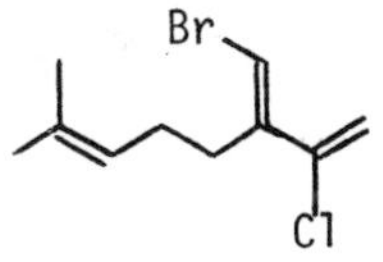 (89)

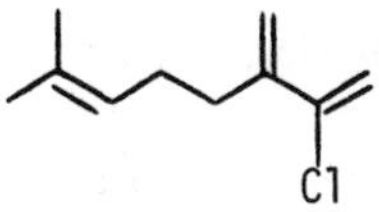

(95) 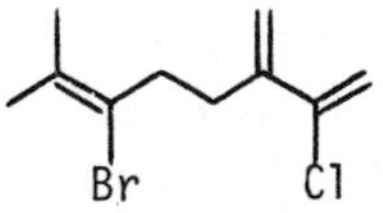 (88)

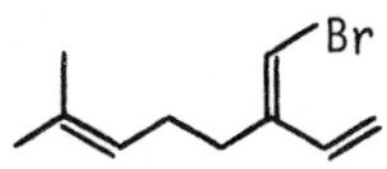

(94) 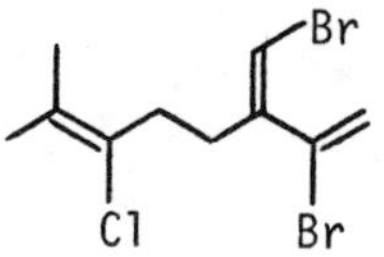 (85)

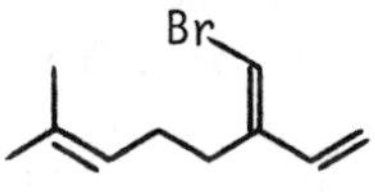

(93) 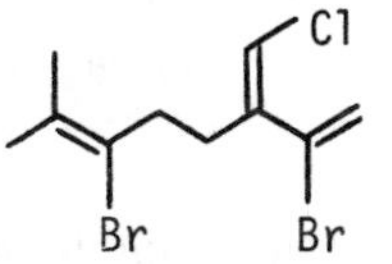 (86)

The major constituent of the fragrant oil was 2-chloromyrcene (95), representing approximately 75% of the volatile fraction.

Mynderse and Faulkner,[220] in reporting the occurrence of violacene (84) as the major constituent of the red alga *Plocamium violaceum*, indicated that another nine polyhalogenated monoterpenes had yet to be identified.

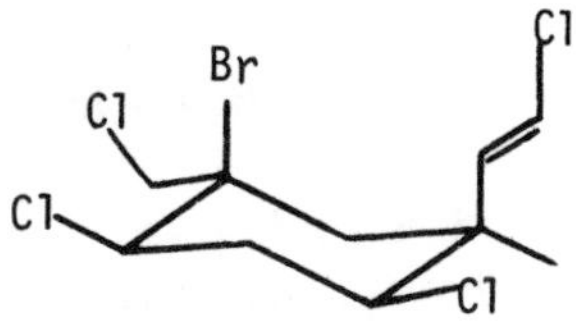

(84) violacene

In the same article Mynderse and Faulkner summarized their findings that the two halogenated monoterpenes (I/146) and (I/157) reported from the sea hare *Aplysia californica* were also obtained from a red alga, *Plocamium pacificum,* on which the sea hare grazes, and that a further nine polyhalogenated monoterpenes were currently being investigated.

Violacene (84) possesses a novel carbon skeleton not previously reported among naturally occurring monoterpenes. Mynderse and Faulkner rationalized the formation of the carbon skeleton as resulting from cyclization of an acyclic monoterpene via a bromonium ion, a process which has been invoked to explain the biosynthesis of many brominated marine natural products.

A further novel odoriferous halogenated monoterpene aldehyde, cartilagineal (79), has been isolated by Crews and Kho[59] from the ether extract of the Pacific Ocean red alga *P. cartilagineum.*

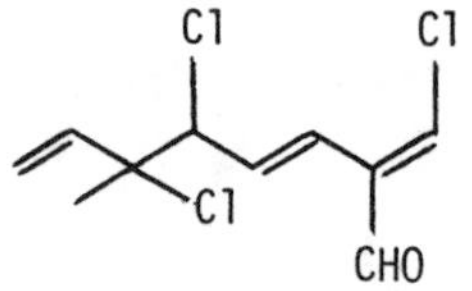

Cl Cl Cl CHO

(79) cartilagineal

The ether-soluble extracts of *P. cartilagineum* are reported to be toxic to goldfish, but no bioassay was reported for cartilagineal. Other uncharacterized polyhalogenated hydrocarbons were stated to be present in the extract, their structure elucidation to be the topic of a subsequent publication.

Thus the literature on this class of polyhalogenated monoterpenes can be expected to expand rapidly and the study of the biosynthesis of these halogenated metabolites should be an interesting area of research.

Sesquiterpenes

Marine organisms have been a rich source of structurally novel sesquiterpenes. Jeffs and Lytle[138] undertook the examination of several species of gorgonians, and, from *Muricea elongata* and *Plexaurella nutans,* reported the known (−)α-curcumene (I/235) and (+)β-bisabolene (I/240) as well as the previously unreported (−)β-curcumene (153) for which they established the absolute configuration:

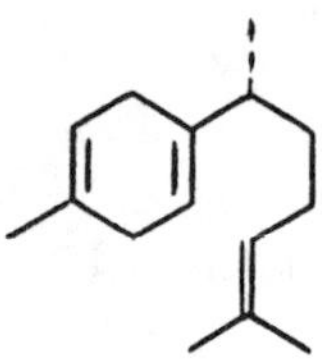

(153) (−)β-curcumene

The absolute configuration of the sesquiterpene zonarene (I/254), previously isolated from the brown alga *Dictyopteris zonarioides,* was described by Iguchi et al.[129] and should be represented as:

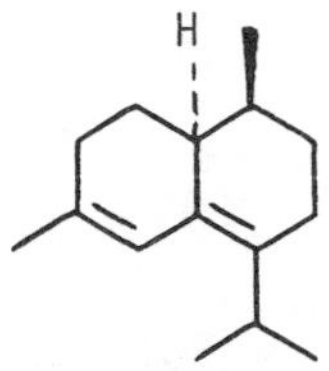

(I/254) zonarene

Oxygenated and halogenated sesquiterpenes have also been reported. Cycloeudesmol (159) from the red alga *Chondria oppositiclada* Dawson was reported, without experimental results, to be strongly antibiotic towards *Staphylococcus aureus* and *Candida albicans.*[81]

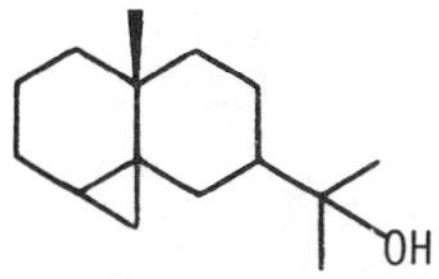

(159) cycloeudesmol

Cycloeudesmol (159) is a cyclopropane sesquiterpene, the formulation of which can be rationalized by postulating the participation of a cation of the type shown below, which has long been considered an intermediate in the biosynthesis of eremophilane sesquiterpenes.

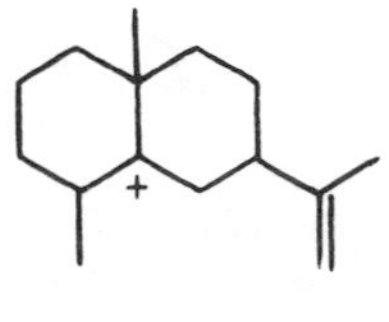

Laurencia species have previously been reported to yield "spiro" sesquiterpenoids, five examples being represented in Volume I.[11] Irie and co-workers[290] have reported on the isolation of spirolaurenone (I/237) as the major constituent of the neutral essential oil of *L. glandulifera.* Analysis of the minor constituents revealed three new isomeric bromine-containing spiro sesquiterpenoids of formula $C_{15}H_{23}BrO$, separated as crystalline substances. Their structures have been assigned as:

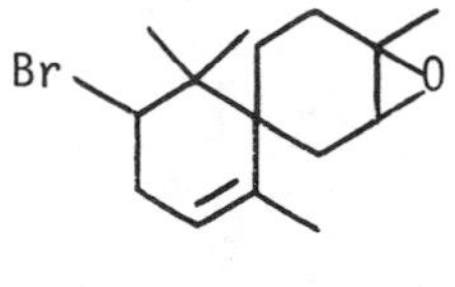

(149) 4-bromo-α-chamigren-8,9-epoxide

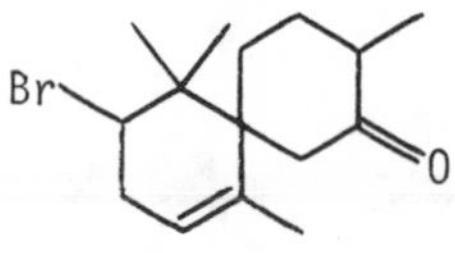

(150) 4-bromo-α-chamigren-8-one

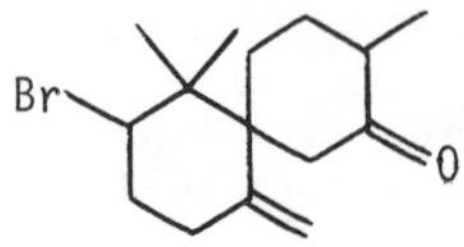

(151) 4-bromo-β-chamigren-8-one

During an antimicrobial screening program on Hawaiian marine algae, Waraszkiewicz and Erickson[270] noted marked activity against *S. aureus* and *Mycobacterium smegmatis* in the methanol/ether extracts of the red alga *Laurencia nidifica*. In the subsequent chemical program, two new halogenated spiro sesquiterpenoids, which they named nidificene (152) and nidifidiene (143), were isolated.

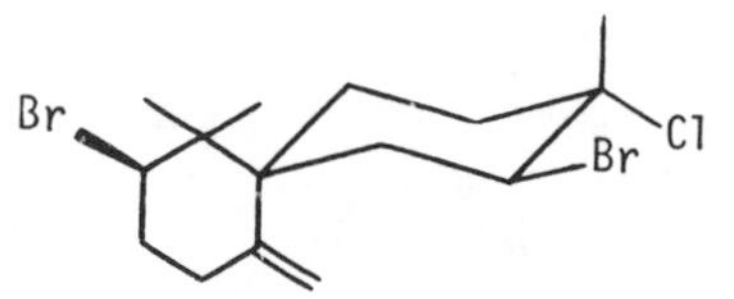

(152) nidificene

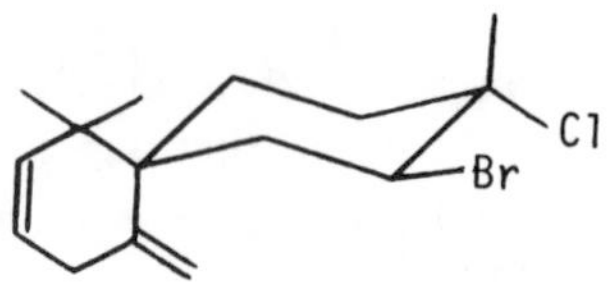

(143) nidifidiene

The antimicrobial activity was not associated with either of these substances, but with the known laurinterol (I/217).[11]

In the hands of Sims et al.[247] *L. elata*, collected from the southeast coast of Australia, yielded another halogenated spiro sesquiterpene (144) which was named elatol. Elatol contains the novel structural feature of a vinyl chloride grouping with the position of substitution of the chlorine atom being one carbon removed from where it is found in other related compounds [cf. prepacifenol (I/231)]. The structure was confirmed by X-ray analysis, and the absolute stereochemistry established as:

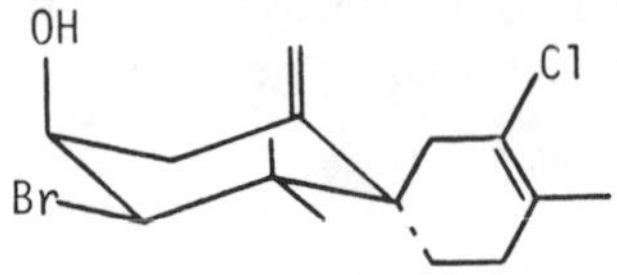

(144) elatol

Faulkner et al. have previously drawn attention to the relationship between digestive gland constituents of certain *Aplysia* species and the constituents of red algae on which they graze.[291,292]

With *Aplysia californica*, these authors extended their previous investigations on the extracts of the digestive glands to isolate a halogenated spiro sesquiterpenoid diepoxide, prepacifenol epoxide (141).[78]

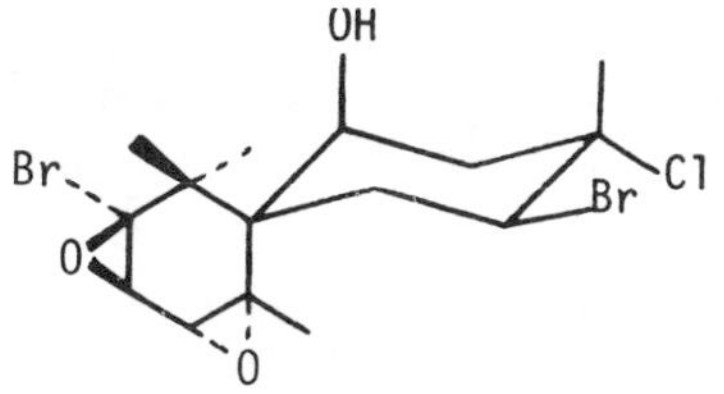

(141) prepacifenol epoxide

The structure of (141) was confirmed by its high yield conversion into johnstonol (I/232).

Johnstonol had previously been isolated from the red alga *L. johnstonii,* and, on investigation of the constituents of this alga, Faulkner and co-workers[78] found that extraction of air-dried material gave johnstonol as the major substance, while extraction of the fresh material with ethyl acetate gave predominantly prepacifenol epoxide (141). It was therefore suggested that johnstonol is most probably an artifact formed from (141), both during methanol extraction of the air-dried alga and also in the digestive glands of the *A. californica.*

The diol (140) was also isolated as a minor constituent of the ether extracts from the digestive glands of *A. californica,* but was not indicated as being derived from *L. johnstonii.*

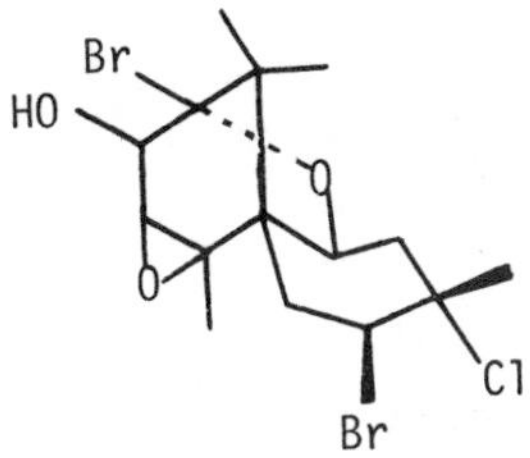

(I/232) johnstonol

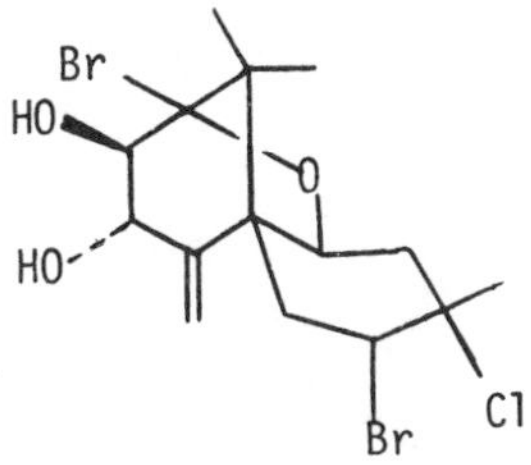

(140)

Pacifenol (139) was isolated from *L. pacifica* and *L. johnstonii* by Sims et al., who subsequently demonstrated with a fresh extract of *L. pacifica* that pacifenol (139) did not occur in the alga but was an artifact generated from the naturally occurring prepacifenol (I/231) during silica gel chromatography.[246] However, pacifenol (139) does exist as a natural product in *L. tasmanica*[246] and the compound has also been reported as a minor constituent of *L. nidifica*[270] and of *A. californica.*[257] In this last case the substance was derived from the algal diet, *L. pacifica,* and is almost certainly an artifact.

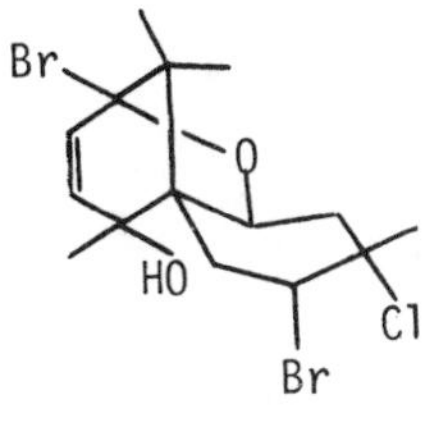

(139) pacifenol

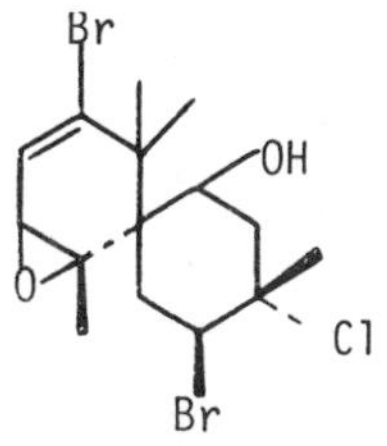

(I/231) prepacifenol

Acetoxyintricatol (174), isolated from *L. intricata* by Hager and co-workers,[187] is but one of 17 new halogenated compounds isolated from this red alga. The structure was elucidated by chemical and physicochemical methods and confirmed by X-ray diffraction studies.

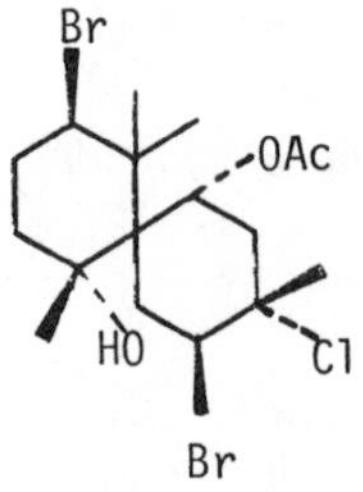

(174) acetoxyintricatol

The sea hare *A. dactylomela* was investigated by Schmitz and McDonald.[243] Using the entire animal, these workers first extracted the sea hares by soaking in aqueous isopropanol, then extracted the air-dried residue with hexane. Following tlc of the material extracted by hexane, three isomeric sesquiterpene ethers have been isolated. One of these, the so-called dactyloxene-B (155) has been characterized but the stereochemistry has not been assigned.

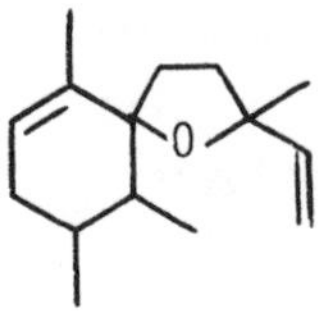

(155) dactyloxene-B

The carbon skeleton of dactyloxene-B can be envisaged as arising biogenetically via single 1:2 methyl migration in a conventional isoprene-derived skeleton.

Following on the correction of the structure of caespitol from the original spiro structure to the structure (I/258)[99,293] isocaespitol (157) was isolated from the same alga, *L. caespitosa* Lamx.,[99,101] and its structure determined by X-ray crystallography.

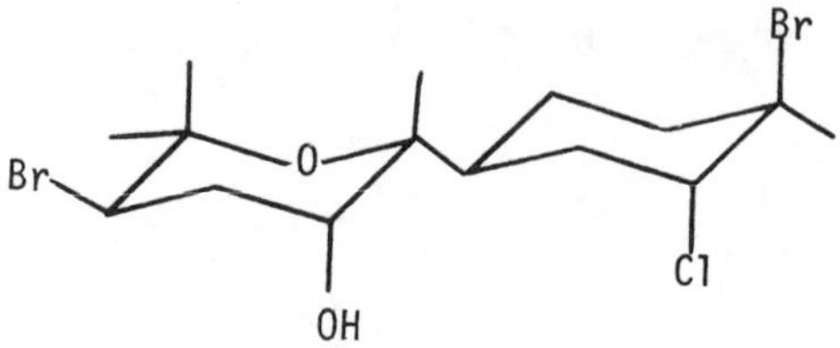

(157) isocaespitol

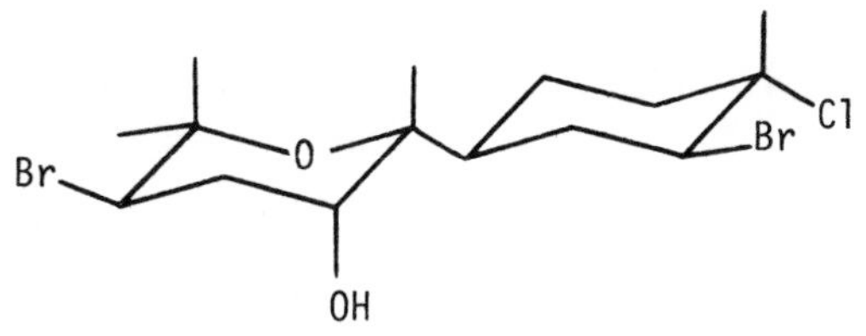

(I/258) caespitol

On melting, isocaespitol (157) is reported to rearrange to caespitol (I/258). When extraction procedures were devised which rigorously excluded rearrangement, the natural occurrence of the isomers in the extract was (157):(I/258)::2:1.

A novel tricyclic sesquiterpene skeleton is a feature of a capnellane derivative (156), isolated from the soft coral *Capnella imbricata* collected in Indonesian waters.[141] The compound (156) is stated as being the major constituent of a mixture containing a "variety of compounds".

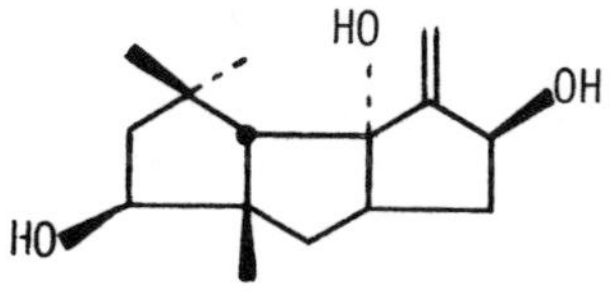

(156)

In addition the chemical and physicochemical elucidation, the structure and absolute configuration of (156) were independently established by X-ray diffraction analysis.

Tursch and colleagues also investigated the occurrence of terpenoids in the alcyonarians (which are closely related to gorgonians) finding them to be a rich source of sesqui- and diterpenes. These authors have characterized a sesquiterpene named africanol (158) from *Lemnalia africana.*[266]

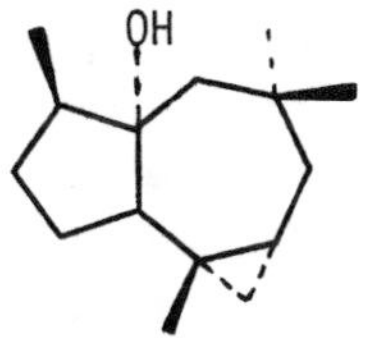

(158) africanol

The relative configuration has been established by single-crystal X-ray diffraction analysis, and the authors believe that the formation of the alcohol can be envisaged as occurring following the rearrangement as shown.[266]

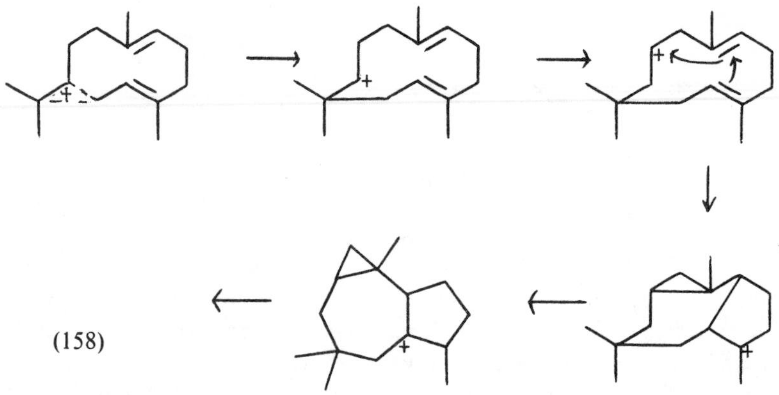

(158)

A number of sesquiterpenoid isocyanides has been isolated from marine sponges. Axisonitrile-1(I/267) was the first sesquiterpenoid monoisocyanide reported, being from the sponge *Axinella cannabina,* collected in the Bay of Taranto.[294] Minale and co-workers, studying a sponge of the same order, *Acanthella acuta,* collected in the Bay of Naples, reported the occurrence of further sesquiterpene isocyanides.[206] The major component acanthellin-1 (164)

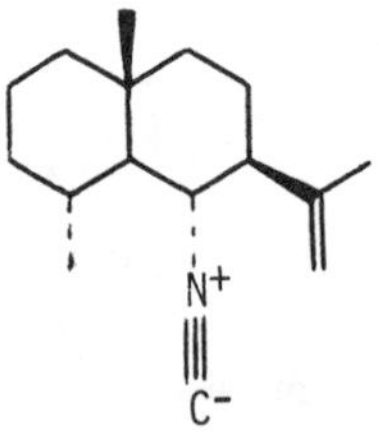

(164) acanthellin-1

showed antibacterial activity (not specified by the authors) and was accompanied by an isocyanide named acanthellin-2, plus at least three isocyanides, the functionality (−N≡C) being assigned on the basis of an i.r. absorption at 2140 $cm.^{-1}$ In this report there was no suggestion of a corresponding isothiocyanate, whereas axisonitrile had co-occurred with the corresponding isothiocyanate (I/268).

Very soon after this report from Minale's group, Fattorusso and co-workers reported the isolation of a different sesquiterpenoid isocyanide during further studies on the sponge *Axinella cannabina.*[77] This isocyanide, axisonitrile-2 (165) possesses an aromadendrane skeleton, which was established by the selenium dehydrogenation of axisonitrile-2 (165) to give guaiazulene in good yield.

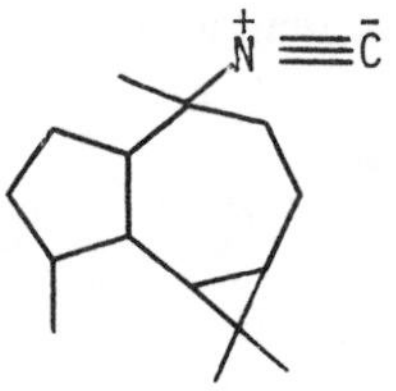

(165) axisonitrile-2

Burreson and Scheuer obtained a sample of a sponge of the genus *Halichondria* which was collected by trawling at 200 meters. This sponge yielded the sesquiterpenoid isocyanide (166), together with the corresponding formamide (169) and an isothiocyanate (167).[36,37]

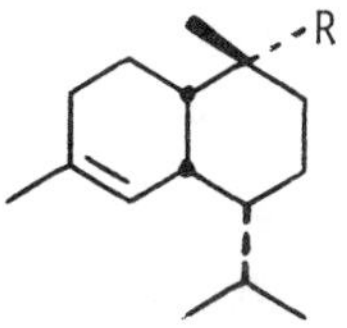

(166) R = $\overset{+}{N}\equiv\overset{-}{C}$
(169) R = NH-CHO
(167) R = N=C=S

Burreson and Scheuer postulate that the formamide (169) is the biogenetic precursor of the isocyanide (166). [The same sponge also yielded a diterpenoid isocyanide (242), together with the corresponding formamide (245) and isothiocyanate (243), which will be discussed at a later stage.][36,37]

A novel furanosesquiterpenoid, pleraplysillin-2 (199) has been isolated from the sponge *Pleraplysilla spinifera,*[48] which previously yielded two simpler furanosesquiterpenes (I/228) and (I/229).[295] Pleraplysillin-2 appears to be derived from a furanosesquiterpenoid and esterified by a hemiterpene alcohol and is therefore correctly illustrated as a sesquiterpenoid rather than as a diterpenoid, which may be suggested by the molecular formula.

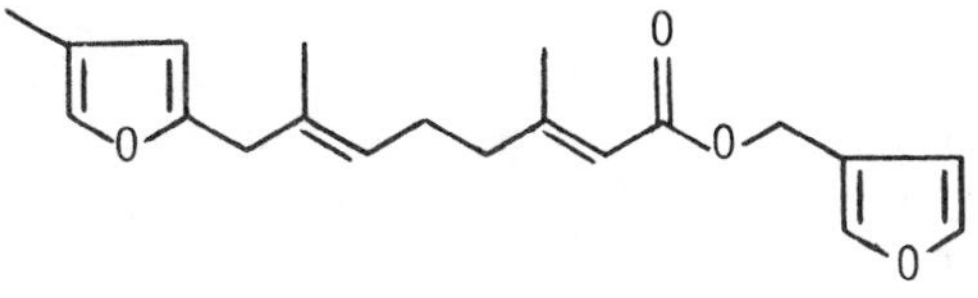

(199) pleraplysillin-2

Sesquiterpenoid moieties in substances of mixed biogenesis have been previously reported as occurring in marine sponges.[11] Minale's group in working on the sponge *Dysidea avara* isolated the hydroquinone avarol (229), and the corresponding quinone avarone (227),[207] both possessing a rearranged drimane skeleton.

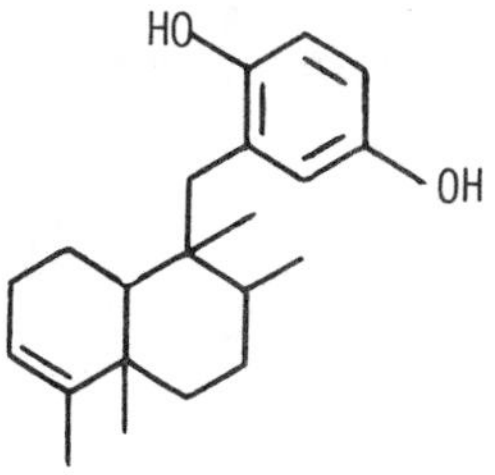

(229) avarol

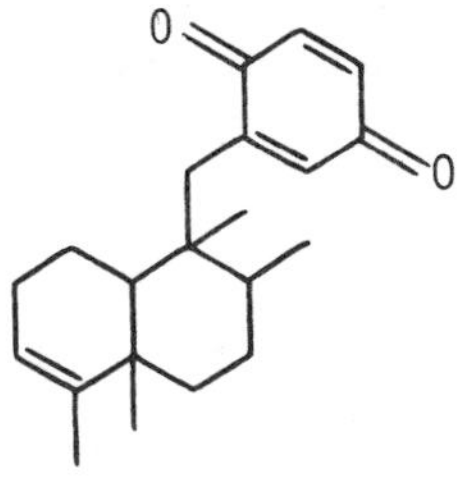

(227) avarone

In examining the less polar fraction of the sponge *Halichondria panicea* which had yielded the paniceins[296] which are triprenyl-derived phenols, from the more polar fractions,[45] Minale and co-workers isolated methyl *trans*-monocyclofarnesate (168) which is a possible precursor for the paniceins.

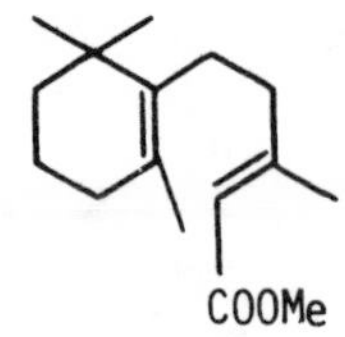

COOMe

(168) methyl *trans*-monocyclofarnesate

Diterpenes

Nagayama and co-workers isolated from the methanol extract of the unsaponifiable material of the brown alga *Hizikia fusiformis* three substances each of which possessed a significant accelerating activity on pig pancreatic lipase.[174] The diterpene phytol (220) was identified as one of the lipase activators, and the other two were not identified except to be described as belonging "to terpenes and sterols respectively".

Phytol (220) was of approximately the same potency as a lipase activator as was farnesol.

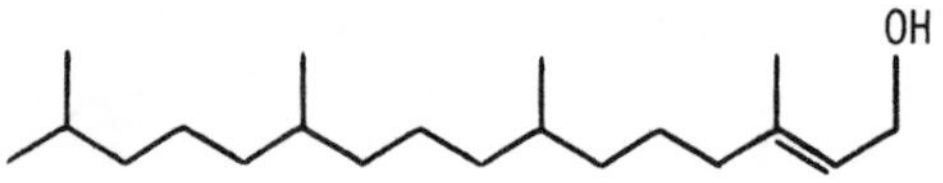

OH

(220) phytol

Another diterpene-derived open chain C_{19} compound, pristane (I/287), which has been previously reported from a brown alga and from shark liver oil,[11] was isolated from the muscle of several species of salmon, from the rainbow trout, from the carp, and from the sea bream.[131]

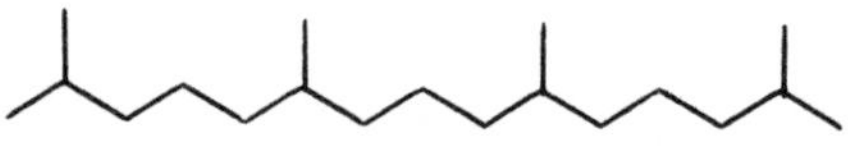

(I/287) pristane

Conventional diterpenoid structures have been found in such compounds as retinol (vitamin A) (208) from crustaceans and from fish,[14,62,86,87,202] and a novel perhydro isomer of retinol, caulerpol (216), has been isolated from a Tasmanian collection of the green alga *Caulerpa brownii*.[21] The corresponding acetate (255) was also present.

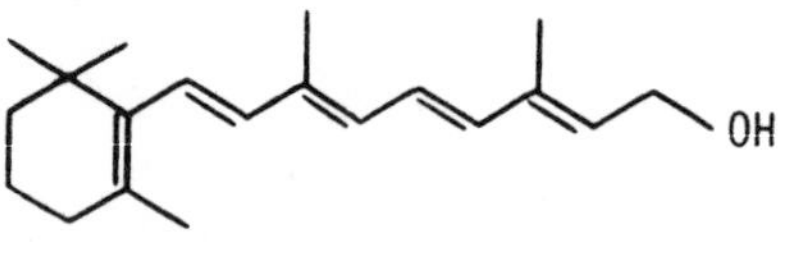

OH

(208) retinol (vitamin A)

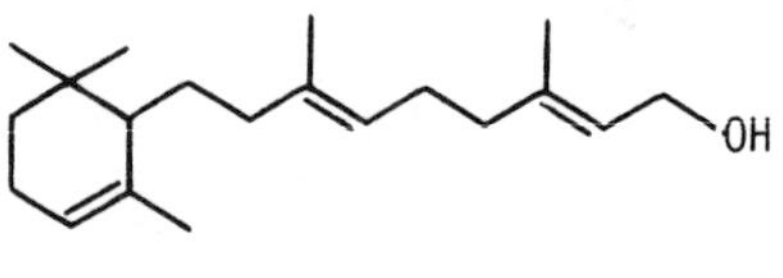

OH

(216) caulerpol

Scientists of the Roche Research Institute of Marine Pharmacology have isolated a series of furans based on a novel tetracyclic diterpene ring system.[162] These are spongiadiol (204), *epi*-spongiadiol (205), spongiatriol (206) and *epi*-spongiatriol (207) and their corresponding diacetates (266, 267) and triacetates (282, 283).

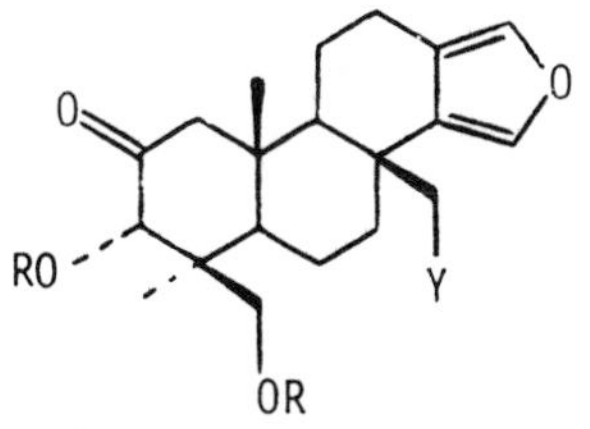

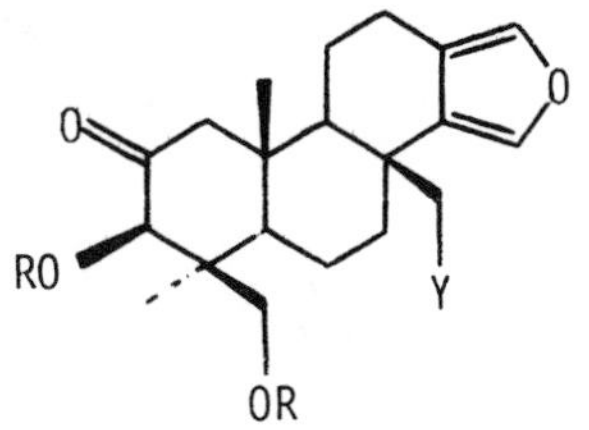

R = H	Y = H	(204) spongiadiol	R = H	Y = H	(205) *epi*-spongiadiol
R = H	Y = OH	(206) spongiatriol	R = H	Y = OH	(207) *epi*-spongiatriol
R = Ac	Y = H	(266) spongiadiol diacetate	R = Ac	Y = H	(267) *epi*-spongiadiol diacetate
R = Ac	Y = OAc	(282) spongiatriol triacetate	R = Ac	Y = OAc	(283) *epi*-spongiatriol triacetate

Isoagatholactone (211), isolated by Minale and co-workers from the sponge *Spongia officinalis,* features a tetracyclic system incorporating a lactone moiety.[50]

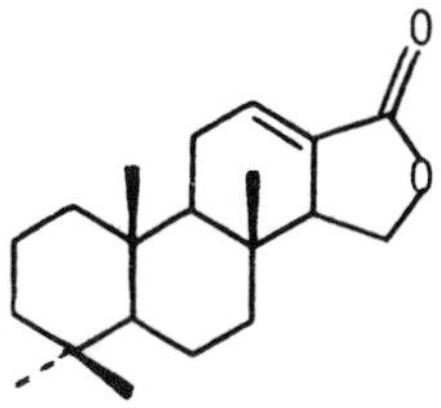

(211) isoagatholactone

Taondiol (I/367) is an example of a diterpenoid moiety in a marine natural product of mixed biogenesis. Gonzales and co-workers, in continuing studies on the brown alga *Taonia atomaria* from which taondiol was obtained, collected the same alga at the same time as the previous collection, but found that taondiol was absent. The only phenolic compound present was an acid $C_{28}H_{42}O_4$ which was named atomaric acid (319).[100]

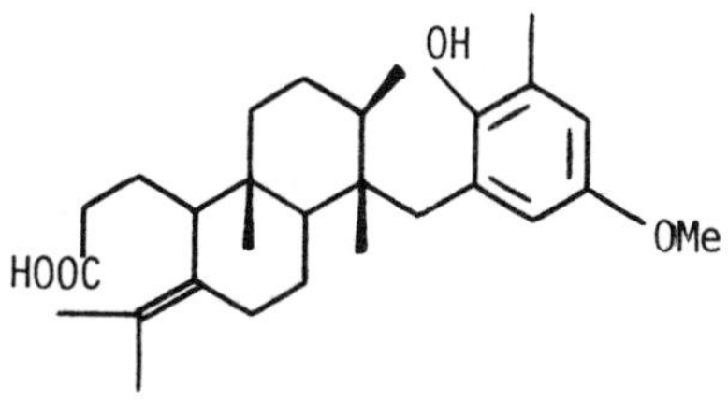

(319) atomaric acid

Atomaric acid was postulated to have been derived from taondiol methyl ether by cationic opening of the chroman ring, followed by apparently concerted hydrogen-methyl shifts.[100] Ring A has been oxidatively cleaved.

More conventional mixed biogenesis compounds containing a diterpene moiety have also been isolated, e.g., δ-tocopherol (310) from several species of algae,[139,202] δ-tocotrienol methyl ether (318) from the brown alga *Cystophora torulosa*,[162] δ-tocotrienol (298) and its epoxide (299) from the brown alga *Sargassum tortile*,[176] the C_{28} homologues β-tocopherol (333),[139,202] λ-tocopherol (334),[139,202] and methyl-δ-tocotrienol methyl ether (337),[162] all being from algae, and the C_{29}-homologue α-tocopherol (vitamin E) (351),[139] also from algae.

Jensen[139] has discussed the significance of Norwegian seaweeds as a potential source of tocopherols.

Alcyonaceans or soft corals are widely distributed in tropical waters of the Indo-Pacific region, both on coral reefs and on island shores. The years 1973 and 1974 have seen significant attention given to these organisms and the vast majority of compounds isolated are derived from the diterpenoid 14-carbon macrocycle for which the name cembrane has been accepted.

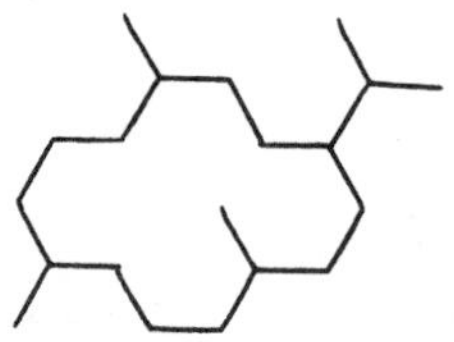

the cembrane ring system

In general, the ring system is unsaturated, and the isopropyl side chain often oxidized and cyclized on to the macrocyclic ring.

Compounds bearing the same fundamental ring system have been reported[11] by several authors as occurring in a related group of coelenterates, the gorgonians. Eunicin (I/290), jeunicin (I/291), crassin acetate (I/328), and eupalmerin acetate (I/329) were all reported as derived from gorgonians.[11]

Eupalmerin acetate (I/329), $C_{22}H_{32}O_5$, was converted to an addition product $C_{22}H_{32}Br_2O_5$ for the purpose of structure determination by X-ray analysis.[268] It was expected that the reaction with bromine would be a simple addition to an isolated double bond, but in fact, the reaction with bromine involved trans-annular participation of the epoxide function, to form an ether bridge and result in bromine at the methyl-substituted carbons of the epoxide ring and double bond, respectively.

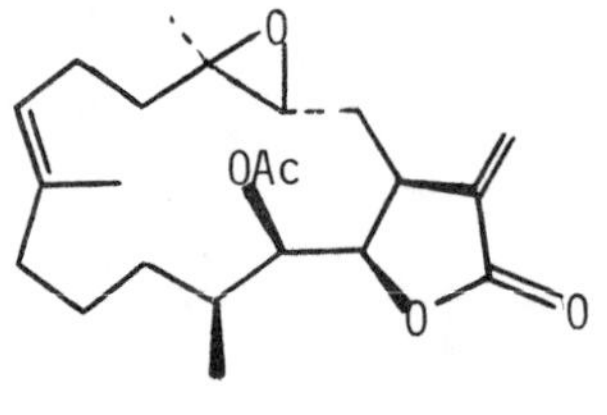

OAc

O

O

O

(I/329) eupalmerin acetate

The rearrangement on bromine addition rendered a direct structure assignment to eupalmerin acetate impossible, and sounded a warning of possible rearrangements in this series.

Early in 1974 Ciereszko and co-workers[242] reported the isolation of epoxynephthenol acetate (256) and of nephthenol (217) from a soft coral of the genus *Nephthea*, collected at Eniwetok in the Marshall Islands.

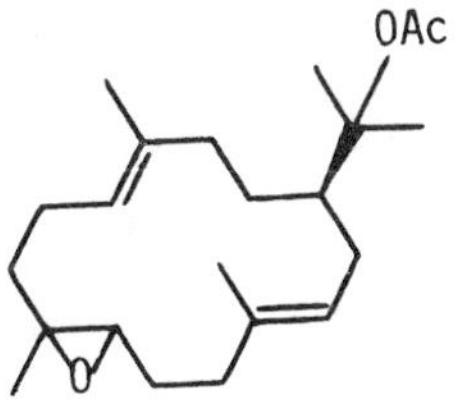

(256) epoxynephthenol acetate

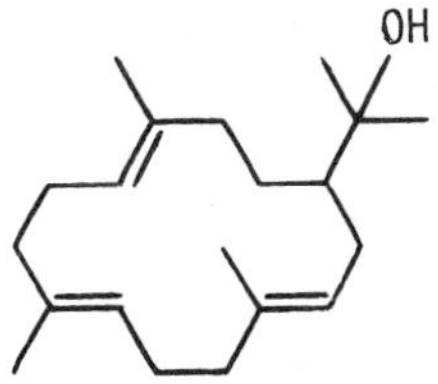

(217) nephthenol

Nephthenol (217) is isomeric with mukulol, which was characterized in 1973 as a constituent of a terrestrial plant.[228]

The related dehydroepoxynephthenol (214) was isolated from a soft coral of the genus *Lobophytum* collected in North Queensland.[51]

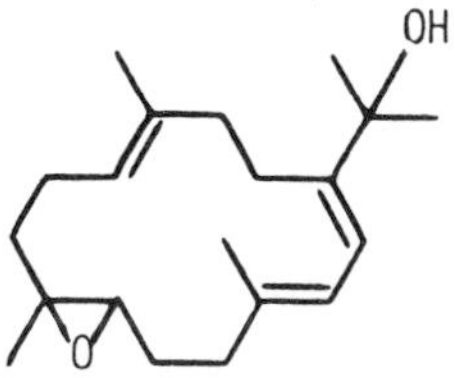

(214) dehydroepoxynephthenol

In general, intact soft corals appear to be remarkably free of predators, although when samples were collected by cutting across the main bulk of the soft coral, the previously unexposed portions of the organism, now exposed, were rapidly attacked by a variety of small fish, which appeared to ignore our presence in their haste to attack this new food source. Similar observations have been made by workers in the Red Sea, where aquarium experiments demonstrated that not only were fish repelled by the intact soft coral, but that they soon died if left in the tank with the soft coral.

On the basis of this type of observation Neeman et al. [222] studied the toxic effects of the substance isolated, sarcophine (203), from the soft coral *Sarcophyton glaucum*. The detailed chemistry of sarcophine was also reported by Kashman and co-workers.[17]

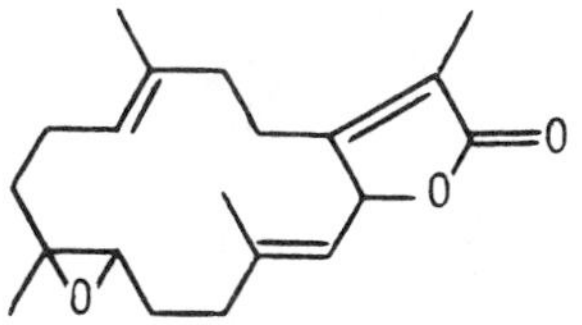

(203) sarcophine

In our studies at RRIMP, the same substance has been isolated from *S. trocheliophorum*.[162]

Lethality tests on the fish *Gambusia affines*[222] showed the LD_{50} of sarcophine to be 3 mg/ℓ after 3 hours. Sarcophine (203) was also found to be toxic to mice, rats, and guinea pigs. Subcutaneous injection in mice at 10 mg sarcophine per kg body weight caused effects similar to those due to atropine — excitation, respiratory increase, and paralytic effects, followed by coma and death. On the isolated guinea pig ileum, sarcophine at a concentration of 0.2 to 0.8 mg/ℓ of solution showed strong anti-acetylcholine action.

Sarcophine was also shown to be a competitive inhibitor of cholinesterase with a Km ≅ 9×10^{-3} M (with inhibitor); Km ≅ 4.5×10^{-3} M (without inhibitor).

In subsequent more detailed studies on *Sarcophyton glaucum* Kashman et al.[150] isolated the simple alcohol (213)

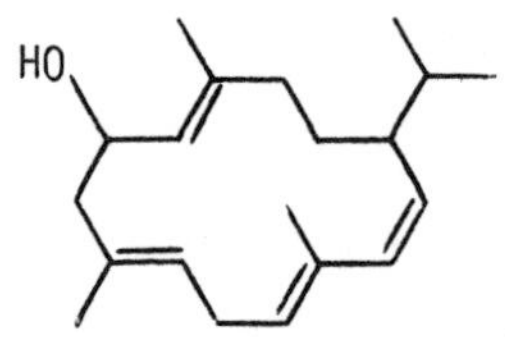

and two compounds featuring a dihydrofuran moiety (209), (210)

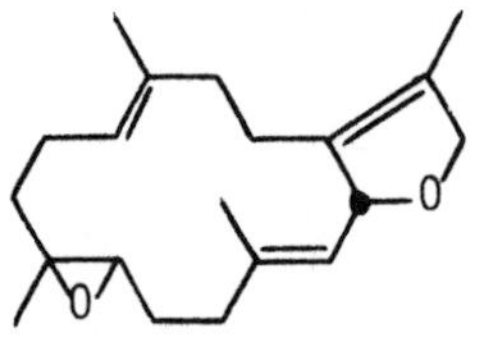

(209)

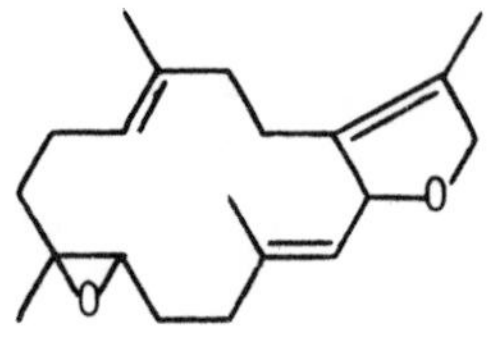

(210)

plus two diastereoisomers (201), (202) of sarcophine (203).

Considerable work is necessary to assign the absolute stereochemistry of the members of this series.

Lobophytolide (200) from the soft coral *Lobophytum cristagalli* collected off Serwaru, Indonesia, was characterized both by chemical means and by X-ray diffraction analysis,[267] and features the lactonization on to the macrocycle in the alternative position to that with sarcophine (203).

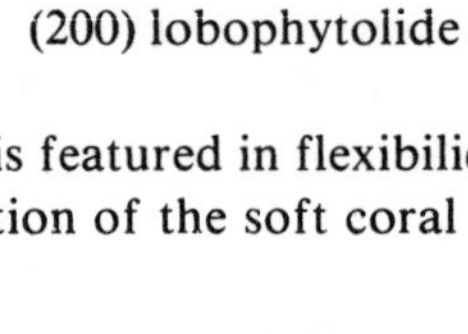

(200) lobophytolide

A six-membered lactone ring is featured in flexibilide (212) and in dihydroflexibilide (215) from an Australian collection of the soft coral *Sinularia flexibilis* for Wells and co-workers.[161]

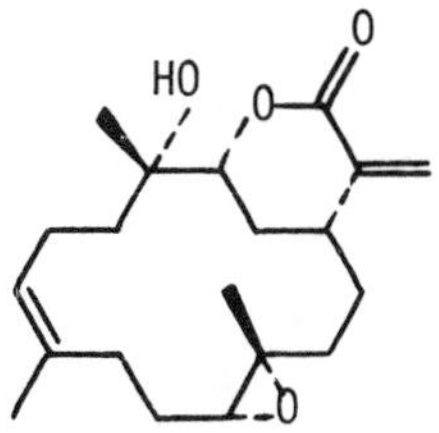

(212) flexibilide

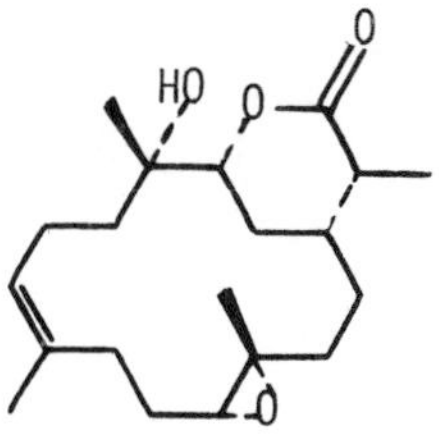

(215) dihydroflexibilide

The interest in soft corals is certain to be maintained and one would hope for increasing attention to the role of the substances obtained in the defence or communication mechanisms of the organisms in the marine ecosystem.

As was the case with the sesquiterpenes, a naturally occurring diterpene isocyanide (242) and the corresponding isothiocyanate (243) and formamide (245) have been found in a marine sponge, in this case of the genus *Halichondria* in a study by Burreson and Scheuer.[36]

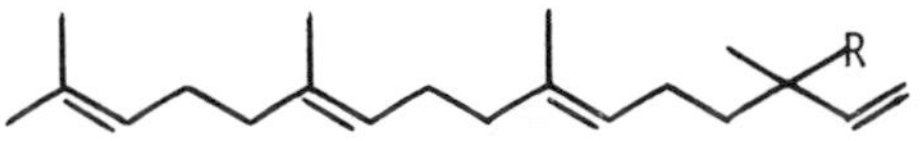

R = $\overset{+}{N}\equiv\overset{-}{C}$ (242)
R = N=C=S (243)
R = NH−CHO (245)

*Carotenoids**

Together with the 31 carotenoids compiled in Volume I, the 22 additional carotenoids included in Volume II make a total of 53 carotenoids reported from marine sources up to the end of 1974. Further up-dating of references prior to 1975 will be attempted in Volume III. How many of the numerous characteristic carotenoids obtained from photosynthetic and other bacteria, yeasts, and blue-green algae should be included remains to be decided.

For additional information about the carotenoids treated in Volumes I and II, the reader is referred to Isler's monograph[133] treating all aspects of carotenoids, and subsequently up-dated by the published main lectures from the IUPAC symposia on carotenoids in 1975 and 1978.[284,285] Entries to the complete original literature for each carotenoid are readily obtained through Straub's list of natural carotenoids[286] covering data published until the middle of 1975. Marine carotenoids have recently been treated in a separate review.[287]

Regarding nomenclature, compound names are cited in Volumes I and II directly from the publications referred to. However, it is pointed out that new, detailed rules for carotenoid nomenclature have now been approved,[288,289] using a double Greek letter prefix to indicate the particular end groups. Thus, for instance, β-carotene (I/483) now is β,β-carotene; α-carotene (I/482) is β,ε-carotene; δ-carotene (I/484; should be Δ^4, not Δ^5) is ε,ψ-carotene; and γ-carotene (II/382) is β,ψ-carotene. The α, γ, and δ prefixes are abandoned in the semirational names. The numbering of the carbon skeleton is shown below:

* This section was contributed by Professor Synnøve Liaaen-Jensen of the Organic Chemistry Laboratories, Norwegian Institute of Technology, University of Trondheim, Norway.

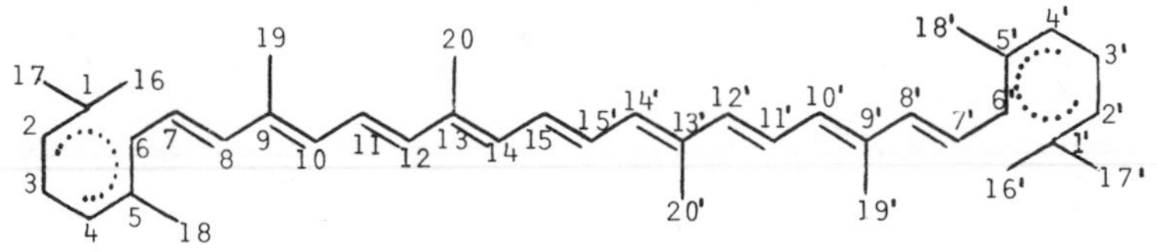

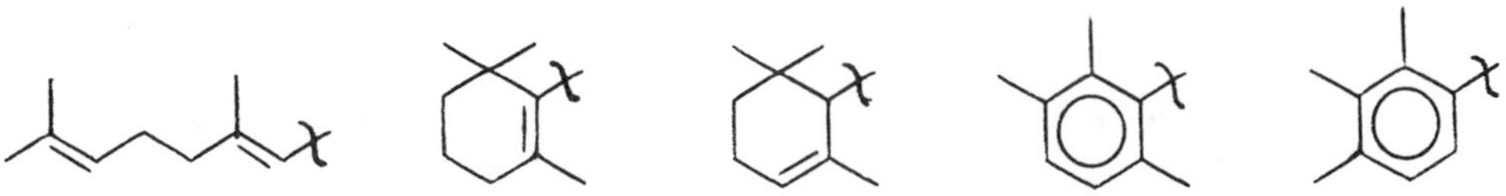

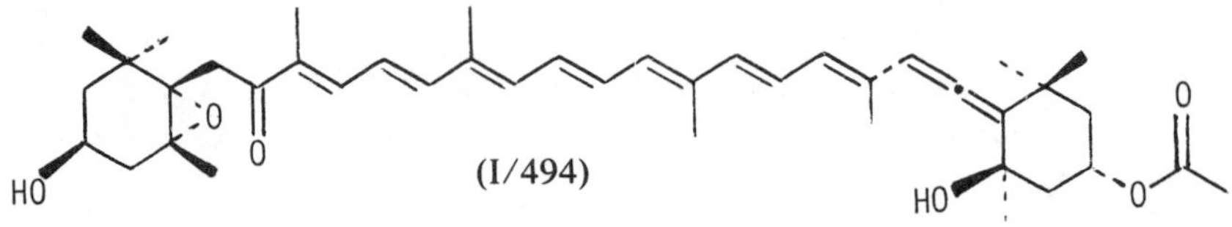

(I/494)

Principles used for naming structurally more complex carotenoids are illustrated by compound I/494 = fucoxanthin = (3*S*,5*R*,6*S*,3′*S*,5′*R*,6′*R*)-5,6-epoxy-3,3′,5′-trihydroxy-6′,7′-didehydro-5,6,7,8,5′,6′-hexahydro-*β*,*β*-caroten-8-one 3′-acetate.

The stereochemistry given in this volume is based on the following assumption: if the chirality of a given carotenoid is established from one source, it is assumed to be the same in other sources until otherwise documented. So far, few examples are known for which this assumption is not valid.

In Volume I the stereochemistry of a particular carotenoid was considered established if it contained a chiral end group stereochemically defined in a different carotenoid. This concerns compounds I/463, 470, 471, 474, 475, 476, 481, 488 and 489. This prediction has since been verified for compounds I/463, 470, 471, 476 and 481 (see Volume III), whereas the situation for I/474, 475, 488 and 489 is not yet settled. It is also pointed out that the chirality of lutein (I/486) has subsequently been revised and is known to be 3′*R*.

The carotenoids tabulated in Volumes I and II are given as claimed by the investigators. Critical evaluation of each identification is beyond the scope of this volume. However, a few general guidelines for evaluation of the evidence are given:

For unequivocal identification of a carotenoid electronic spectrum (visible region), mass spectrum and co-chromatography with the authentic compound are a minimum requirement. Identifications based on R_F-values and electronic spectra alone should be considered only as tentative.

When saponification is included in the isolation procedure *α*-ketols (a) readily rearrange to the corresponding diosphenols (b), a reaction caused by alkali in the presence of traces of oxygen. Thus compounds I/474 and II/373 are likely artifacts of II/379 and I/476, respectively.

Evidence for the 4-hydroxy-*β*-end group (c) is frequently weak, and should be considered satisfactory only if allylic oxidation to d is achieved or allylic dehydration demonstrated, cf. compounds II/377, 385, 387.

5,6-Epoxides (e) readily rearrange to furanoid 5,8-epoxides (f_1 and f_2) in the presence of acids, etc., during the isolation procedure. Consequently 5,8-epoxides are usually artifacts, cf. flavoxanthin (II/389). Isolation of both C8 epimeric furanoxides supports a chemical rearrangement.

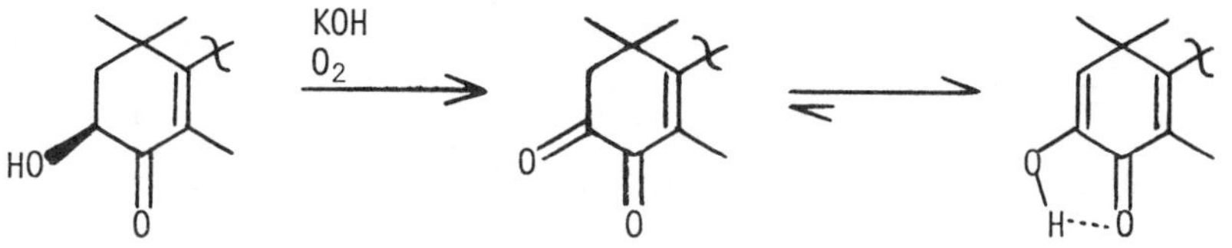

KOH
O_2

a

b

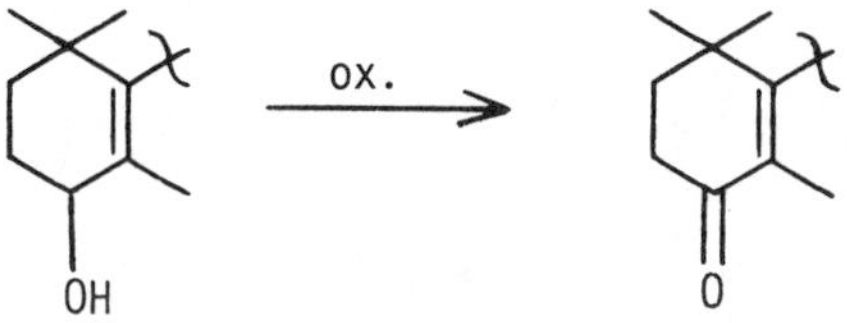

ox.

c

d

H^+

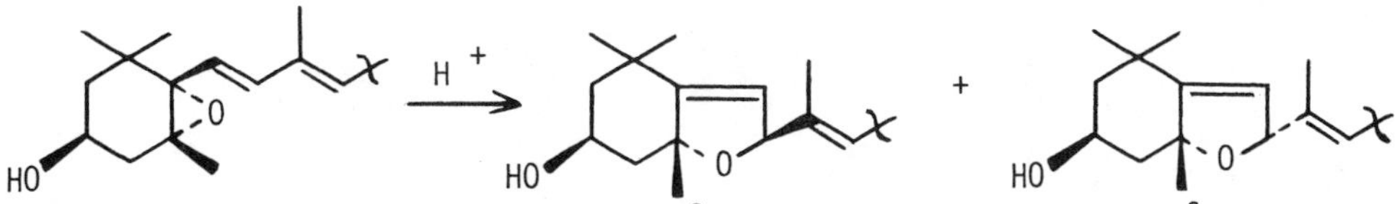

e

f_1

f_2

The carotenoids included in this volume are classified below by the same system as used for Volume I:

Table 11. Derivatives of β-carotene (β,β-carotene): 15 compounds
Table 12. Derivatives of α-carotene (β,ε-carotene): 5 compounds
Table 13. Allenic and aromatic carotenoids: 2 compounds

Where additional stereochemical evidence is available after 1974, this is indicated by footnotes.

Whereas allenic and aryl groups are also encountered in carotenoids from nonmarine sources, acetylenic carotenoids (I/465, 466, 470, 471, 472, 475, 477, 479, 480, 481, and II/ 380, 381, 391) are hitherto restricted to marine organisms.

Although tunaxanthin (II/386) is a common marine carotenoid, its structure is not satisfactorily documented. Natural occurrence of the allylic ethyl ether II/394 needs verification.

Table 11
COMPOUNDS RELATED TO β-CAROTENE

Cpd. no.	Formula	Source	References
382	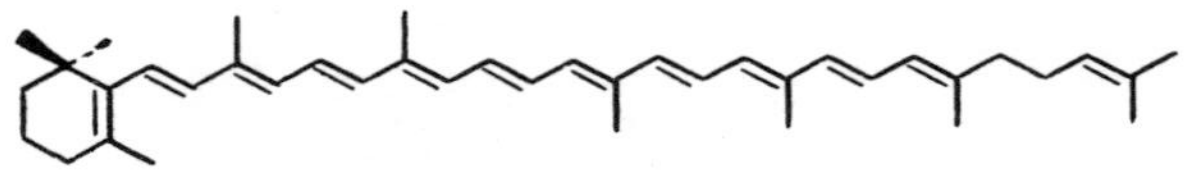	Crustaceans Sponges	61,65,102,133 153,202
385	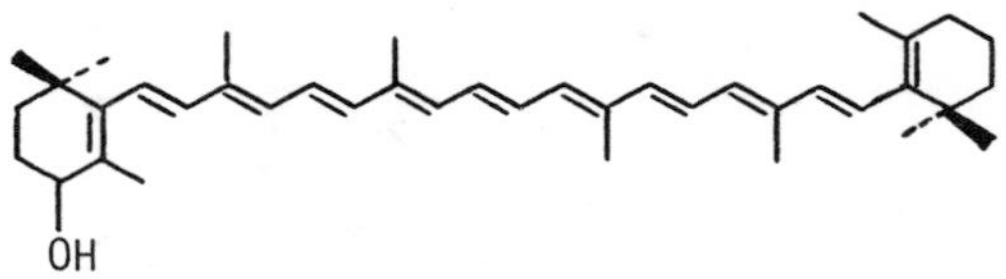	Crustaceans Sea worm	63,133 151,156
383	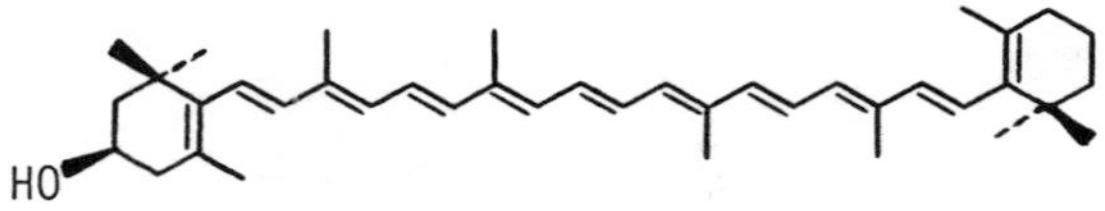	Crustaceans Fish Molluscs Starfish	63,65 103,133 151,198 202
387	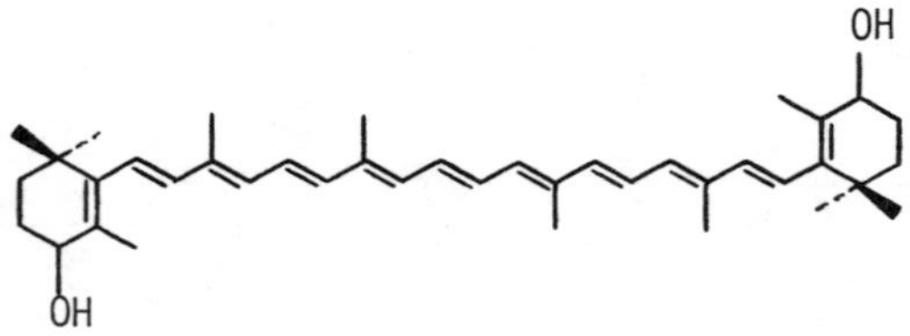	Crustaceans Fish Sponge	61,63 64,65 133,151
376	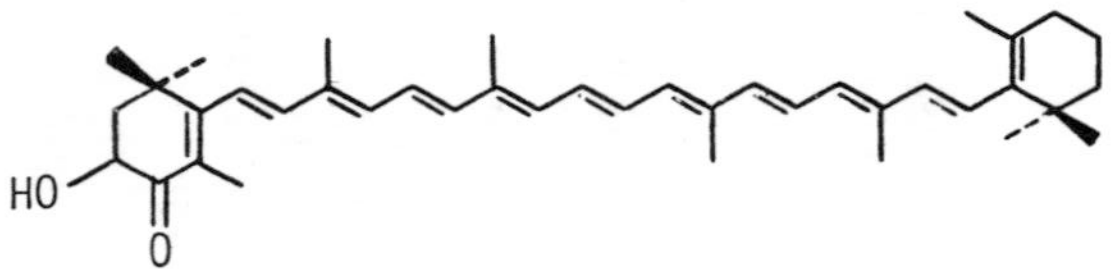	Crustacean Starfish	133 151

377	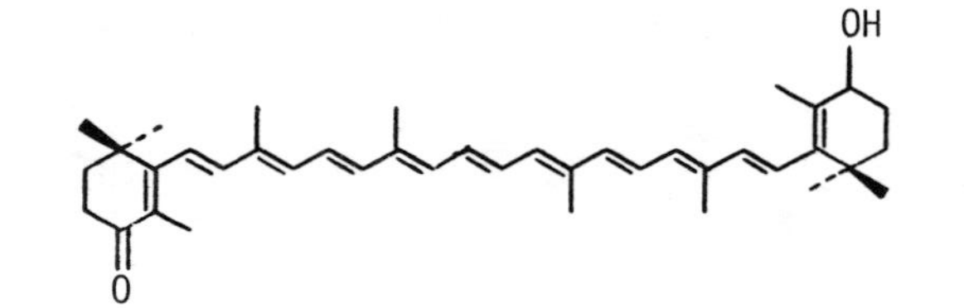 Crustaceans Fish	64,65,133 151,156
378	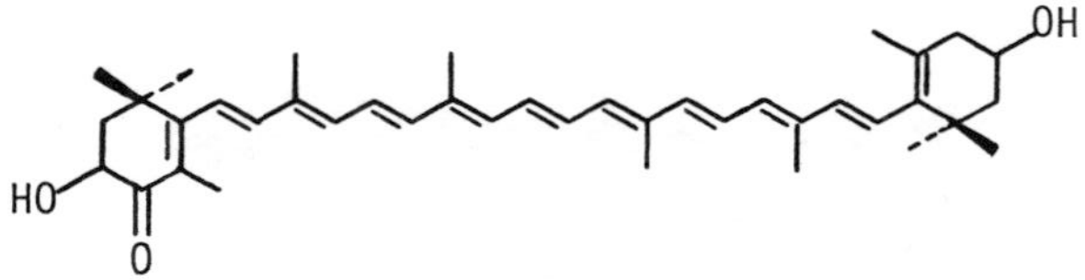 Crustaceans Starfish	133,151 196,199
374	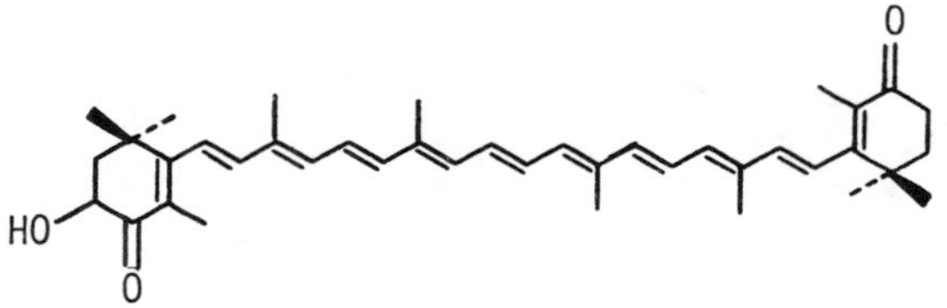 Crustaceans Fish Sea cucum- ber	34,133,153 156,157,158 196,199
390	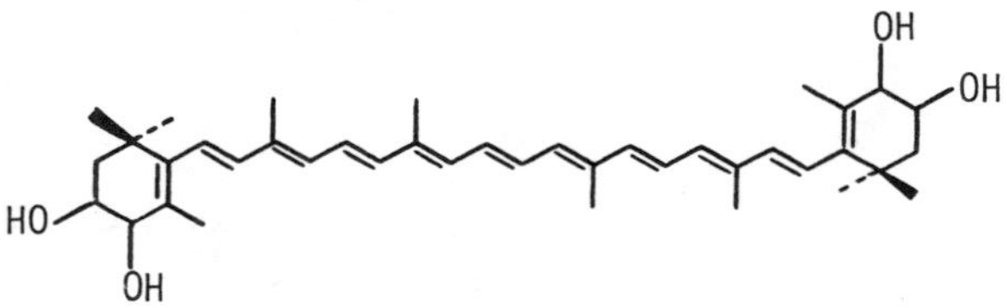 Crustaceans	133,151,153
393	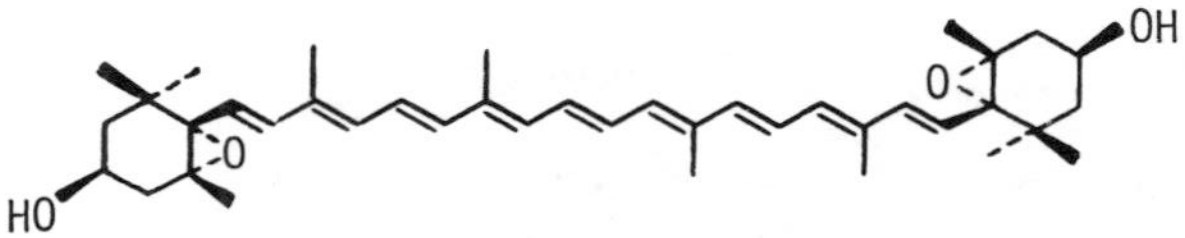 Sponge	61,133,202

Table 11 (continued)
COMPOUNDS RELATED TO β-CAROTENE

Cpd. no.	Formula	Source	References
391[a]	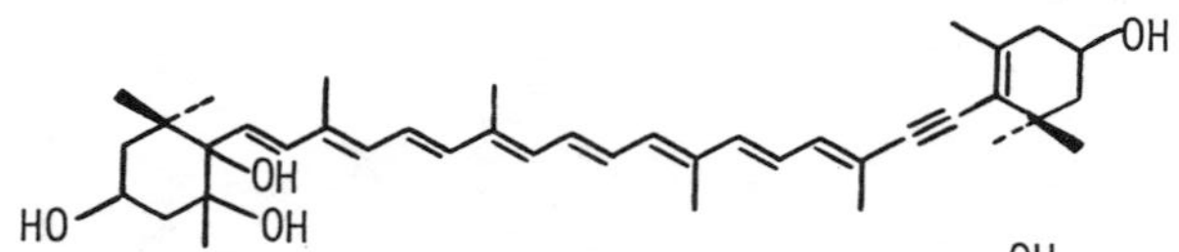	Alga	133,225
381	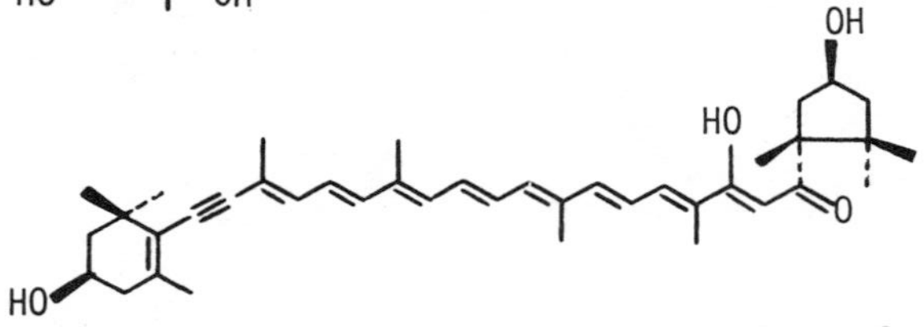	Mussels	103,165
380	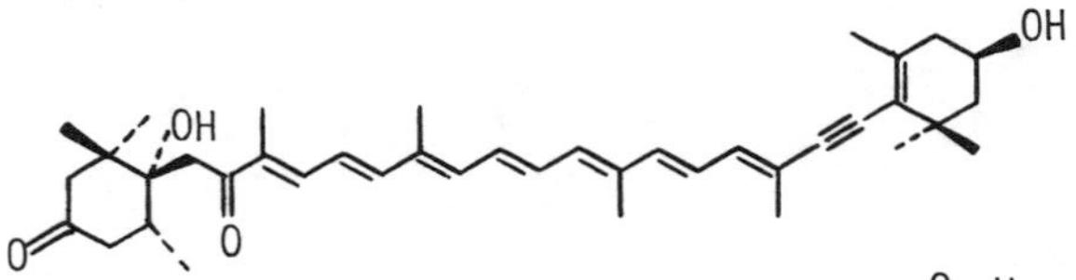	Mussel	165
373	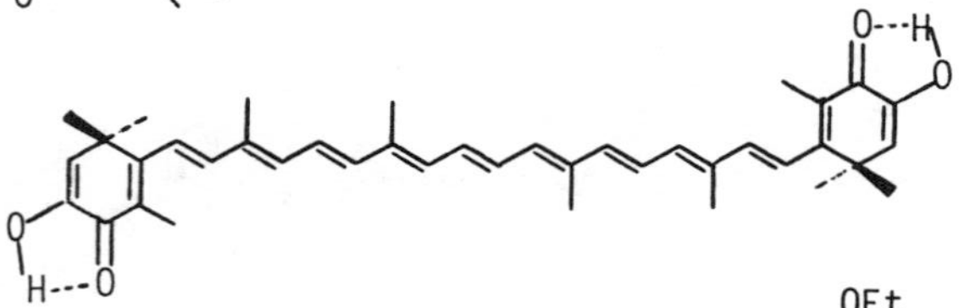	Crustaceans Fish Sponge	61,62,63 64,65,133 198,200
394	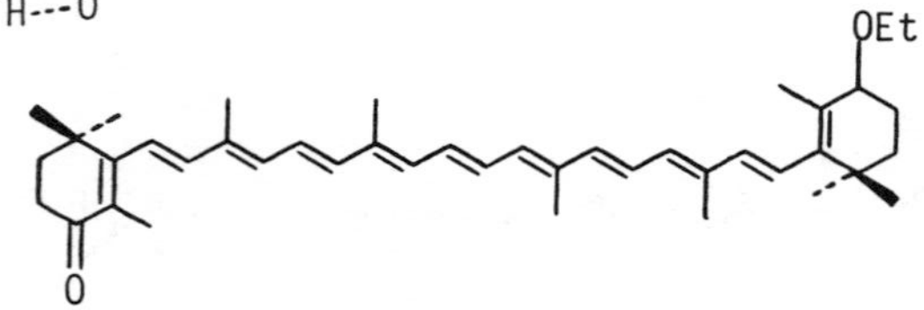	Crustacean	65

[a] Stereochemistry later established (see Volume III).

Table 12
COMPOUNDS RELATED TO α-CAROTENE

Cpd. no.	Formula	Source	References
384	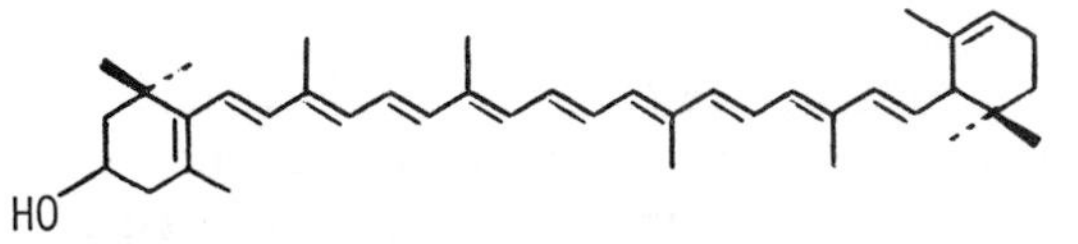	Fish	133,157 158,198
386	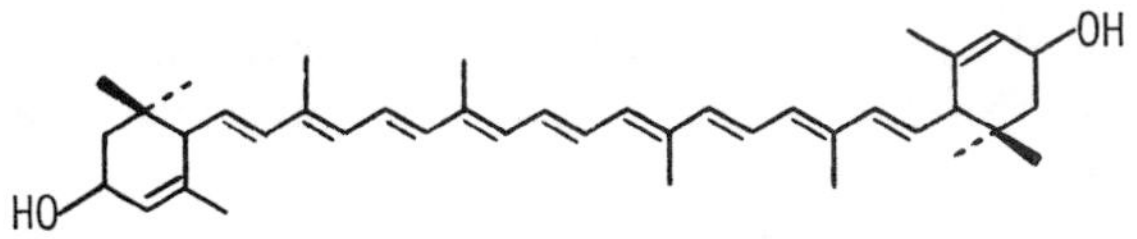	Fish Prawn	60,62,64,133 152,154,155 157,158,198
388[a]	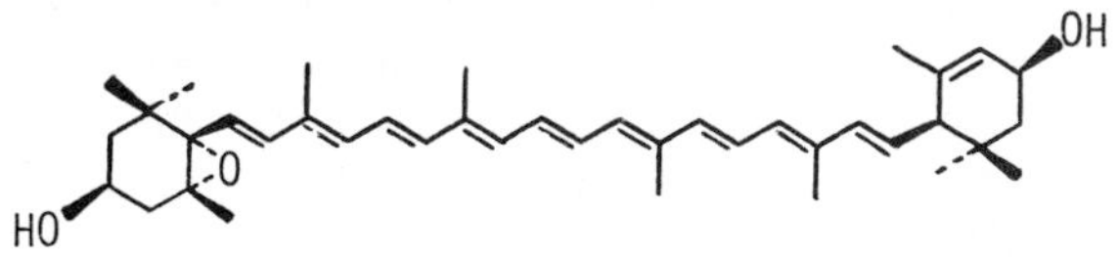	Crustaceans Fish	62,63,64 65,133
389[b]	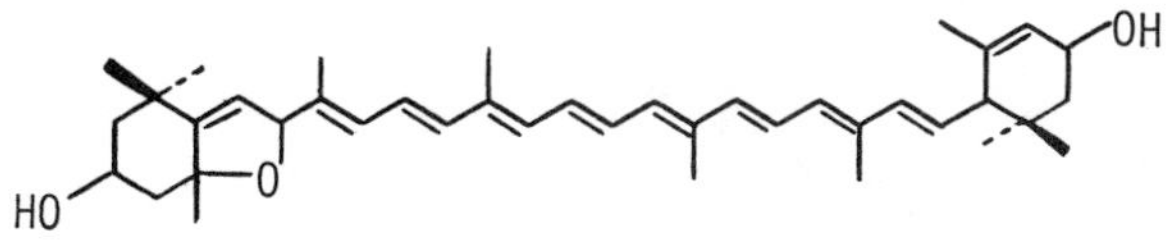	Crustaceans Mollusc	103,133,151 200,202
379	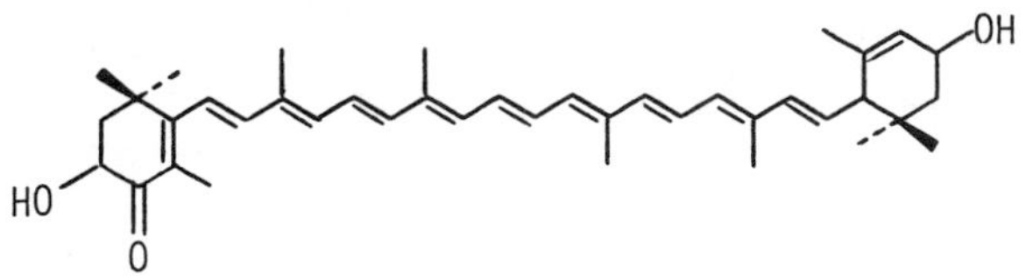	Sea bream	133,157

[a] Stereochemistry later revised to 3′*R* (see Volume III).
[b] Stereochemistry later established (see Volume III).

Table 13
ALLENIC AND AROMATIC CAROTENOIDS

Cpd. no.	Formula	Source	References
392	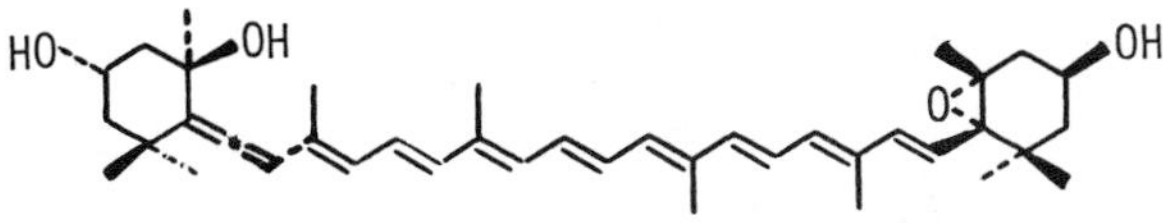	Brown alga Sponge	61,133,223
375[a]	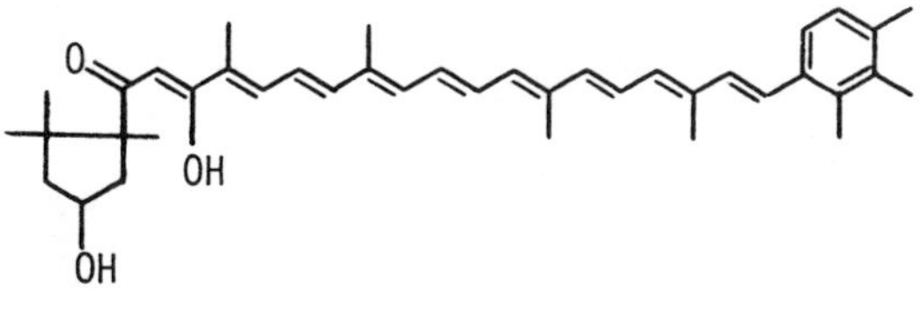	Sponge	5

[a] Stereochemistry later revised to 3′ *R* (see Volume III).

Sterols

Sterols are widely distributed in marine organisms and interest is maintained in the possibility that within different species in certain phyla, the specific sterol composition may provide a basis for a chemotaxonomic classification. To this time, this potential application is limited by an incomplete set of data and often imprecise classical taxonomy of species concerned.

Within the sterols themselves, greater care is still needed by authors in specifying the stereochemistry at the different chiral centers. In our tabulation of sterols we have not attempted to predict a stereochemistry if the author has not specified that stereochemistry.

Because of the implied importance of sterol content, the practice in Volume I of this series of tabulating sterols according to different fundamental ring systems is continued in Tables 14 to 26 (at end of chapter).

Sterols are distributed in many different marine phyla. In the algae, the sterol (326) was isolated from the diatom *Phaeodactylum tricornutum*; it had previously been reported from a brittle star, *Ophiocomina nigra*.[238] The suggestion has been made[238] that these marine invertebrates may derive the sterol via the food chain from the diatoms, and extensive studies are required to clarify such postulates as this.

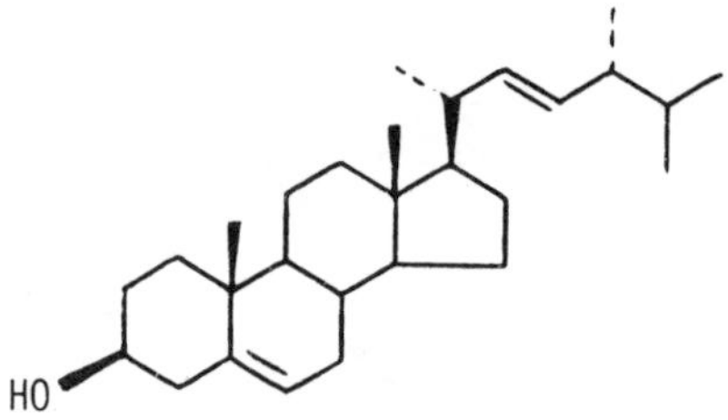

HO

(326)

When *P. tricornutum* was cultured in the presence of [CD_3]-methionine, two deuterium atoms were incorporated, indicating the production of a 24-methylene intermediate in the C24 alkylation mechanism.[238]

The green alga *Codium fragile* was found[237] to contain a sterol (323) isomeric with (326) as well as the homologue (339).

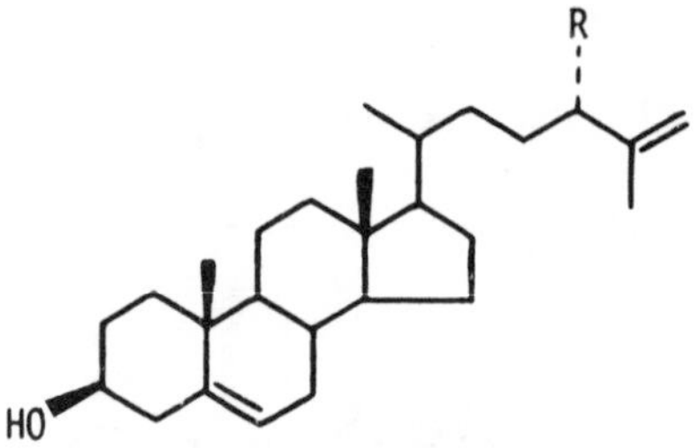

R

HO

(323) R = Me codisterol
(339) R = CH_2Me clerosterol

Clerosterol (339) has been reported as occurring in high vascular plants, whereas codisterol has not been previously reported.

Fucosterol (I/409), which has lipase-activating activity, is regarded as the principal sterol of the brown algae.

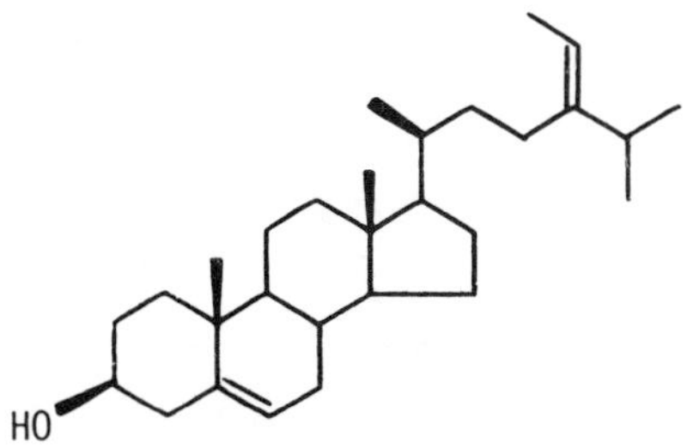

HO

(I/409) fucosterol

Studies by Smith et al.[250] confirm this theory for the warm water *Sargassum fluitans* which contained, as well, at least nine other sterols of which cholesterol (I/380), 24-methylenecholesterol (I/396), and *trans*-22-dehydrocholesterol (I/374) were positively identified.

Fucosterol (I/409) and 24-methylenecholesterol (I/396) were also identified in the brown alga *Hizikia fusiformis.*[173] In the Norwegian brown alga *Pelvetia canaliculata,* fucosterol (I/409) was again the principal sterol, and 24-oxocholesterol (306) was also characterized.[216] This ketone was also found to be present in the

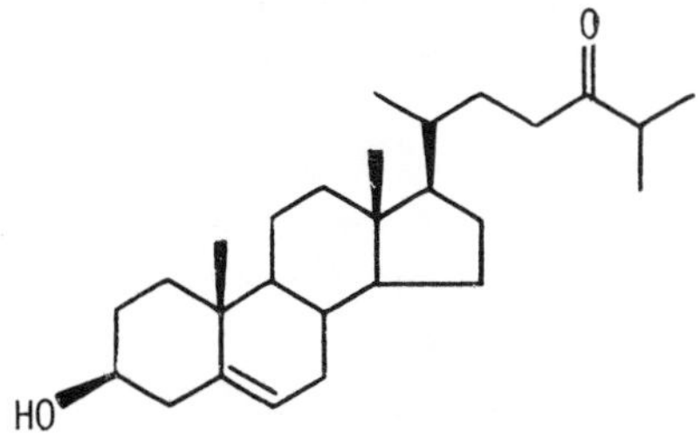

O

HO

(306) 24-oxocholesterol

brown algae *Ascophyllum nodosum* and *Laminaria saccharina.*[240] 24-Oxocholesterol (306) had first reported in an extract of dried *A. nodosum* and was considered likely to be an artifact produced by aerial oxidation of fucosterol (I/409). However, the two separate reports of the occurrence of 24-oxocholesterol (306) in three different species of freshly collected brown algae indicate that the compound (306) is a true algal metabolite.[240]

Safe et al.[240] also reported minor traces of fucosterol (I/409) in extracts of the red alga *Furcellaria fastigiata,* in which cholesterol (I/380) represented more than 99% of the sterol fraction. Other authors have reported cholesterol content at a much lower level in the same red alga, but significant seasonal variation in sterol content has been shown in other species of red algae[240] and such variations may, in the future, prove valuable indicators of the role of certain sterols in the life cycle of the algae.

From the red algae *Meristotheca papulosa*[143] and *Gracilaria textorii,*[144] the keto-steroid cholest-4-en-3-one (304), was isolated.

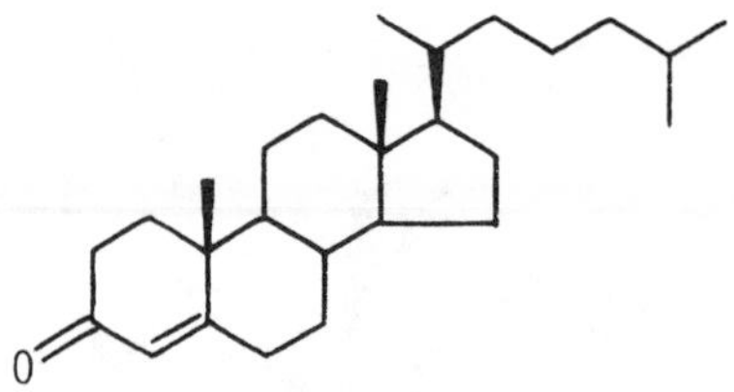

(304) cholest-4-en-3-one

Using cholesterol-4-^{14}C, Kanazawa and Yoshioka[143] ascertained that the cholest-4-en-3-one (304) was formed as a metabolite of cholesterol (I/380) and suggested that the oxidation takes place because of the absence of steroid C20—C22 lyase in the red alga. It is known that the Δ^4-3-keto-steroids occurwidely in the animal kingdom and play an important function as sexual and adrenocortical hormones. The role of such compounds in the plant kingdom is not as well studied, and the question of whether cholest-4-en-3-one (304) possesses a physiological function in red algae has not been reported.

In a separate study Kanazawa and co-workers reported that the red algae *Meristotheca papulosa* and *Gracilaria textorii* contained only cholesterol (I/380) whereas *Porphyridium cruentum,* which is a unicellular red alga, can produce sterols when cultured on a chemically defined medium. However the previous report that *P. cruentum* does not produce sterols in the natural environment has not been proven incorrect, and the question requires further study.[145]

The red alga *Rhodymenia palmata* was found[84] to contain several known sterols (or sterol precursors) and also the novel 31-norcycloartanol (350) and cycloartanol (357) which were concurrently reported in the starfish *Asterias rubens* by Goad and co-workers[249] who suggested that these sterols (350) and (357) were probably derived from a dietary source.

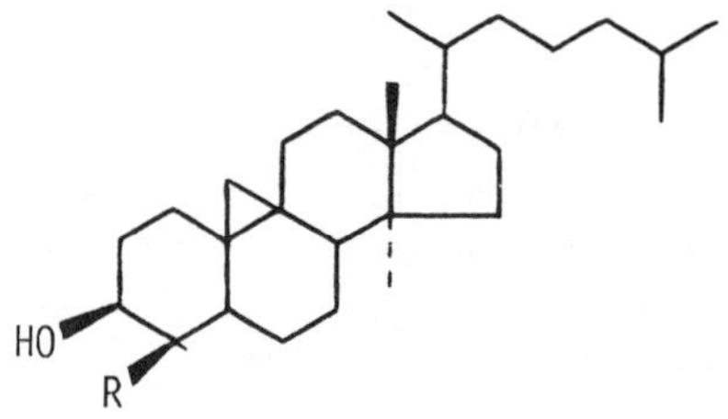

HO
R

(350) R = H 31-norcycloartanol
(357) R = Me cycloartanol

A different sterol containing a cyclopropane ring system was isolated from a different species of starfish, the *Acanthaster planci.*[148] This was the novel C_{30} sterol gorgostanol (358).

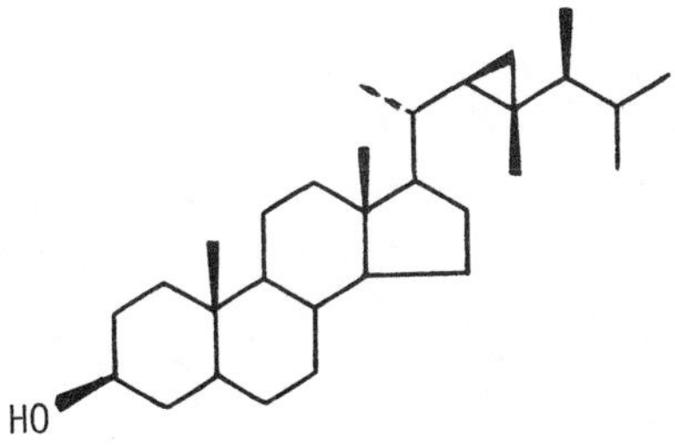

HO

(358) gorgostanol

On coral reefs *A. planci* feeds on the corals which contain gorgosterol (I/437) having the same carbon skeleton as gorgastanol, but possessing the Δ^5 double bond.

Bergmann and co-workers[297,298] had demonstrated that marine invertebrates in many cases contain complex sterol mixtures consisting of C_{27}, C_{28}, and C_{29} sterols of varying degrees of unsaturation. Their studies suggested that echinoderms can be divided into two groups on the basis of the types of sterols they contain:

1. Crinoidea, Ophiuroidea, and Echinoidea contain sterols with a Δ^5 double bond.
2. Asteroidea and Holothuroidea contain sterols with a Δ^7 double bond.

Bergmann[298] further suggested that the sterols found in all echinoderms were possibly derived from the diet and that the Asteroidea and the Holothuroidea may be able to modify Δ^5 sterols and give predominantly the Δ^7 sterols.[98] The work of Goad and co-workers which resulted in the isolation and identification of 22 sterols from the starfish *Asterias rubens* supports this postulate,[249] as did the studies of Kanazawa and co-workers on the starfish *Leiaster leachii*[146,261] and *Coscinasterias acutispina*,[146] and the studies of Kobayashi and Mitsuhashi on *Asterias amurensis* and several other asteroids.[171]

However, cholesteryl sulfate (311) was reported as

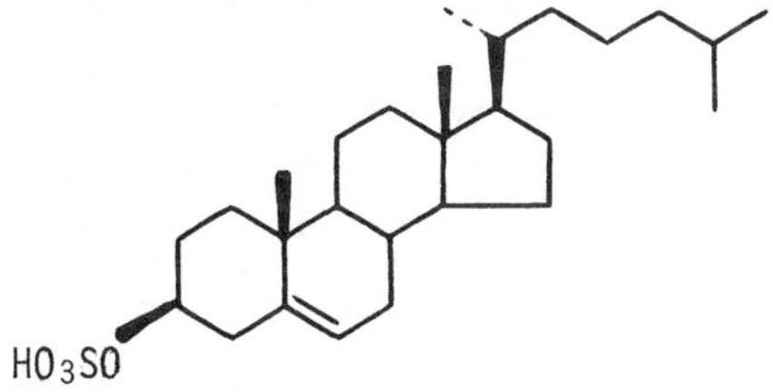

HO_3SO

(311) cholesteryl sulfate

the principal sterol of *Asterias rubens* in the study by Björkman and co-workers.[20] This study looked specifically at membrane lipids, and no sterol derivatives other than that of cholesterol were reported. Analysis of the saponified extracts of the whole starfish *A. rubens* gave cholesterol (I/380) as the only Δ^5 sterol in the 22 sterols isolated by Smith et al.[249] Cholesteryl sulfate (311) has also been isolated from the eggs of the sea urchin *Anthocidaris crassispina*.[282]

Asterias amurensis, which initially yielded the sterol amuresterol (302),[171] also yielded the asterosaponins A and B, both saponins containing the same aglycone, which has been characterized as (314),[142] the saponins differing on the basis of the number of molecules of sugars attached.

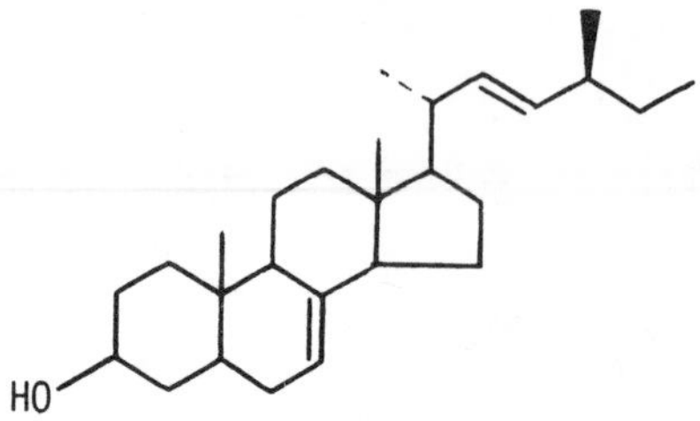

(302) amuresterol

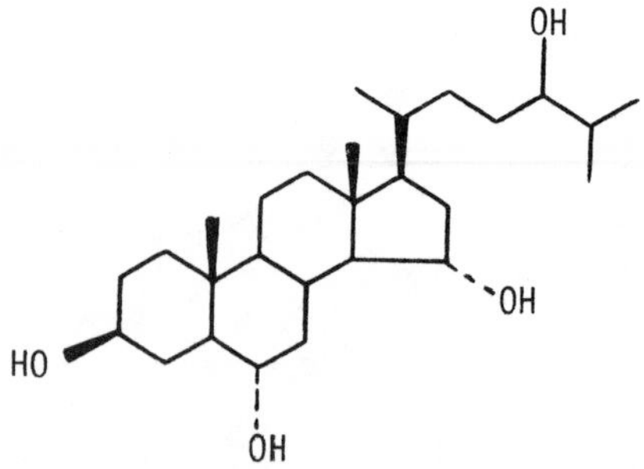

(314)

Ovaries of the starfish *Pisaster ochraceous* were extracted by acetone to give after purification two fractions possessing estrogenic activity.[31] One fraction gave rise to material which suggested the presence of progesterone (230) and the other fraction gave a substance characterized as β-estradiol (183). Estrone (182) was not detected.

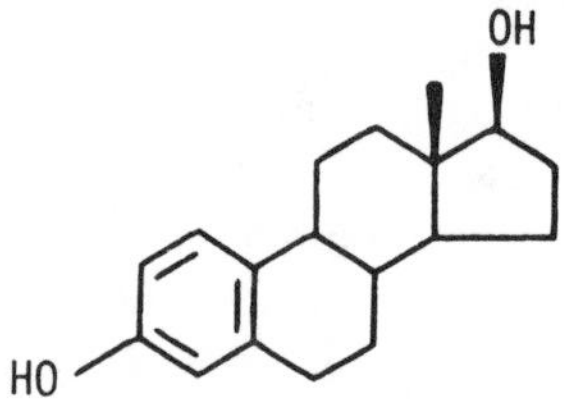

(183) β-estradiol

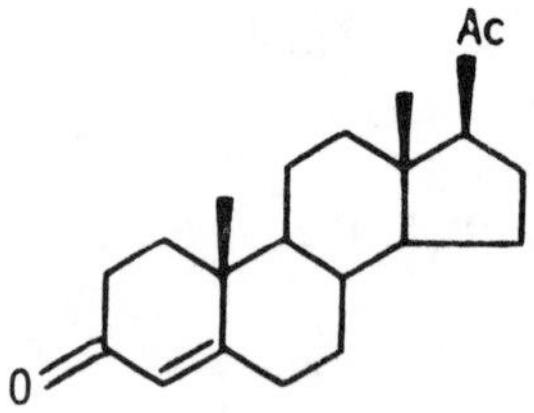

(230) progesterone

The same authors who worked on *P. ochraceous*[31] studied the estrogenic steroids of the sea urchin *Strongylocentrotus franciscanus*,[32] and confirmed the presence of β-estradiol (183) while tentatively assigning a minor component as progesterone (230). β-Estradiol (183), estrone (182), and progesterone (230) have been detected in the ovaries of the dogfish *Squalus suckleyi*.[273]

Holothurians have been noted for the presence of antifungal and toxic glycosides in some species. For *Stichopus japonicus* the antifungal glycoside has been characterized as (397)[167] and the aglycone as (354),[168] the structure being corrected from that shown in Volume I[11] as (I/431).

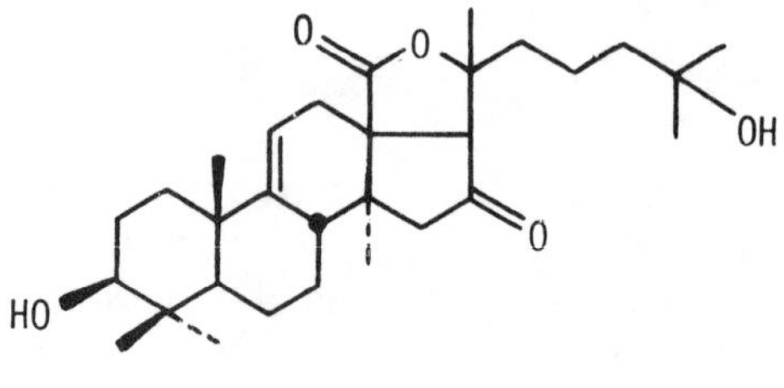

(354) stichopogenin A_4

Sponges are believed to contain the greatest diversity of sterols. In reviewing the sterols from sponges Minale and co-workers[68] noted that all sponge sterols reported to 1972 contain the conventional tetracyclic nucleus with a C_8 chain at C17, which may be modified by the addition of one or two carbon atoms at C24. All other differences could be accounted for by nuclear and/or side-chain unsaturation and/or differences in stereochemistry at C20 and C24.

During 1972 de Luca et al.[299] reported that the sponge *Verongia aerophoba* contained two new sterols, aplysterol (I/416) and 24,28-didehydroaplysterol (I/407), which showed not only the normal alkylation at C24 but also an additional methyl substitution at C26. Minale and co-workers[68] believe that this unusual type of substitution is confined to sterols of the family Verongidae.

As with other species of marine organisms, the authors have noted that in the sponges, the data on sterol content are too limited and fragmented to permit their correlation for taxonomic purposes. Improved separation techniques and a better understanding of symbiotic relationships in sponges will almost certainly lead to further attempts to develop a system of chemotaxonomy based on sponge sterols.

Studies on sponges of the genus *Axinella* have led to the characterization of sterols with a modified tetracyclic ring system in *Axinella verrucosa*.[205] These unique stanols contain a 3β-hydroxymethyl-A-nor-5α-cholestane nucleus, typified by (307).

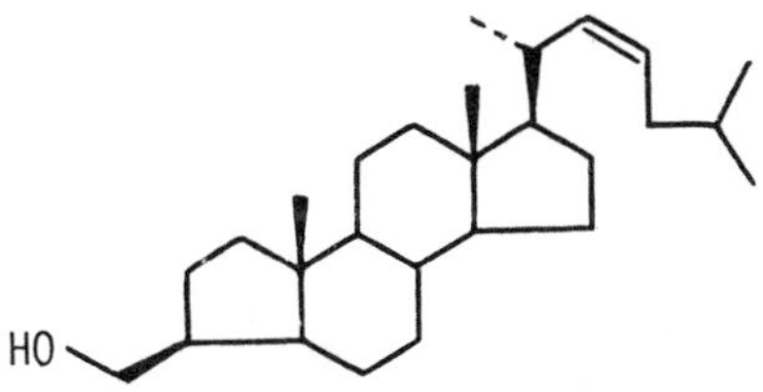

HO

(307)

Other stanols isolated from this sponge are (312), (329), (336), (349), and (352).

Not all species of *Axinella* give sterols with this modified ring system. *Axinella cannabina* gave rise to the 5,8-peroxides (301) and (322).[76]

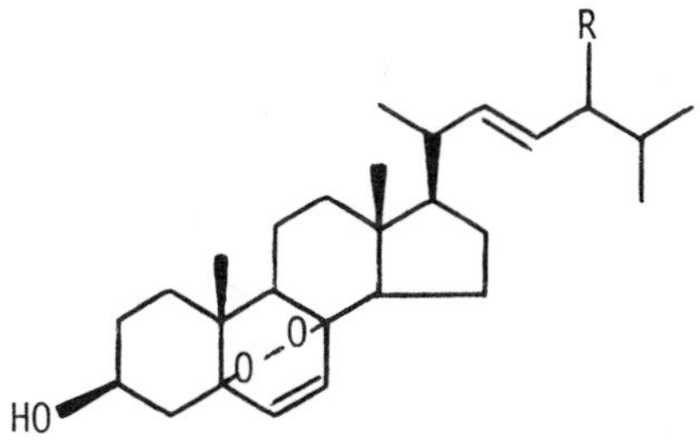

R
O
O
HO

(301) R = H
(322) R = Me

Extracts of *Axinella polypoides* have given rise to a number of 19-nor-stanols, in which the major component is 19-nor-5α,10β-ergost-22-en-3β-ol (309).[204]

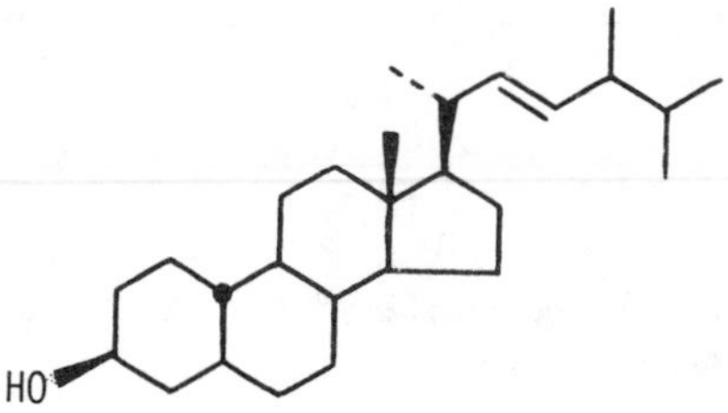

HO

(309)

Other stanols in which the C19 methyl group is absent are (278), (293), (294), (308), (313), (328), and (335).

A new C_{27} sterol has been isolated from an annelid, *Pseudopotamilla occelata,*[172] and assigned the name occelasterol (305), with the C22—C23 double bond being proven to be *trans*. Subsequently, glc studies have shown the sterol (305) to be distributed in various amounts in most of the marine invertebrates.

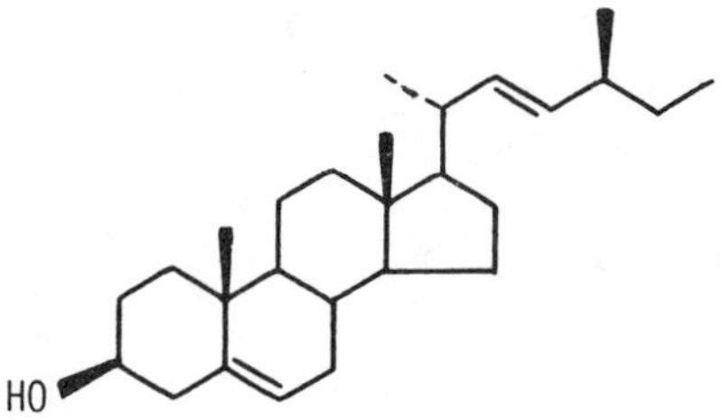

HO

(305) occelasterol

Molluscs have been extensively examined for sterol content. A new sterol, 22-*trans*-24-norcholesta-5,22-dien-3β-ol(I/360), featuring a hitherto uncharacterized C_7 side chain first isolated from the scallop *Placopecten magellanicus,* is now believed, on the basis of gas chromatographic studies, to be a component of many marine sterol mixtures.[127]

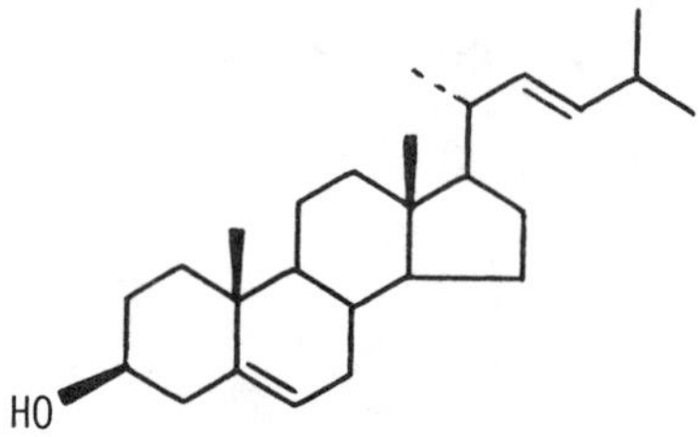

HO

(I/360)

Sterol biosynthesis in molluscs and in echinoderms has been undertaken,[75] and it appears that the evolutionary trend is towards cholesterol (I/380), which becomes the major sterol in most advanced forms.

Table 14

Ring system	Cpd. no.	C_2	C_3	C_5	C_6	C_7	C_9	C_{11}	C_{12}	C_{14}	C_{15}	C_{16}	C_{17}*	Source	References
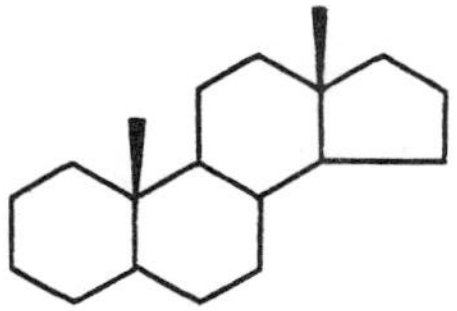	268		α-OH			α-OH			α-OH				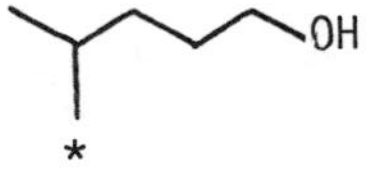 * OH	Fish	258
	289		α-OH			α-OH			α-OH				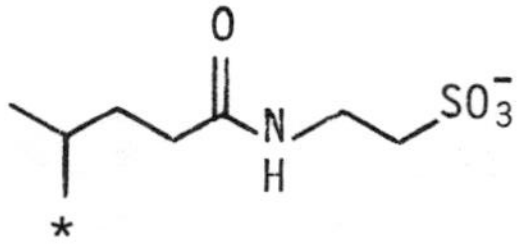 * O N H SO_3^-	Fish	258
	314		β-OH		α-OH						α-OH		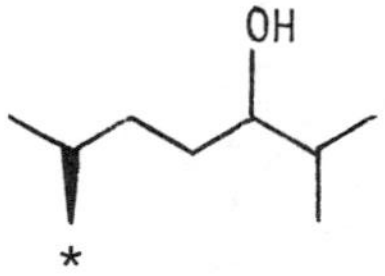 * OH	Starfish	142
	315		α-OH			α-OH			α-OH				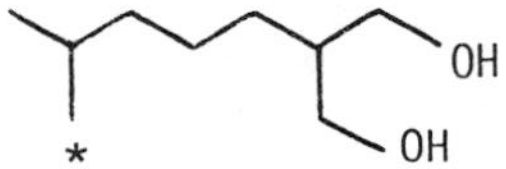 * OH OH	Fish	258
	317		β-OH			α-OH			α-OH				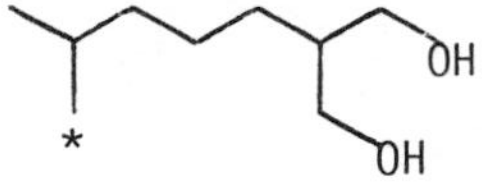 * OH OH	Fish	258

Table 14 (continued)

Ring system	Cpd. no.	C_2	C_3	C_5	C_6	C_7	C_9	C_{11}	C_{12}	C_{14}	C_{15}	C_{16}	C_{17}*	Source	References
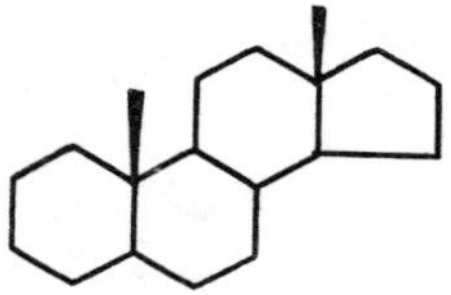	355	α-OAc	α-OH					β-OH					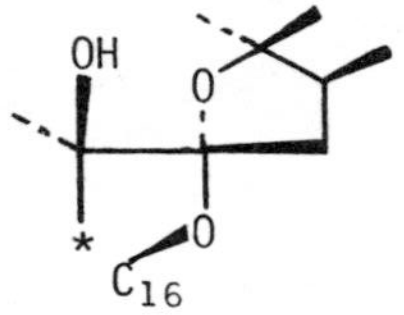	Gorgonian	162
	358		β-OH										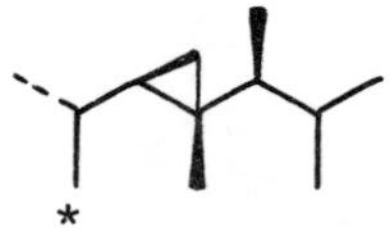	Starfish	148
	360		β-OH	α-OH	β-OH								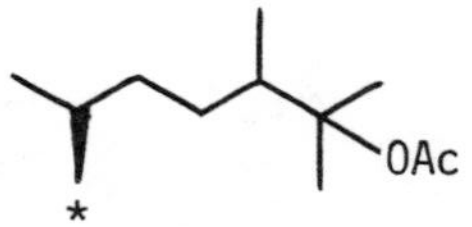	Soft coral	209

Table 15

Ring system	Cpd. no.	C_2	C_3	C_6	C_7	C_9	C_{11}	C_{12}	C_{14}	C_{16}	C_{17}*	Source	References
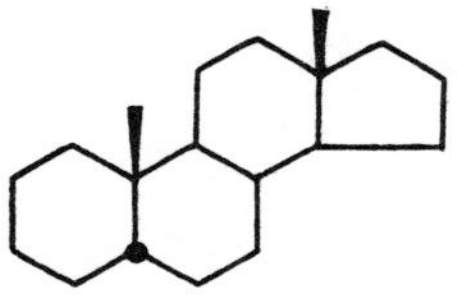	286		α-OH		α-OH						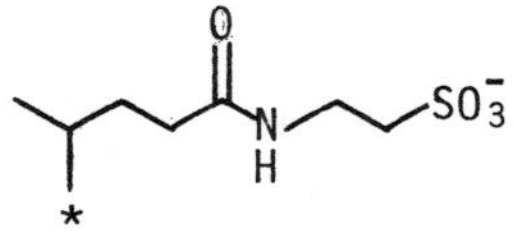	Fish bile	258
	287		α-OH					α-OH			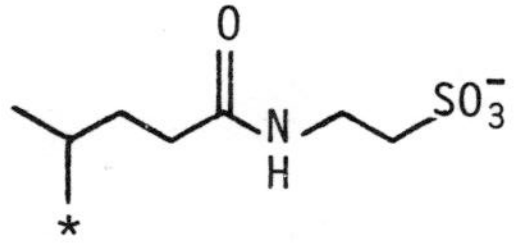	Fish	258
	288		α-OH		β-OH						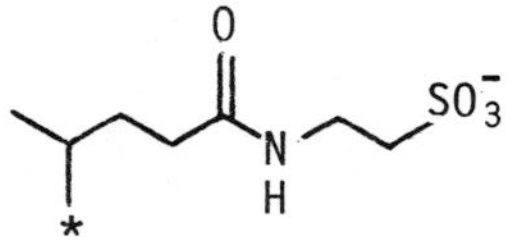	Fish	258
	290		α-OH		α-OH			α-OH			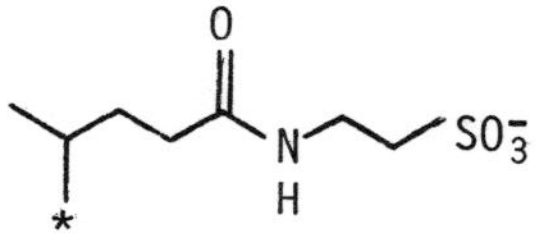	Fish bile	258
	291		α-OH		α-OH						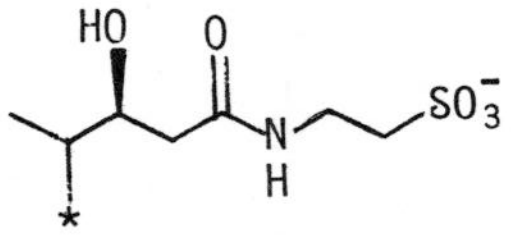 HO	Fish	258

Table 15 (continued)

Ring system	Cpd. no.	C_2	C_3	C_6	C_7	C_9	C_{11}	C_{12}	C_{14}	C_{16}	C_{17}*	Source	References
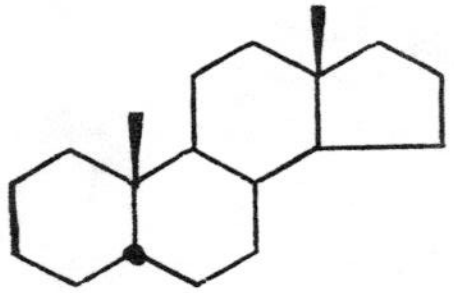	292		α-OH	β-OH			α-OH				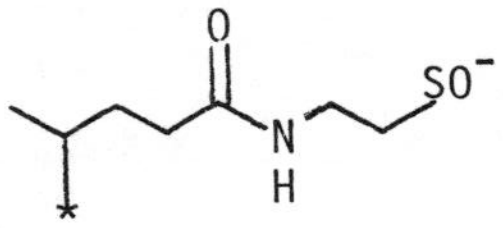	Eel Fish	258
	316		α-OH	α-OH			α-OH				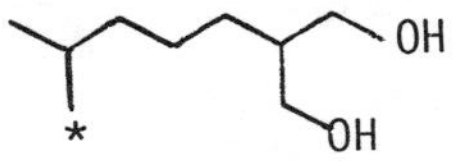	Eels	258

Table 16

Ring system	Cpd. no.	C_2	C_3	C_6	C_7	C_9	C_{11}	C_{12}	C_{14}	C_{16}	C_{17}*	Source	References
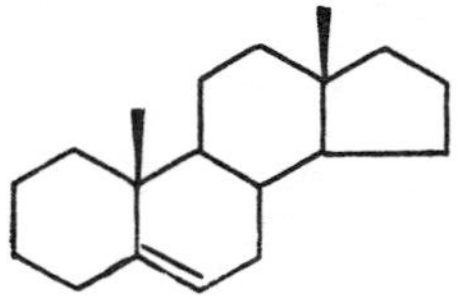	285		β-OH								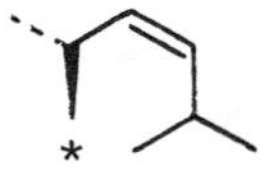	Sponges	68
	305		β-OH								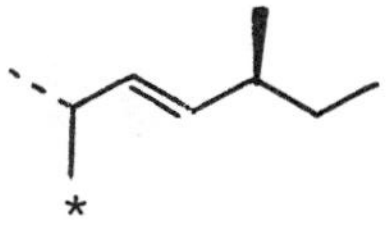	Crustacean Molluscs Echinoderms Tunicate	172
	306		β-OH								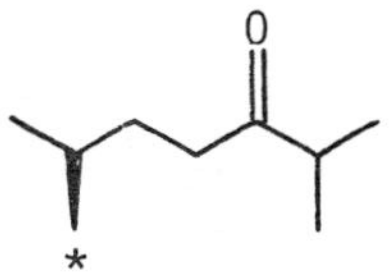	Brown algae	216 240
	311		β-OSO_2OH								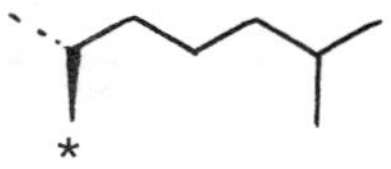	Sea urchin Starfish	20 282
	323		β-OH								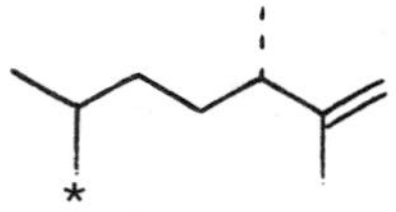	Green alga	237

Table 16 (continued)

Ring system	Cpd. no.	C_2	C_3	C_6	C_7	C_9	C_{11}	C_{12}	C_{14}	C_{16}	C_{17}*	Source	References
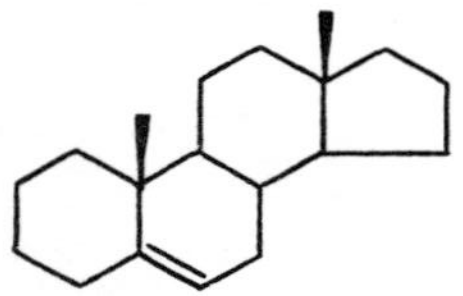	325		β-OH								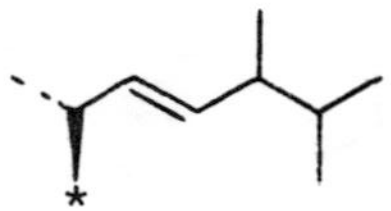	Brown alga	250
	326		β-OH								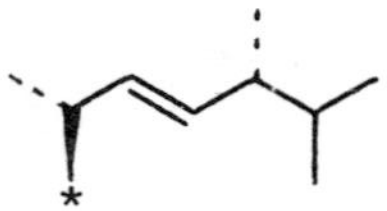	Brittle star Diatom	238
	331		β-OH								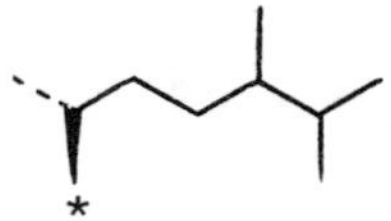	Brown alga Gorgonians Jellyfish Sea anemone Sea urchins Sea worm	22, 169, 172, 250, 275, 276
	339		β-OH								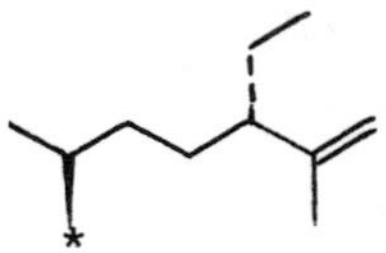	Green alga	237

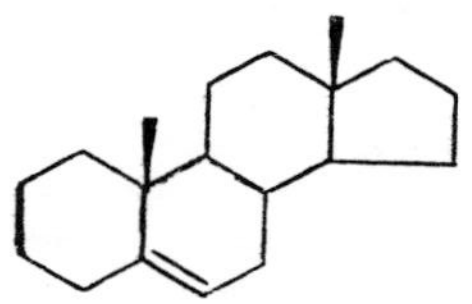

	342	β-OH	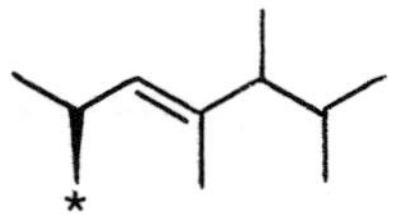	Soft coral	147
	343	β-OH	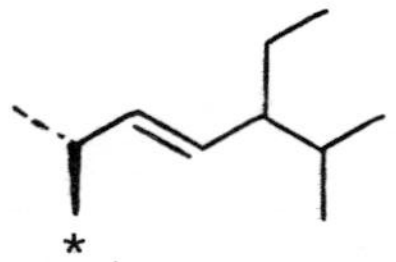	Brown alga Sea worm	169, 172, 250
	348	β-OH	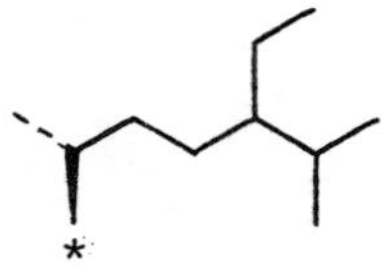	Brown alga Jellyfish Sea anemone Sea urchins Sea worm	169, 172, 250, 275, 276
	356	β-OH	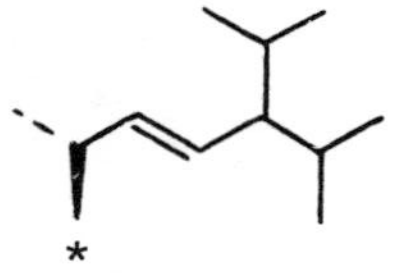	Sponge	120
	359	β-OH	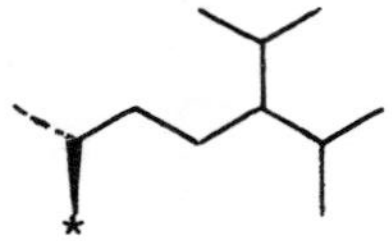	Sponge	120

Table 17

Ring system	Cpd. no.	C_2	C_3	C_6	C_7	C_9	C_{11}	C_{12}	C_{14}	C_{16}	C_{17}*	Source	References
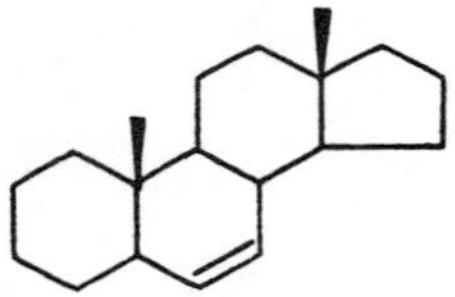	301		β-OH	(C_{5-8} peroxy linkage)							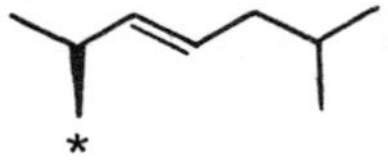	Sponge	76
	322		β-OH	(C_{5-8} peroxy linkage)							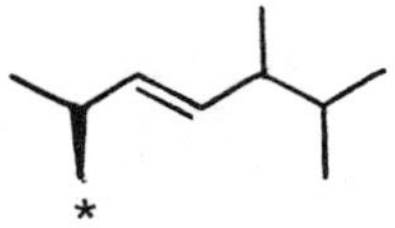	Sponge	13, 76

Table 18

Ring system	Cpd. no.	C_2	C_3	C_4	C_6	C_7	C_9	C_{11}	C_{12}	C_{14}	C_{16}	C_{17}*	Source	References
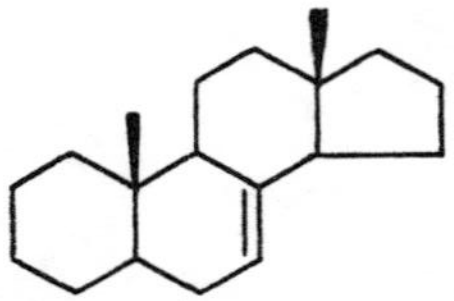	302		OH									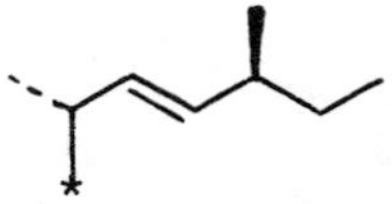	Starfish	171
	303		β-OH									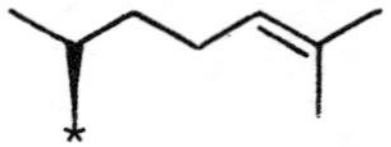	Starfish	249
	321		β-OH									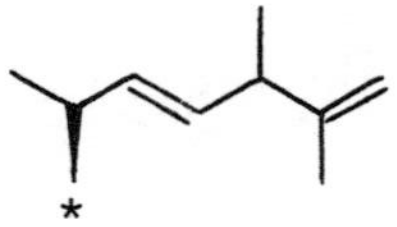	Starfish	261

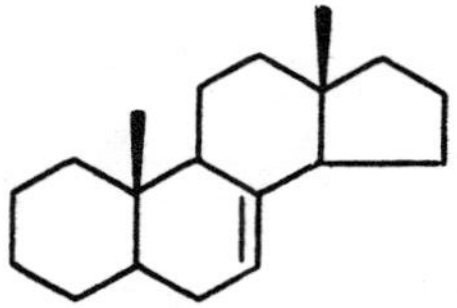 324	β-OH	α-Me	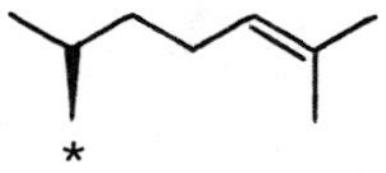	Starfish	249
327	β-OH		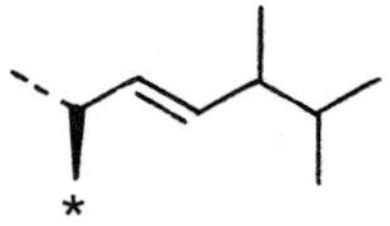	Starfish	261
330	β-OH	α-Me	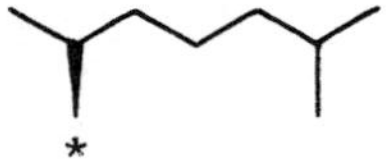	Starfish	249
332	β-OH		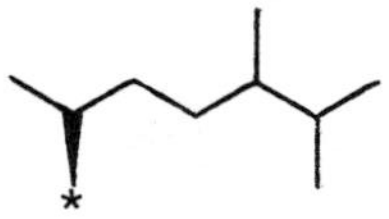	Starfish	146,249
344	β-OH		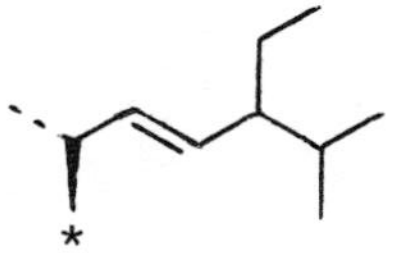	Starfish	146,249,261
345	β-OH		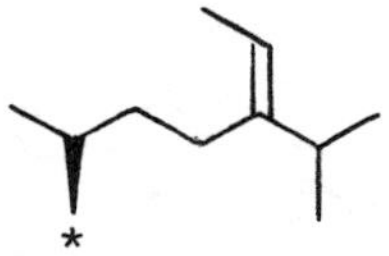	Starfish	146,249
346	β-OH		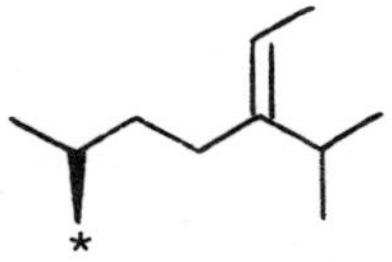	Starfish	249

Table 19

Ring system	Cpd. no.	C_2	C_3	C_6	C_7	C_9	C_{11}	C_{12}	C_{14}	C_{16}	C_{17}*	Source	References
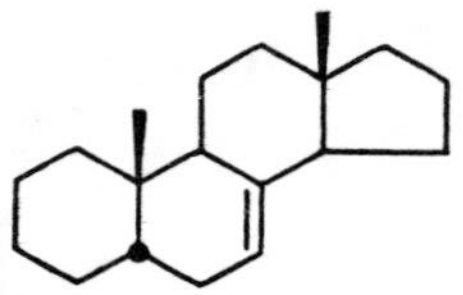	284		α-OH					α-OH			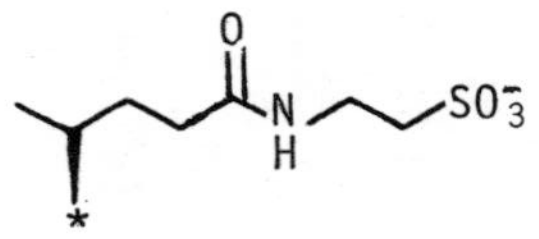	Fish	258

Table 20

Ring system	Cpd. no.	C_2	C_3	C_6	C_7	C_9	C_{11}	C_{12}	C_{14}	C_{16}	C_{17}*	Source	References
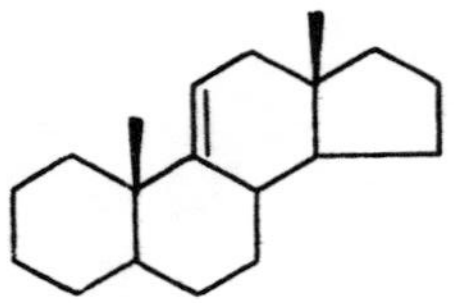	366		β-OH	α-O-$C_6H_{11}O_5$							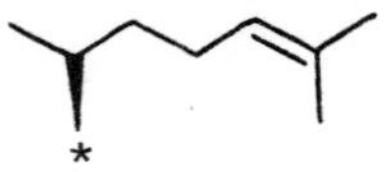	Starfish	265

Table 21

Ring system	Cpd. no.	C_2	C_3	C_6	C_7	C_9	C_{11}	C_{12}	C_{14}	C_{16}	C_{17}*	Source	References
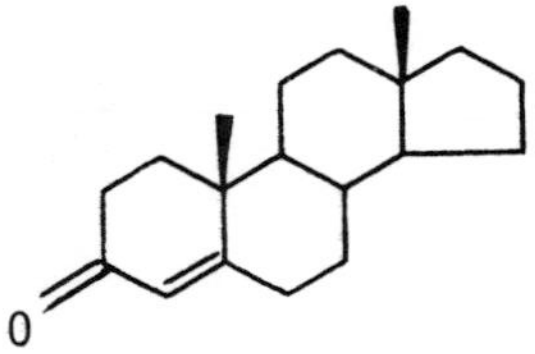 O	230										 Ac	Dogfish Mollusc Sea urchin Starfish	31 32 202 273
	240										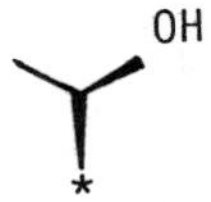 OH	Mollusc Sea urchin	32
	304										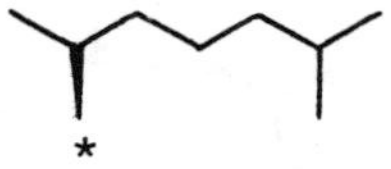	Red algae	143, 144

Table 22

Ring system	Cpd. no.	C_2	C_3	C_6	C_7	C_9	C_{11}	C_{12}	C_{14}	C_{16}	C_{17}*	Source	References
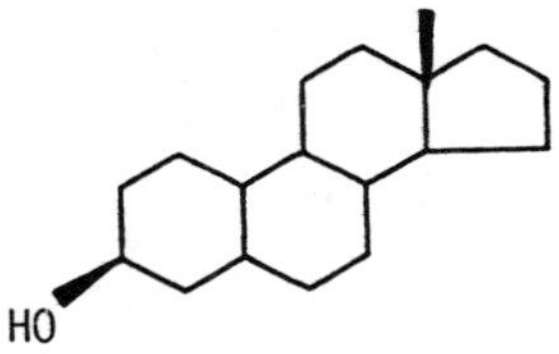 HO	278		β-OH								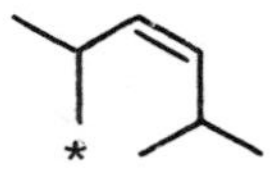	Sponge	204
	293		β-OH								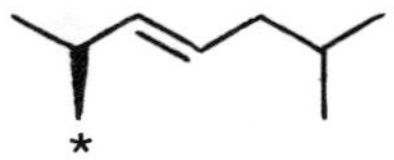	Sponge	204

Table 22 (continued)

Ring system	Cpd. no.	C_2	C_3	C_6	C_7	C_9	C_{11}	C_{12}	C_{14}	C_{16}	C_{17}*	Source	References
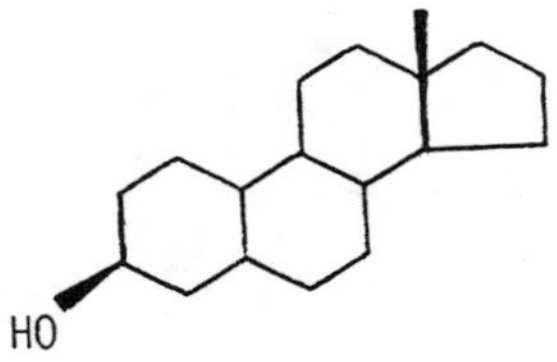 HO	294		β-OH								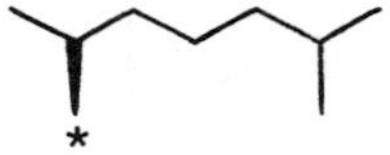	Sponge	204
	308		β-OH								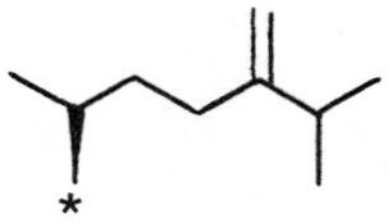	Sponge	204
	313		β-OH								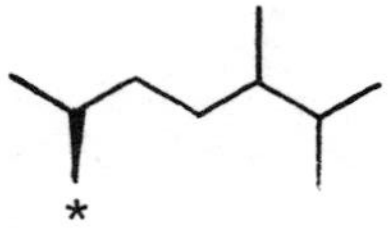	Sponge	204
	328		β-OH								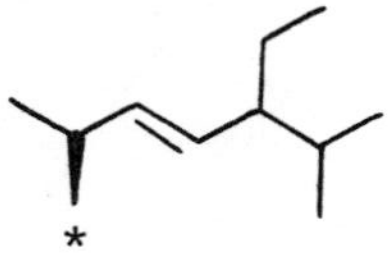	Sponge	204
	335		β-OH								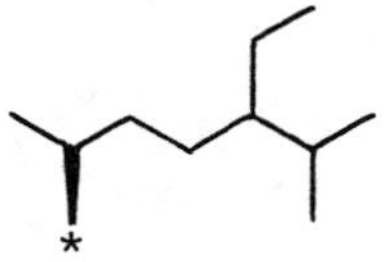	Sponge	204

Table 23

Ring system	Cpd. no.	C_2	C_3	C_6	C_7	C_9	C_{11}	C_{12}	C_{14}	C_{16}	C_{17}*	Source	References
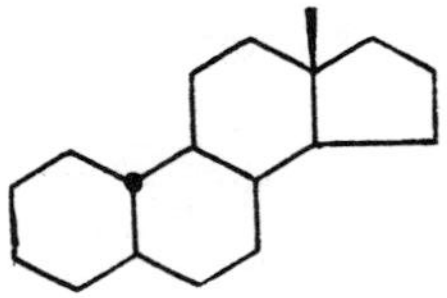	309		β-OH								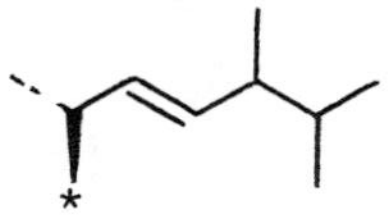	Sponge	204

Table 24

Ring system	Cpd. no.	C_2	C_3	C_6	C_7	C_9	C_{11}	C_{12}	C_{14}	C_{16}	C_{17}*	Source	References
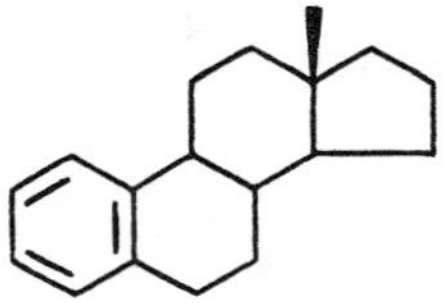	182		OH								 O	Dogfish Mollusc	32, 202 273
	183		OH								 OH	Dogfish Mollusc Sea urchin Starfish	31 32 202 273

Table 25

Ring system	Cpd. no.	C_2	C_3	C_4	C_6	C_7	C_9	C_{11}	C_{12}	C_{14}	C_{16}	C_{17}*	Source	References
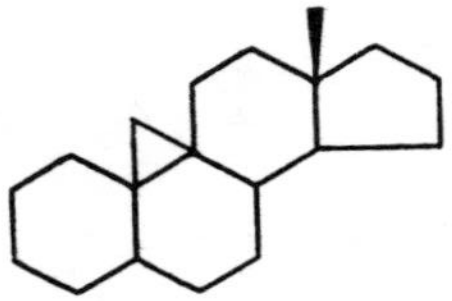	350		β-OH	α-Me						α-Me		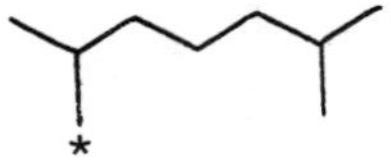	Red alga Starfish	84 249

Table 26

Ring system	Cpd. no.	C_2	C_3	C_6	C_7	C_9	C_{11}	C_{12}	C_{14}	C_{16}	C_{17}*	Source	References
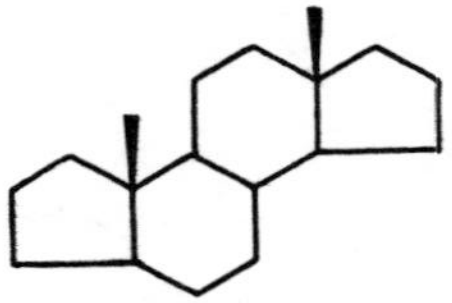	307		β-CH_2OH								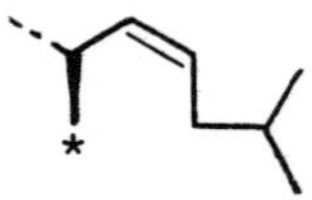	Sponge	205
	312		β-CH_2OH								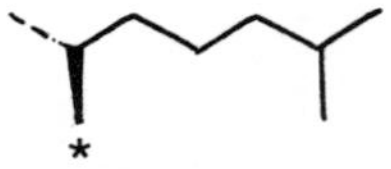	Sponge	205
	329		β-CH_2OH								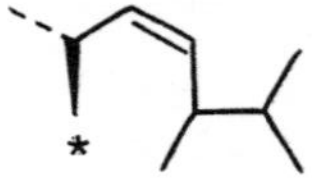	Sponge	205

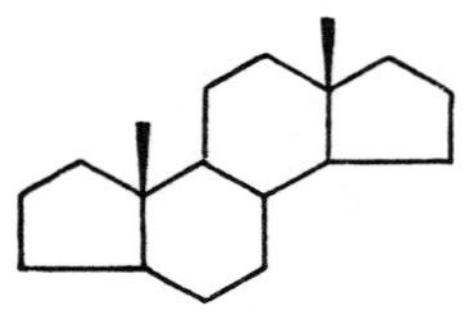

336	β-CH_2OH	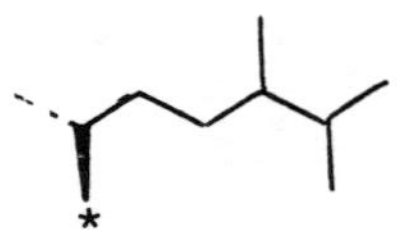 Sponge	205
349	β-CH_2OH	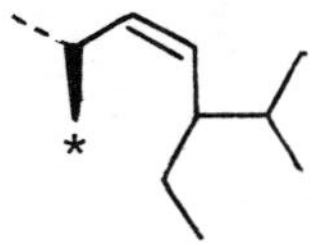 Sponge	205
352	β-CH_2OH	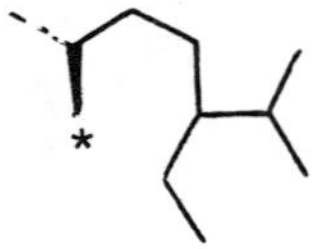 Sponge	205

Compounds Containing Nitrogen

Simple Amines

Trimethylamine (I/20) and trimethylamine oxide (16) are known to be widely distributed in marine fish muscle, with only trace amounts of trimethylamine being present in freshwater fish muscle, in general. It has been suggested that trimethylamine oxide (16) is an end-product of nitrogen metabolism in marine teleosts and elasmobranchs.[92] There were few reports on its occurrence in marine algae prior to the paper by Fujiwara-Araski and Mino, in which simultaneous determination of trimethylamine and trimethylamine oxide was reported for 20 species of marine algae.[92] Two species of green algae, *Enteromorpha linza* and *Ulva reticulata*, showed zero content of trimethylamine oxide and one species of brown alga, *Sargassum thunbergii*, showed a trace amount. In general, relative to the trimethylamine content, trimethylamine oxide concentraction was low in the brown algae, intermediate in the green algae, and high in the red algae. The authors suggest that, as is the case for marine fishes, trimethylamine oxide may be an end product in the metabolism of trimethylamine in algae.

The study must be regarded as preliminary in the field of nitrogen metabolism in marine algae.

Harada and Yamada[116] studied the concentration of trimethylamine oxide in 62 species of fish and made some preliminary correlations between levels of trimethylamine oxide in different groups of fishes. Their results confirm earlier reports that in freshwater fishes the trimethylamine oxide level in muscle was almost undetectable.

The amine tyramine (59) has been isolated from the posterior salivary gland of the blue-ringed octopus *Hapalochlaena maculosa*,[125,202] and the corresponding trimethyl ammonium salt candicine chloride (113) from the mollusc *Turbo argyrostoma*.[279] This mollusc also contains toxic (cholinomimetic) amines (52)[278] and (64).[277,281] Octopamine (I/112) has been isolated from several species of octopus.

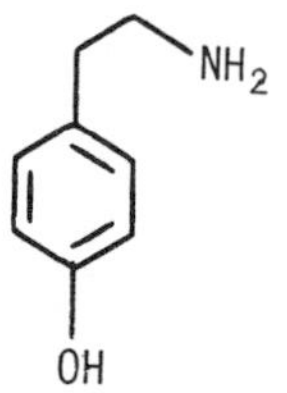

NH_2 — OH — $\overset{+}{N}Me_3$ Cl^- — OH — HO — NH_2 — OH

(59) tyramine 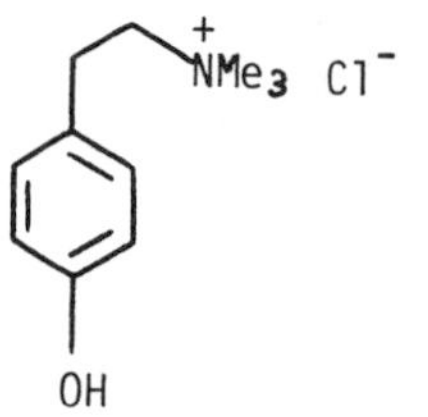 (113) candicine chloride 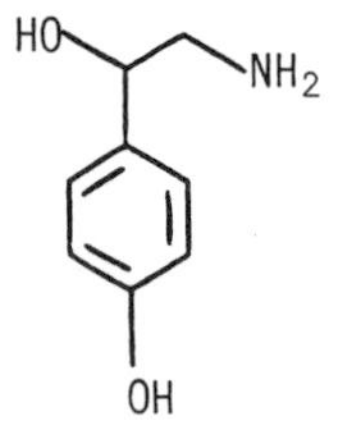 (I/112) octopamine

Octopamine has also been detected in individual neurons of the mollusc *Aplysia californica*, these nerve cells being devoid of dopamine and norepinephrine, and it has been suggested that octopamine may function as a neurotransmitter in *Aplysia*.[239]

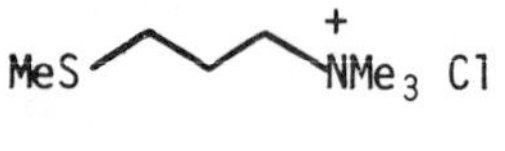

MeS — $\overset{+}{N}Me_3$ Cl^- — $Me_2\overset{+}{S}$ — $\overset{+}{N}Me_3$ $2Cl^-$

(52) [3-(Methylthio) propyl] trimethylammonium chloride

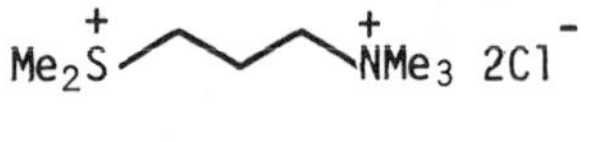

(64) [3-(Dimethylsulfonio) propyl]-trimethylammonium dichloride

Further studies are proceeding on these two toxins.

Imidazole Derivatives

Apart from the simple acyclic amines, there are numerous examples of imidazole derivatives from different marine species.

imidazole

2-Aminoimidazole (14) has been reported from the sponge *Reniera cratera*.[46]

(14) 2-aminoimidazole

No definite conclusion has been reached concerning the biosynthesis of this amine, but the authors noted its isolation from a bacterial culture in which the accumulation of 2-aminoimidazole was enhanced by the addition of arginine.

The 2-aminoimidazole moiety and its methylated derivatives appear in the fluorescent tetrazacyclopentazulene pigments termed the zoanthoxanthins (104, 117, 121, 125) which were isolated from the Mediterranean zoanthid *Epizoanthus arenaceus* by Prota and co-workers.[40,41,42] Two of these zoanthoxanthins (121) and (125) have the same skeleton as the previously described parazoanthoxanthin D (I/202)[300,301] and the other two (104) and (117) represent an isomeric tetrazacyclopentazulene system, as illustrated below.

(121) epizoanthoxanthin A

(104) 3-norpseudozoanthoxanthin

Histamine (I/53) and other 4-substituted imidazoles are also widely distributed in marine organisms.[203] In a study by Das et al.[66] the sponge *Suberites inconstans* was found to contain histamine and lesser amounts of five other compounds which were not characterized, but which behaved like histamine when tested on the guinea-pig ileum. This sponge *S. inconstans* was selected because it produced itching and slight swelling of the fingers when handled.

(I/53) histamine

N-Urocanylhistamine (105)[236] and *N*(4-imidazolepropionyl)histamine (106)[235] have been found in the soft tissues of the gastropod mollusc *Drupa concatenata* Lam.

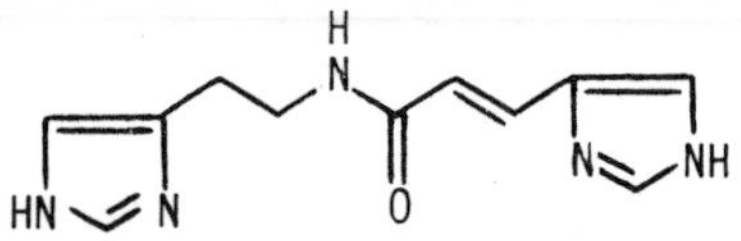

(105) *N*-urocanylhistamine

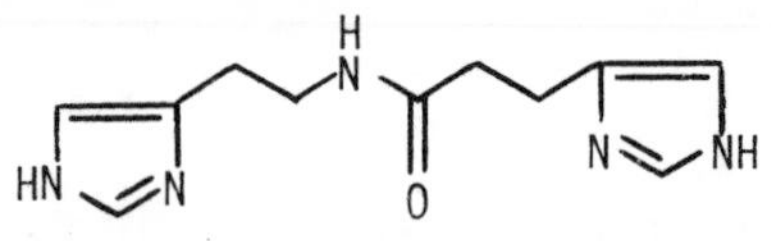

(106) *N*(4-imidazolepropionyl)histamine

These two imidazole derivatives were previously unrecorded from natural sources, and previously, the only known *N*-acylated histamine derivative found in nature was *N*-acetylhistamine.[235]

The studies on *Drupa concatenata* also revealed the presence of choline, imidazolepropionic acid, methyl imidazolepropionate, histamine, and an unspecified histamine derivative, and are being continued to determine if the imidazolepropionylhistamine (106) is used by the carnivorous mollusc in the capture of its prey.[235]

It is interesting to note that carnivorous gastropod molluscs of the family Muricidae contain urocanyl choline (murexine) (I/177), another 4-substituted imidazole,[164] in their hypobranchial glands. *Thais haemastoma* hypobranchial glands have yielded choline (I/71), senecioylcholine (I/158), and dihydromurexine (I/179), as well as murexine.[234] Murexine stimulates autonomic ganglia and blocks neuromuscular transmission. The unsubstituted choline acrylate (I/114) which occurs in the whelk *Buccinum undatum*[302] and in the dinoflagellate *Amphidinium carteri*[260] has little blocking action and primarily stimulates smooth muscle and autonomic ganglia.

Choline Derivatives

Several choline esters have been reported[233,234] in hypobranchial glands of gastropod molluscs, and a new ester *N*-methylmurexine (118) has been isolated from *Nucella emarginata.*[15] This derivative produced contraction of the frog rectus abdominus muscle, different in character from that produced by acetylcholine, the contraction being much slower.

Glycerylphosphorylcholine (65) has been isolated from the sperm-storing organ, the spermatophore, of the cephalopod mollusc *Octopus dofleini martini,* along with L-carnitine (49),[33] the latter having been previously reported in many classes of marine organisms,[89,202] and also known as Vitamin B_T.

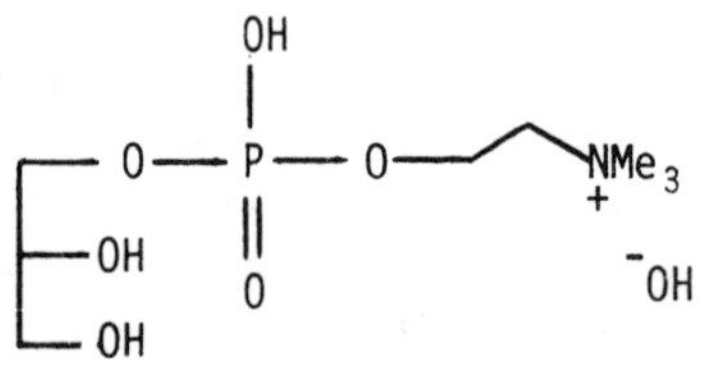

(65) glycerylphosphorylcholine

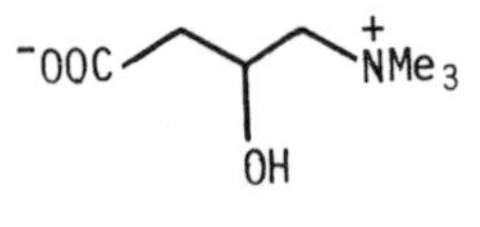

(49) L-carnitine

Catecholamines

Dopamine (60) is recognized as the most important catecholamine in the invertebrate central nervous system. It is widely distributed in marine invertebrates[58,163,202] and there is usually at least 10 times more dopamine than noradrenaline (61) in the invertebrate central nervous system, except for the octopus *Octopus vulgaris,* where the ratio is approximately 3:1::(60):(61).

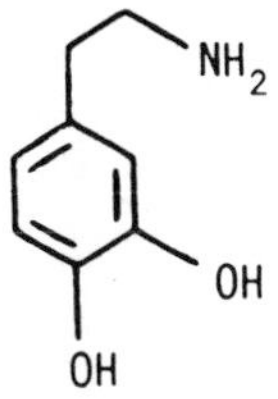

(60) dopamine

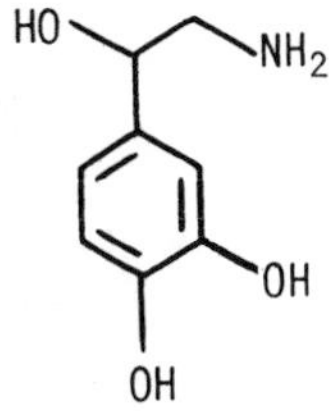

(61) 1-noradrenaline

Amino Acids

Amino acids from marine organisms have been extremely varied in structural type. Studies on Australian algae by Madgwick and Ralph[190] involved 16 brown, 12 red, and 6 green marine algae. In the brown algae glutamic acid (I/52), aspartic acid (I/26) and alanine (I/11) were consistently the most significant representatives. Aspartic acid (I/26) was, in general, more common in the brown algae than in the red algae, and alanine (I/11) tended to be of higher concentration in brown and green algae than in the red algae.

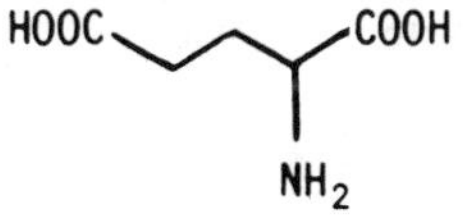

(I/52) glutamic acid 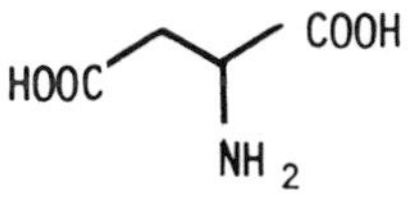 (I/26) aspartic acid 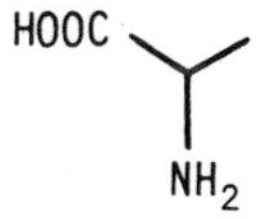 (I/11) alanine

Some relatively rare amino acids were isolated in the survey. These were L-baikiain (I/80) from the red alga *Corallina officinalis,* chondrine (I/50) from the brown alga *Zonaria sinclairii,* 1-methylhistidine (46) from the brown alga *Phyllospora comosa,* and 1,3-dimethylhistidine (I/113)[189,190] from the red alga *Gracilaria secundata.*

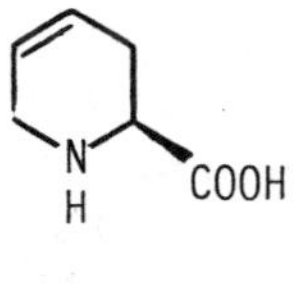

(I/80) L-baikiain

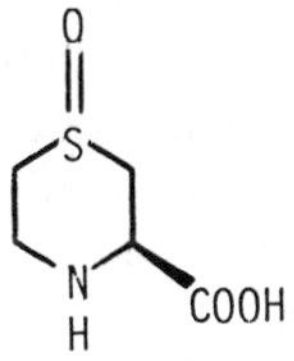

(I/50) chondrine

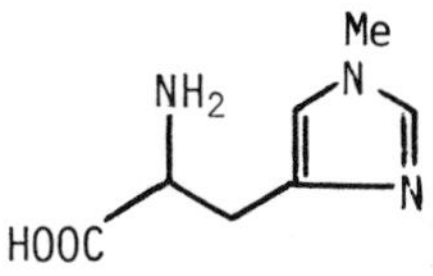

(46) 1-methylhistidine

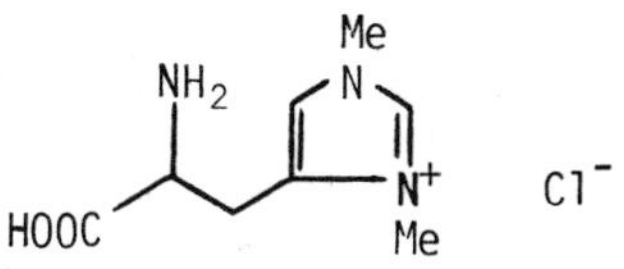

(I/113) 1,3-dimethylhistidine (chloride)

Porphyra umbilicalis is one of the species of "nori", the edible seaweed extensively used in its natural state. Its amino acid content was principally alanine (I/11), glutamic acid (I/52) and glycine (I/3).[186]

Some sulfur-containing amino acids have been isolated from red algae, e.g., (−)*S*-hydroxymethyl-L-homocysteine (25) from *Chondrus ocellatus*[257] and L-methionine-1-sulfoxide (26) and *N*-methylmethionine sulfoxide (34) from *Grateloupia turuturu.*[208]

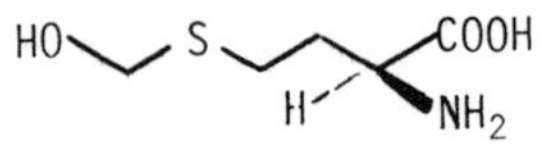

(25) (−)S-hydroxymethyl-L-homocysteine

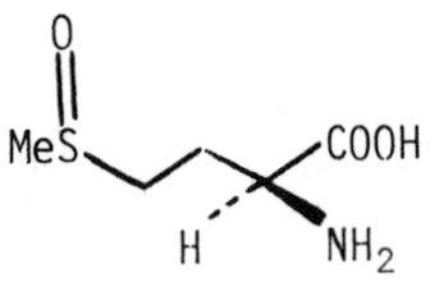

(26) L-methionine-1-sulfoxide

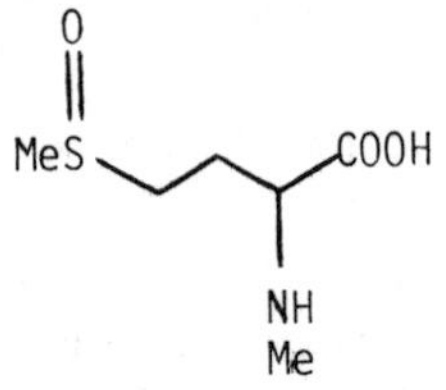

(34) *N*-methylmethionine sulfoxide

Betaines have been isolated from a green alga and from the ovary of a shellfish. Homoserine betaine (50) was found in the hydrolyzate of an uncharacterized base from the green alga *Monostroma nitidum*[2] and by ion-exchange chromatography followed by paper chromatography from the ovary of the mollusc *Callista brevisiphonata*.[280] This latter procedure also yielded L-valine betaine (63) from the mollusc.

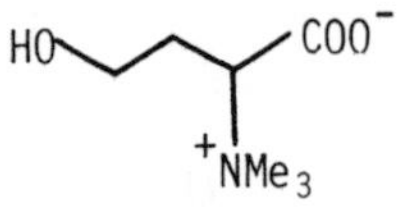

(50) homoserine betaine

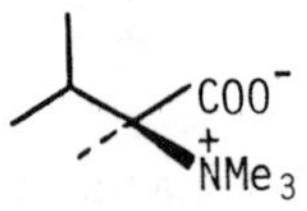

(63) L-valine betaine

The ovary of the mollusc *C. brevisiphonata* had been reported to be poisonous when ingested during the spawning season. The betaines were isolated while seeking the choline ester involved, and no report on pharmacological activities was given. Homoserine betaine (50) was reported to occur in the turban shell *Turbo argyrostoma*, and suggested as likely to occur in other marine organisms.[280]

Suyama and Yoshizawa have observed that the usual cation-exchange column chromatographic method used for separation of amino acids led to unreliable results for free histidine and lysine in the muscles of fish.[254] In their modified separation techniques on 13 species of migratory fish, a high level of the amino acid histidine (I/81) was confirmed in fish muscle, as well as appreciable levels of carnosine (76) and anserine (98).[254]

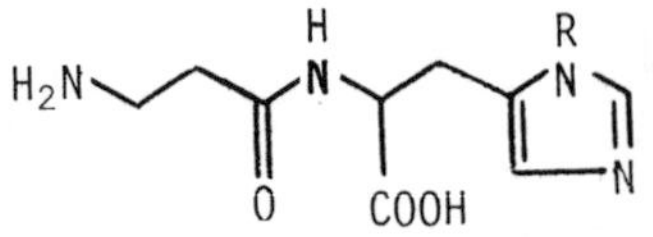

(76)	carnosine	R = H
(98)	anserine	R = Me

Extensive studies on nitrogenous constituents in eight species of fish have come from Shimizu and co-workers.[175] Free amino acid nitrogen varied between 7 to 45% of the total extractive nitrogen and they analyzed for the distribution of nearly 90% of the extractive nitrogen. Taurine (I/8) was the only amino acid found at a fairly high level throughout the species investigated — red sea bream, stone flounder, flathead flounder, flounder, puffer, angler, common mackerel, and jack mackerel.

Glycine (I/3) and glutamic acid (I/52) have been found in the nervous system of two fish species and in some invertebrates. High glycine content was detected in the nervous system of echinoderms and arthropods and low glycine content in that of molluscs, annelids, and the fish species. Glutamic acid was present in low concentration in the nervous systems of echinoderms, moderate concentration in those of molluscs and annelids, and high concentration in those of arthropods and the fish species.[227]

3-Hydroxy-L-kynurenine (81), as well as 15 common amino acids, have been isolated from three common European gorgonians, *Eunicella cavolini, E. verrucosa,* and *E. stricta.*[39]

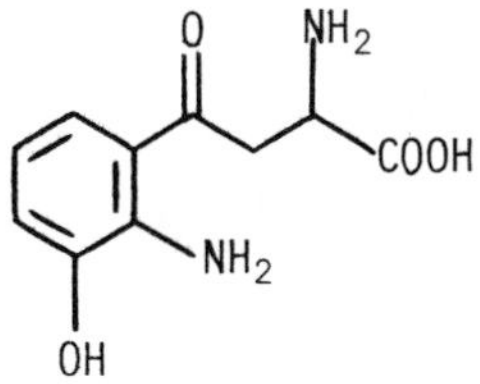

(81) 3-hydroxy-L-kynurenine

Porphyrins and Chlorophylls

The structure, distribution, and metabolism of porphyrins in marine organisms have been reviewed by Rimington and Kennedy[231] and by Goodwin,[103] the latter specifically in the molluscs.

Dougherty et al.[71] have isolated chlorophyll c from the marine diatom *Nitzschia closterium* and Jeffrey[137] has isolated and purified the compound from *Sargassum flavicans.* The magnesium-containing pigment chlorophyll d has been shown to be present in red algae, along with chlorophyll a, and was isolated in highest yield by extraction of *Gigartina agardhii.* Chlorophyll d was unstable at room temperature, being isomerized to a mixture of three chlorophylls.

Table 27

Molecular formula II/—	Structural formula	Name	Source of compound	Available by synthesis	Available nonmarine sources	Activity	References
C_1							
1. $CHBr_3$		Bromoform	Red alga: *Asparagopsis taxiformis*				85
C_2							
2. $C_2H_8NO_4P$		Phosphoethanolamine	Pufferfish: *Fugu vermiculare porphyreum*				175
C_3							
3. $C_3H_2Br_3ClO$		3-Chloro-1,1,3-tribromoacetone	Red alga: *Asparagopsis taxiformis*	+		Antimicrobial	82 85
4. $C_3H_2Br_4O$		1,1,3,3-Tetrabromoacetone	Red alga: *Asparagopsis taxiformis*	+		Antimicrobial	82 85
5. $C_3H_3Br_2ClO$		1-Chloro-1,3-dibromoacetone	Red alga: *Asparagopsis taxiformis*	+		Antimicrobial	82 85

Table 27 (continued)

Molecular formula II/—	Structural formula	Name	Source of compound	Available by synthesis	Available nonmarine sources	Activity	References
C_3							
6. $C_3H_3Br_2ClO$		3-Chloro-1,1-dibromo-acetone	Red alga: *Asparagopsis taxiformis*	+		Antimicrobial	82 85
7. $C_3H_3Br_2IO$		1,1-Dibromo-3-iodoacetone	Red alga: *Asparagopsis taxiformis*				85
8. $C_3H_3Br_3O$		1,1,3-Tribromoacetone	Red alga: *Asparagopsis taxiformis*	+		Antimicrobial	82 85
9. C_3H_4BrClO		1-Bromo-3-chloroacetone	Red alga: *Asparagopsis taxiformis*	+		Antimicrobial	82
10. $C_3H_4Br_2O$		1,3-Dibromoacetone	Red alga: *Asparagopsis taxiformis*	+		Antimicrobial	82
11. $C_3H_4O_2$		Acrylic acid	Yellow-brown alga: *Phaeocystis* sp.	+		Antibiotic	202 245

Formula	Structure	Name	Source			Activity	References
12. C_3H_5BrO	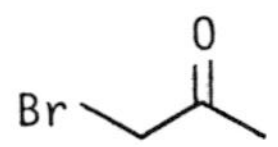	Bromoacetone	Red alga: *Asparagopsis taxiformis*				85
13. C_3H_5IO	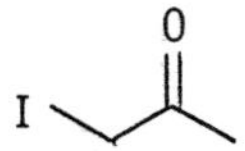	Iodoacetone	Red alga: *Asparagopsis taxiformis*				85
14. $C_3H_5N_3$	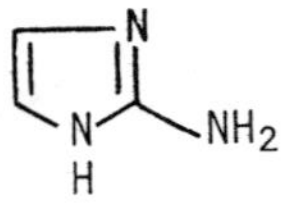	2-Aminoimidazole	Sponge: *Reniera cratera*				46
15. $C_3H_6O_3$	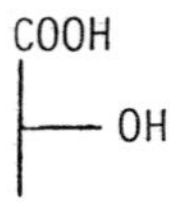	D(−)-Lactic acid	Octopus: *Octopus dofleini martini*	+			191 202
16. C_3H_9NO	$Me_3N \longrightarrow O$	Trimethylamine oxide	Algae Fish				92,116 175,202
C_4							
17. C_4HBr_4N	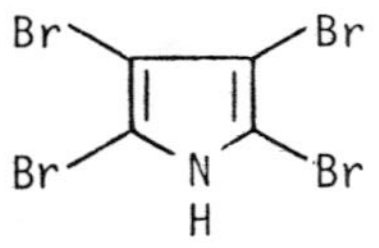	Tetrabromopyrrole	Bacterium: *Chromobacterium* sp.	+		Antibiotic	6
18. $C_4H_4O_4$	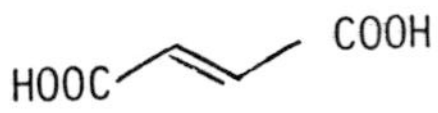	Fumaric acid	Clam: *Tapes japonica*	+	+		117 202
19. $C_4H_6O_4$	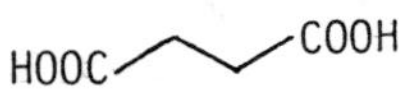	Succinic acid	Clam: *Tapes japonica* Sea worm: *Arenicola marina*	+	+	Laxative	3,117 202

Table 27 (continued)

Molecular formula II/—	Structural formula	Name	Source of compound	Available by synthesis	Available nonmarine sources	Activity	References
C_4							
20. $C_4H_6O_5$		Malic acid	Clam: *Tapes japonica*	+	+		117 202
21. $C_4H_7N_3O$		Creatinine	Fish	+	+		175 202 254
22. $C_4H_9NO_2$		α-Amino-*n*-butyric acid	Fish: *Chrysophrys major*, *Trachurus japonicus*	+			175 202
23. $C_4H_9NO_2$		γ-Amino-*n*-butyric acid	Algae Pufferfish: *Fugu vermiculare porphyreum*	+			175 190 202
24. $C_4H_9NO_3$		Homoserine	Gorgonians: *Eunicella cavolini*, *E. stricta*, *E. verrucosa*	+			39 202
C_5							
25. $C_5H_{11}NO_3S$		(−)-*S*-Hydroxymethyl-L-homocysteine	Red alga: *Chondrus ocellatus*				257

Formula	Name	Source			References
26. $C_5H_{11}NO_3S$ 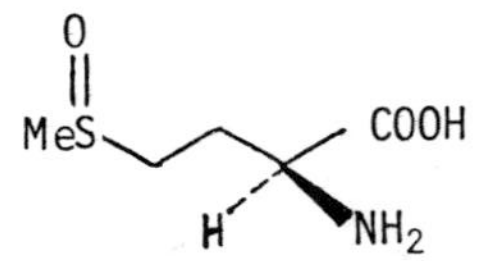	L(−)-Methionine-*l*-sulfoxide	Red alga: *Grateloupia turu-turu*	+		208
27. $C_5H_{12}N_2O_2$	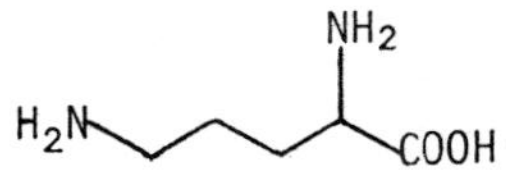 Ornithine	Algae Fish	+	+	175,190 202,208
C_6					
28. $C_6H_5N_5O_2$	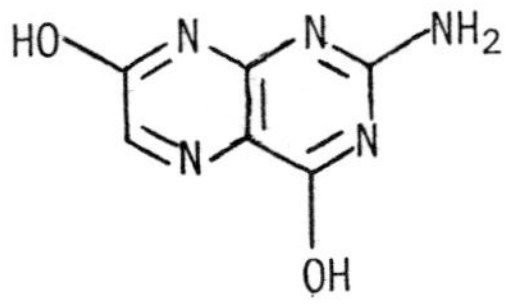 Isoxanthopterin	Copepods			210
29. $C_6H_6O_3$	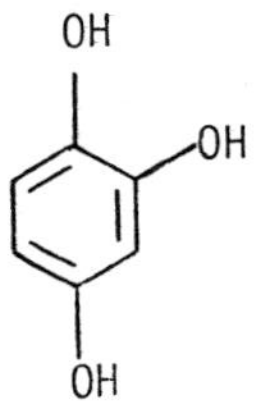 Hydroxyhydroquinone 1,2,4-Benzenetriol	Sponge: *Axinella polypoides*	+		46 202
30. $C_6H_6O_3$	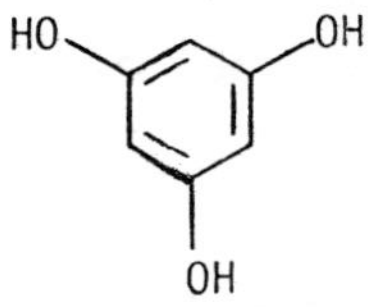 Phloroglucinol	Brown algae	+		94 202

Table 27 (continued)

Molecular formula II/—	Structural formula	Name	Source of compound	Available by synthesis	Available nonmarine sources	Activity	References
C_6							
31. $C_6H_7N_5O_2$	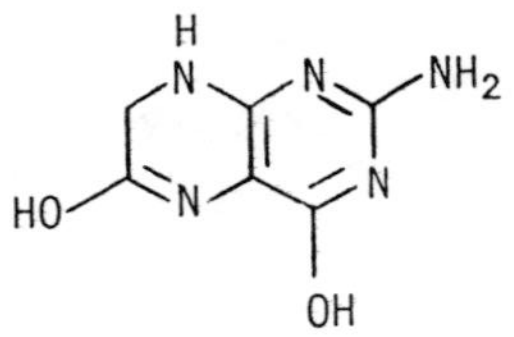	7,8-Dihydroxanthopterin	Salmon: *Oncorhynchus kisutch*				179
32. $C_6H_{12}N_2O_4S_2$	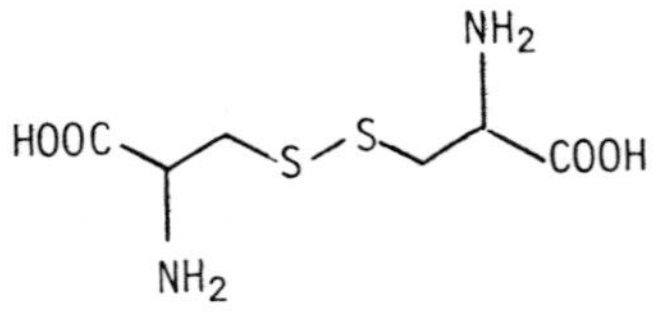	Cystine	Algae Clam: *Tapes japonica*	+	+		117 190 202
33. $C_6H_{12}O_8S$	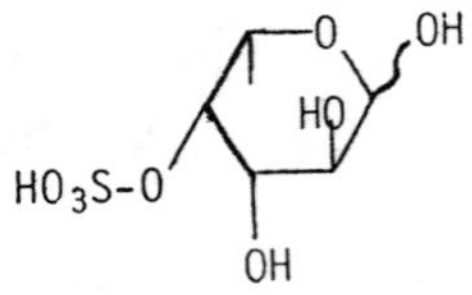	L-Fucose-4-sulfate	Sea urchins: *Hemicentrotus pulcherrimus* *Pseudocentrotus depressus*				132
34. $C_6H_{13}NO_3S$	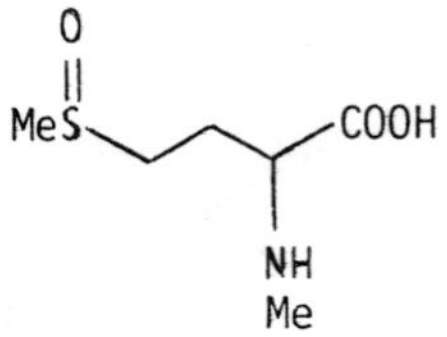	*N*-Methylmethionine sulfoxide	Red alga: *Grateloupia turuturu*	+			208

Formula	Structure	Name	Source	+	+	Activity	Ref.
35. $C_6H_{13}N_3O_3$	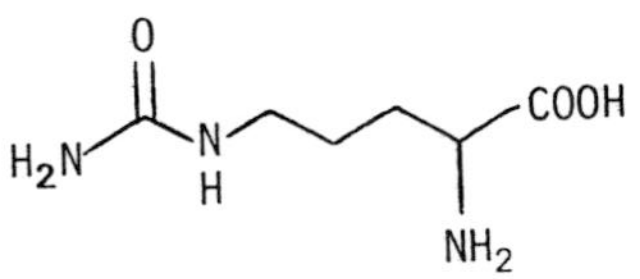	Citrulline	Algae	+	+		190 202 208
C_7 36. $C_7H_4Br_2O_2$	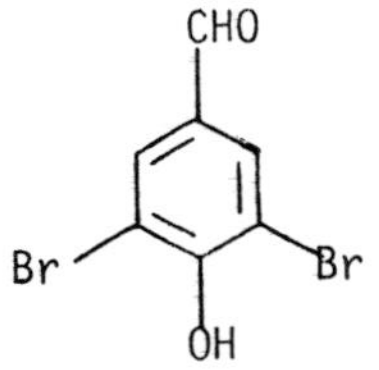	3,5-Dibromo-4-hydroxy-benzaldehyde	Sea worm: *Thelepus setosus*	+			119
37. $C_7H_5Br_3O_3$	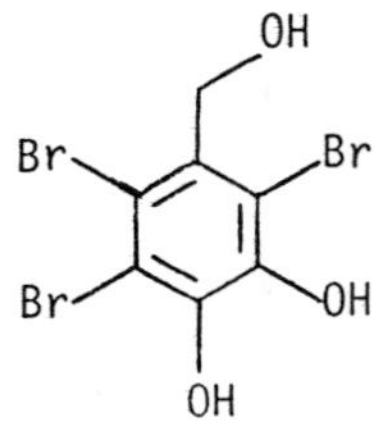	2,3,6-Tribromo-4,5-dihydroxybenzyl alcohol	Red algae: *Polysiphonia lanosa* *Rhodomela subfusca*			Antibiotic	97
38. $C_7H_5N_5O_3$	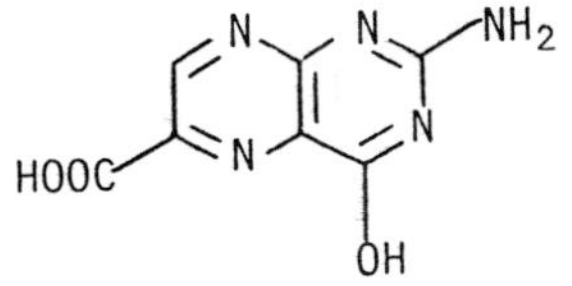	Pterin-6-carboxylic acid	Salmon: *Oncorhynchus kisutch*				179
39. $C_7H_6O_2$	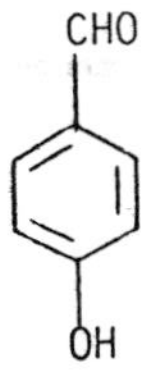	4-Hydroxybenzaldehyde	Bacterium: *Chromobacterium* sp.	+	+	Antibiotic	6 202

Table 27 (continued)

Molecular formula II/—	Structural formula	Name	Source of compound	Available by synthesis	Available nonmarine sources	Activity	References
C_7							
40. $C_7H_7BrO_2$	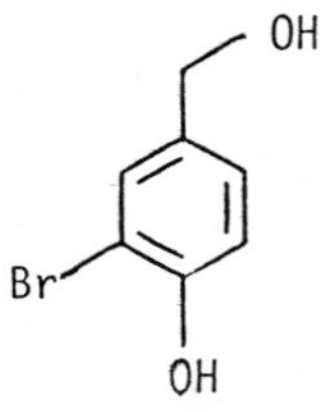	3-Bromo-4-hydroxybenzyl alcohol	Red algae: *Halopithys incurvus*, *Polysiphonia brodiaei*, *P. fruticulosa*, *P. nigra*, *P. urceolata*			Antibiotic	97
41. $C_7H_7BrO_3$	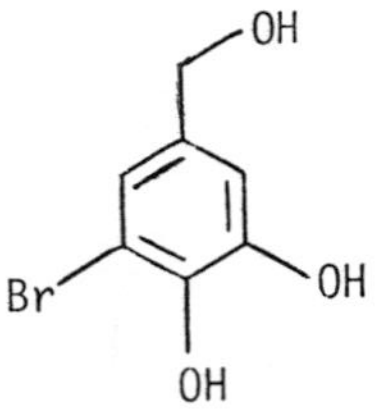	3-Bromo-4,5-dihydroxybenzyl alcohol	Red algae: *Halopithys incurvus*, *Polysiphonia lanosa*, *P. nigrescens*, *P. urceolata*	+		Antibiotic	97 252
42. $C_7H_{10}Br_4O$	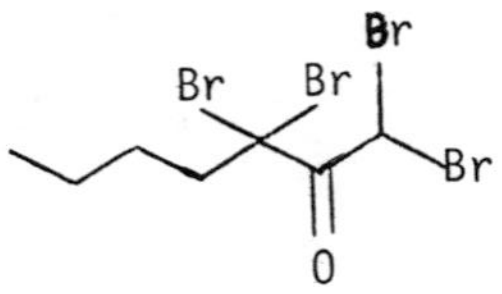	1,1,3,3-Tetrabromo-2-heptanone	Red alga: *Bonnemaisonia hamifera*	+			232 248

No.	Formula	Structure	Compound	Source		References
43.	$C_7H_{11}Br_2IO$	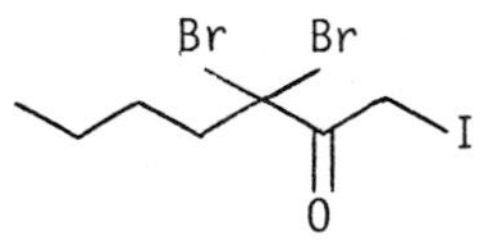	1-Iodo-3,3-dibromo-2-heptanone	Red alga: *Bonnemaisonia hamifera*		232 248
44.	$C_7H_{11}Br_3O$	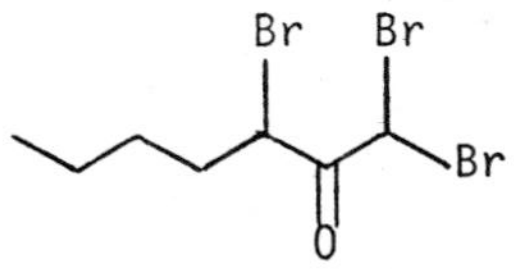	1,1,3-Tribromo-2-heptanone	Red alga: *Bonnemaisonia hamifera*		232 248
45.	$C_7H_{11}Br_3O$	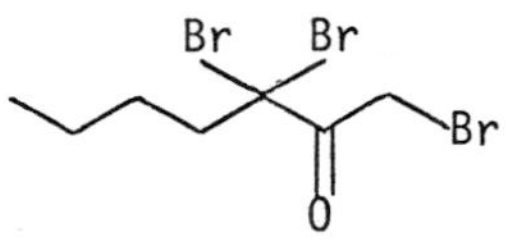	1,3,3-Tribromo-2-heptanone	Red alga: *Bonnemaisonia hamifera*		232 248
46.	$C_7H_{11}N_3O_2$	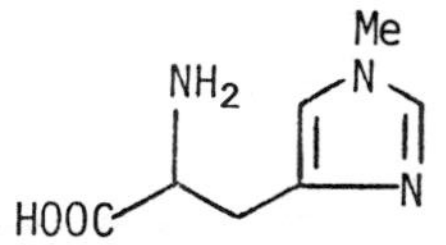	1-Methylhistidine	Brown algae: *Phyllospora comosa*, *Zonaria turneriana* Red algae: *Corallina officinalis*, *Gracilaria secundata*	+	189 190
47.	$C_7H_{11}N_3O_2$	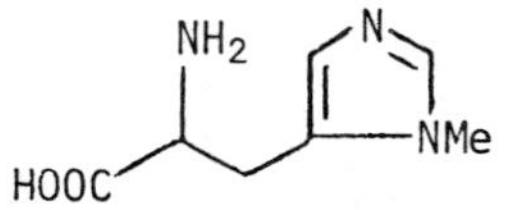	3-Methylhistidine	Pufferfish: *Fugu vermiculare porphyreum*		175
48.	$C_7H_{12}Br_2O$	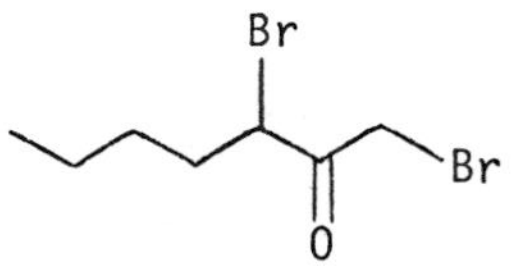	1,3-Dibromo-2-heptanone	Red alga: *Bonnemaisonia hamifera*		232 248

Table 27 (continued)

Molecular formula II/—	Structural formula	Name	Source of compound	Available by synthesis	Available nonmarine sources	Activity	References
C_7							
49. $C_7H_{15}NO_3$	^-OOC, OH, $\overset{+}{N}Me_3$	L-Carnitine Vitamin B_T	Coelenterates Crustaceans Echinoderms Fish Molluscs Sea worms Sponge	+	+		33 89 202
50. $C_7H_{15}NO_3$	HO, COO^-, $^+NMe_3$	Homoserine betaine	Green alga: *Monostroma nitidum* Molluscs: *Callista brevisiphonata* *Turbo argyrostoma*	+			2 280
51. $C_7H_{17}NO_3$	O, O, $\overset{+}{N}Me_3$, ^-OH	Acetylcholine	Lobster: *Homarus americanus* Molluscs Starfish: *Asterias rubens*	+	+		58 69 164
52. $C_7H_{18}ClNS$	MeS, $\overset{+}{N}Me_3$, Cl^-	[3-(Methylthio)propyl] trimethylammonium chloride	Mollusc: *Turbo argyrostoma*	+		Cholinomimetic	277 278
C_8							
53. $C_8H_2Br_6N_2$	Br, Br, Br, Br, Br, Br, N, H, N, H	Hexabromo-2,2′-bipyrrole	Bacterium: *Chromobacterium* sp.	+		Antibiotic	6

54. $C_8H_6Br_2O_3$	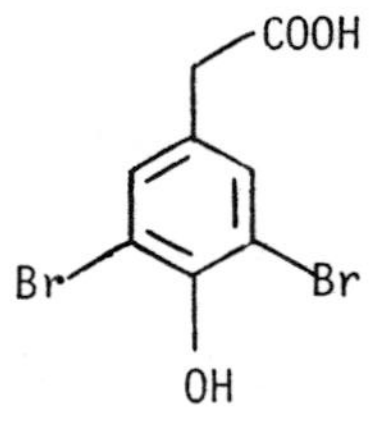 3,5-Dibromo-4-hydroxyphenylacetic acid	Red alga: *Halopithys incurvus*		43	
55. $C_8H_8O_2$	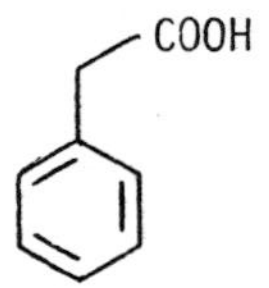 Phenylacetic acid	Brown alga: *Undaria pinnatifida*	+		1
56. $C_8H_8O_3$	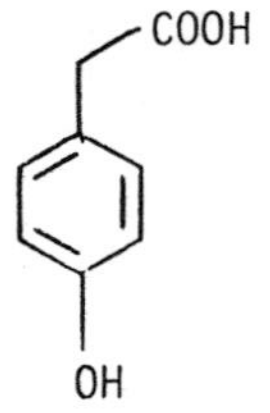 *p*-Hydroxyphenylacetic acid	Brown alga: *Undaria pinnatifida*	+		1
57. $C_8H_8O_4$	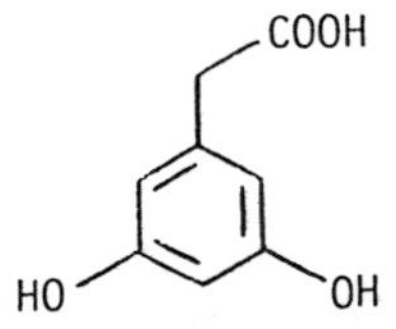 3,5-Dihydroxyphenylacetic acid	Octopus: *Octopus vulgaris* Squid: *Loligo vulgaris*	+		163
58. $C_8H_{10}N_2O_2$	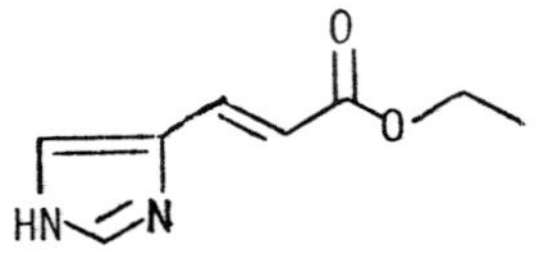 Ethyl urocanate	Molluscs: *Concholepas concholepas* *Murex trunculus*			233

Table 27 (continued)

Molecular formula II/—	Structural formula	Name	Source of compound	Available by synthesis	Available nonmarine sources	Activity	References
C_8							
59. $C_8H_{11}NO$		Tyramine	Octopus: *Hapalochlaena maculosa*	+	+	Sympathomimetic	125 202
60. $C_8H_{11}NO_2$		Dopamine	Crab Lobster Molluscs Sea urchin Starfish	+	+	Investigative antihypotensive agent	58 163 202
61. $C_8H_{11}NO_3$		*l*-Noradrenaline	Clam Octopuses Sea urchin Squids Starfish	+	+	Sympathomimetic; Vasopressor	58 163 202
62. C_8H_{12}		Fucoserratene	Brown alga: *Fucus serratus*	+			135 218

Formula	Structure	Name	Source			Activity	Ref.
63. $C_8H_{17}NO_2$	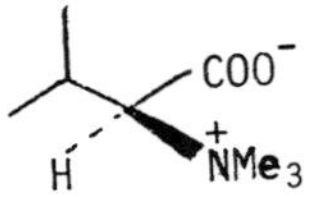	L-Valine betaine	Mollusc: *Callista brevisiphonata*	+			280
64. $C_8H_{21}Cl_2NS$	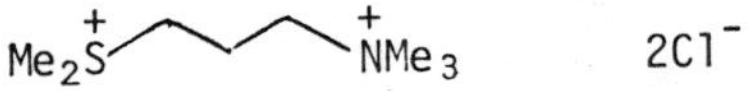	[3-(Dimethylsulfonio) propyl]-trimethylammonium dichloride	Mollusc: *Turbo argyrostoma*	+		Cholinomimetic	277 281
65. $C_8H_{22}NO_7P$	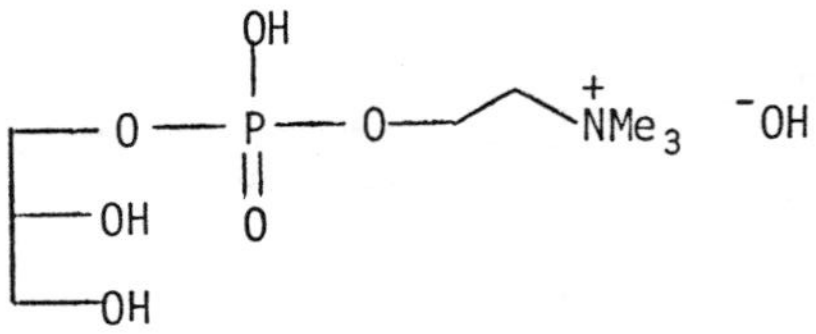	Glycerylphosphorylcholine	Octopus: *Octopus dofleini martini*		+		33
C_9 66. $C_9H_6Br_2O_4$	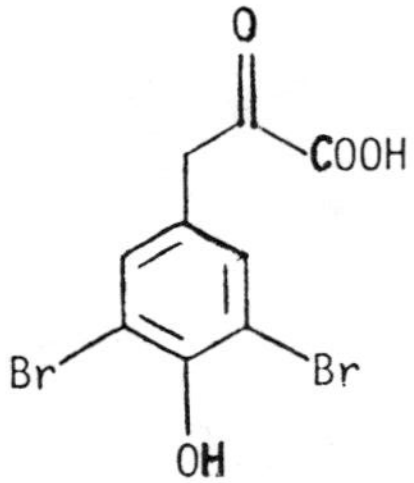	3,5-Dibromo-4-hydroxyphenylpyruvic acid	Red alga: *Halopithys incurvus*				43
67. $C_9H_7N_5O_5$	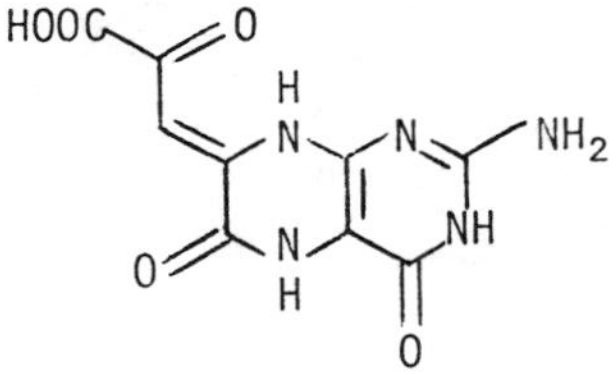	Erythropterin	Copepods	+	+		202 210

Table 27 (continued)

Molecular formula II/—	Structural formula	Name	Source of compound	Available by synthesis	Available nonmarine sources	Activity	References
C_9							
68. $C_9H_9Br_3O_2$	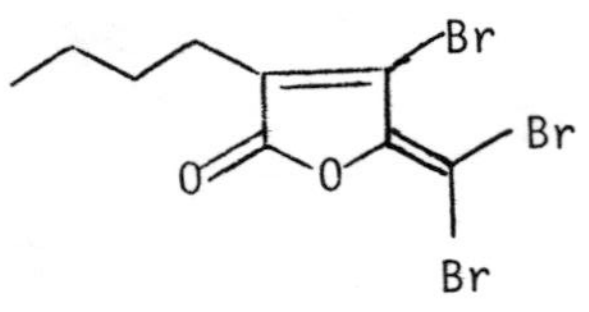	Fimbrolide C	Red alga: *Delisea fimbriata*				162
69. $C_9H_9Br_3O_3$	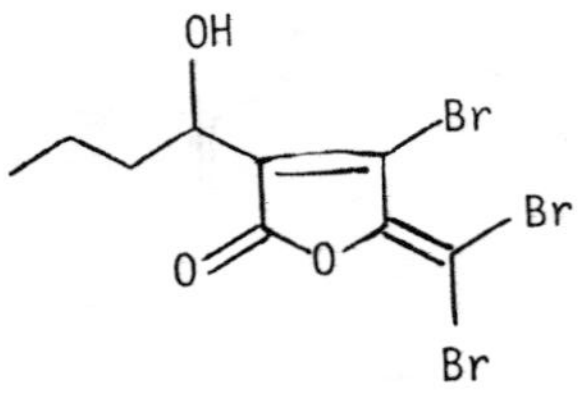	Hydroxyfimbrolide C	Red alga: *Delisea fimbriata*				162
70. $C_9H_{10}Br_2O_2$	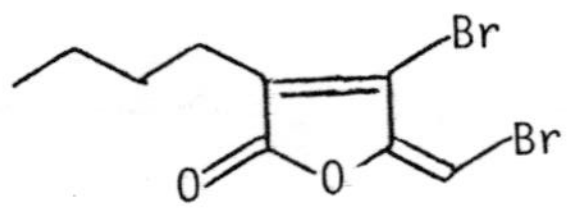	Fimbrolide A	Red alga: *Delisea fimbriata*				162
71. $C_9H_{10}Br_2O_2$	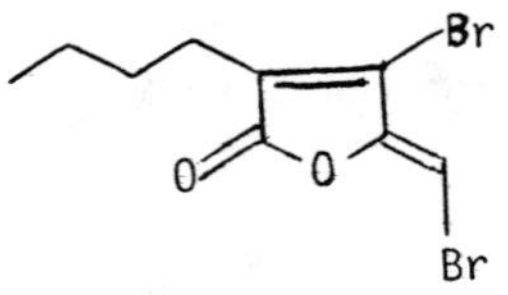	Fimbrolide B	Red alga: *Delisea fimbriata*				162

Formula	Name	Source			Ref.
72. $C_9H_{10}Br_2O_3$	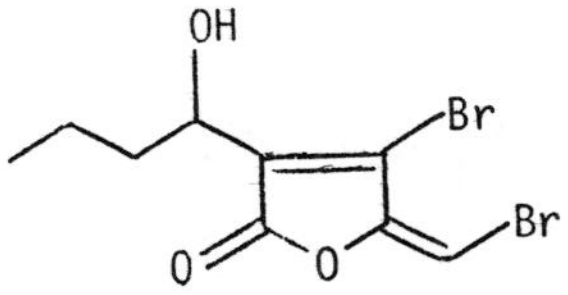 Hydroxyfimbrolide A	Red alga: *Delisea fimbriata*			162
73. $C_9H_{10}Br_2O_3$	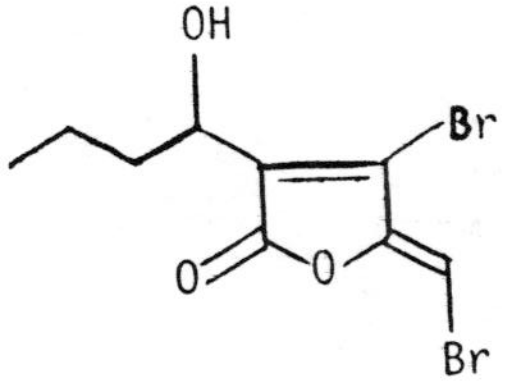 Hydroxyfimbrolide B	Red alga: *Delisea fimbriata*			162
74. $C_9H_{11}N_5O_4$	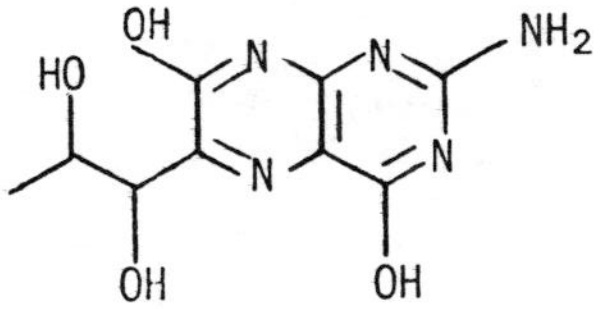 Ichthyopterin	Salmon: *Oncorhynchus kis-utch*		+	179 202
75. $C_9H_{13}N_2O_9P$	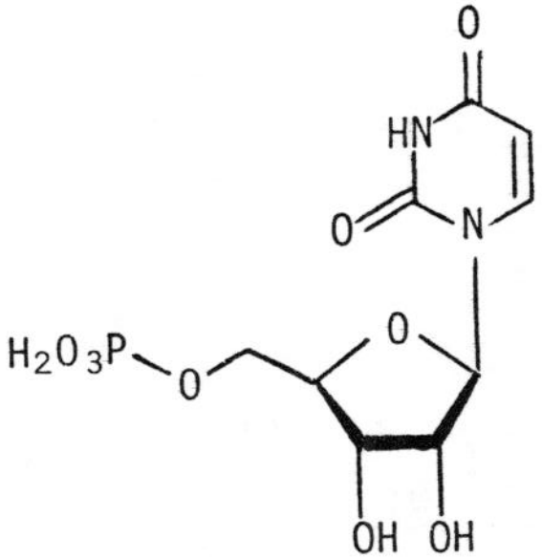 5′-Uridylic acid	Clam: *Tapes japonica*	+	+	117 202

Table 27 (continued)

Molecular formula II/—	Structural formula	Name	Source of compound	Available by synthesis	Available nonmarine sources	Activity	References
C_9							
76. $C_9H_{14}N_4O_3$	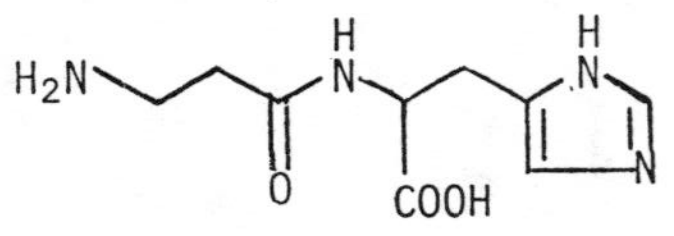	Carnosine	Fish	+	+		202 254
77. $C_9H_{18}O_8$	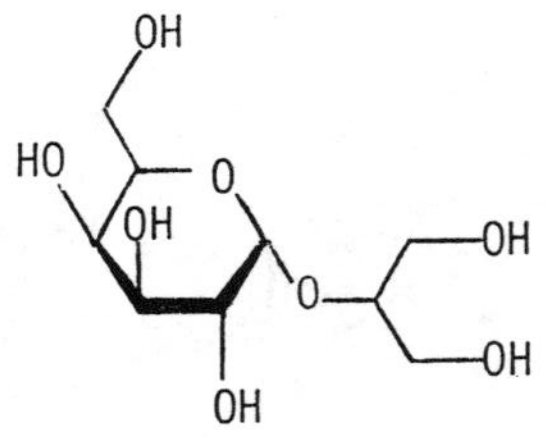	Floridoside	Red alga: *Plocamium costatum*				162
C_{10}							
78. $C_{10}H_{10}N_6$	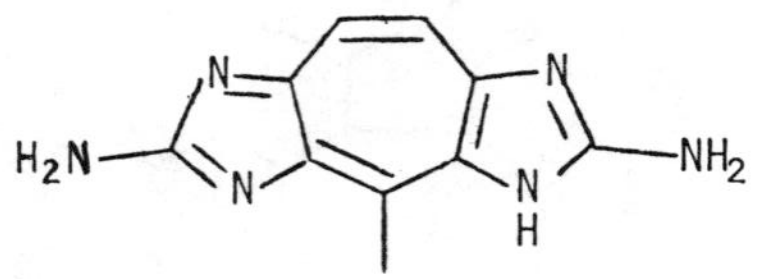	Parazoanthoxanthin A	Zoanthid: *Parazoanthus axinellae*				40
79. $C_{10}H_{11}Cl_3O$	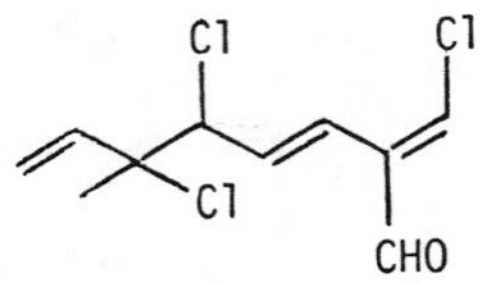	Cartilagineal	Red alga: *Plocamium cartilagineum*				59

80. $C_{10}H_{12}Br_2O_3$	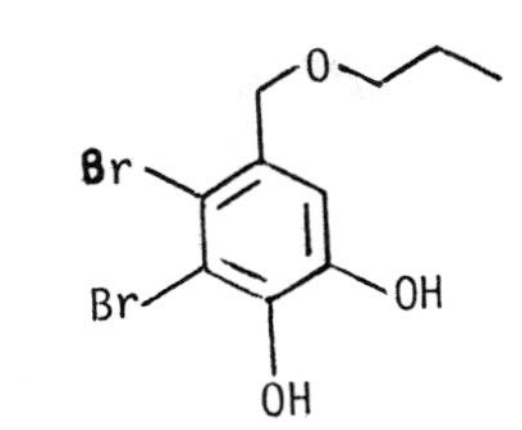 2,3-Dibromo-4,5-dihydroxybenzyl-*n*-propyl ether	Red algae: *Polysiphonia lanosa* *P. nigrescens*			Antibiotic	97
81. $C_{10}H_{12}N_2O_4$	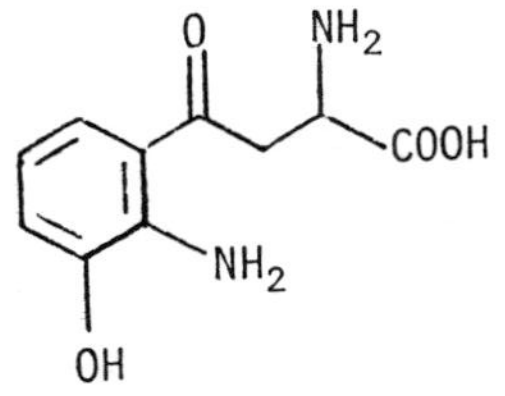 3-Hydroxy-L-kynurenine	Gorgonians: *Eunicella cavolini* *E. stricta* *E. verrucosa*		+		39 69
82. $C_{10}H_{12}N_4O_5$	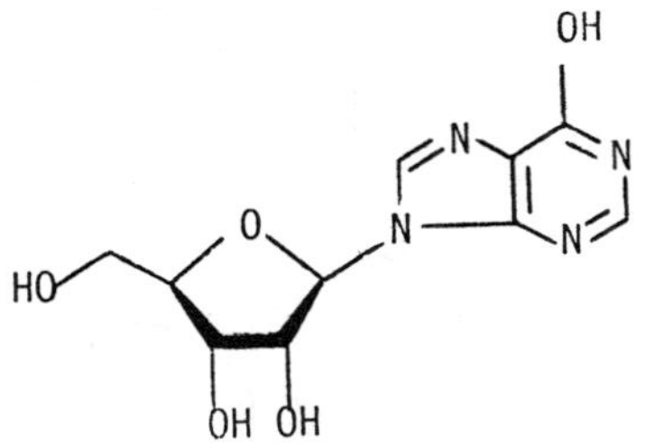 Inosine	Clam: *Tapes japonica* Fish	+	+		117 175 202
83. $C_{10}H_{12}O_3$	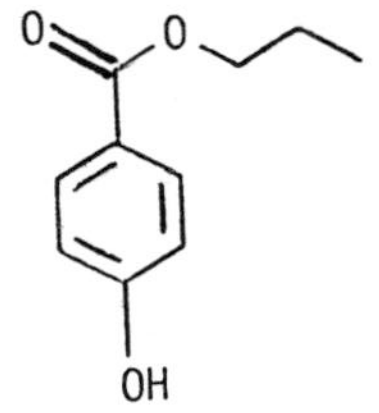 *n*-Propyl-4-hydroxybenzoate	Bacterium: *Chromobacterium* sp.	+		Antibiotic	6

Table 27 (continued)

Molecular formula II/—	Structural formula	Name	Source of compound	Available by synthesis	Available nonmarine sources	Activity	References
C_{10}							
84. $C_{10}H_{13}BrCl_4$	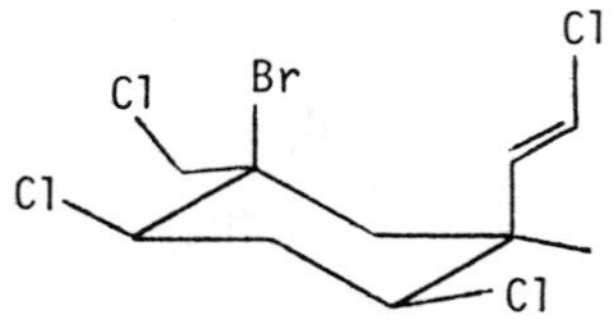 Cl, Br, Cl, Cl, Cl	Violacene	Red alga: *Plocamium violaceum*				220
85. $C_{10}H_{13}Br_2Cl$	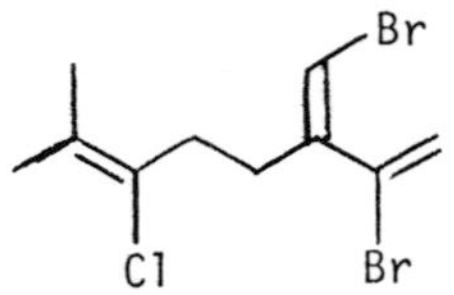 Br, Cl, Br	6-Chloro-2,(Z)-9-dibromomyrcene	Red alga: *Desmia hornemanni*				126
86. $C_{10}H_{13}Br_2Cl$	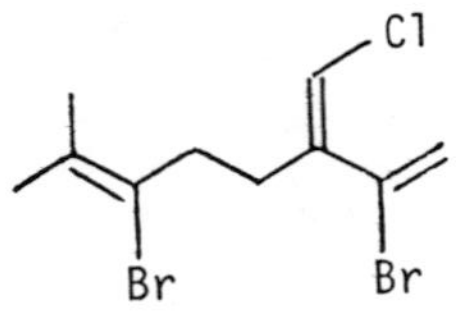 Cl, Br, Br	(Z)-9-Chloro-2,6-dibromomyrcene	Red alga: *Desmia hornemanni*				126
87. $C_{10}H_{13}N_4O_8P$	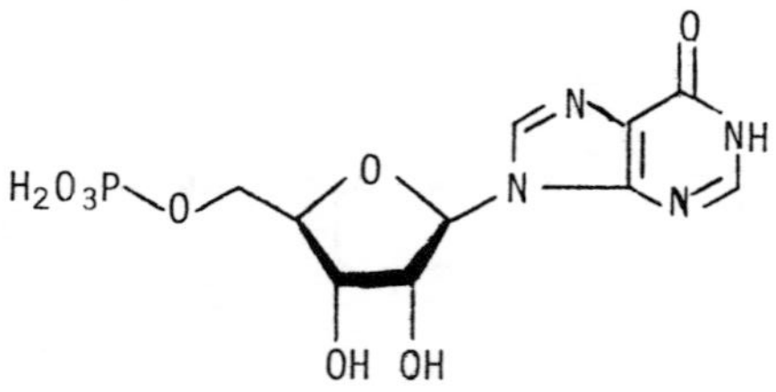 H_2O_3P, O, O, N, N, O, NH, N, OH, OH	Inosinic acid	Clam: *Tapes japonica* Fish	+	+		117 175 179 202

88. $C_{10}H_{14}BrCl$	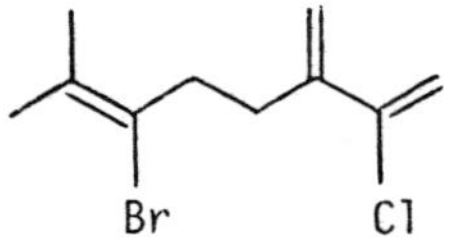 6-Bromo-2-chloromyrcene	Red alga: *Desmia hornemanni*			126
89. $C_{10}H_{14}BrCl$	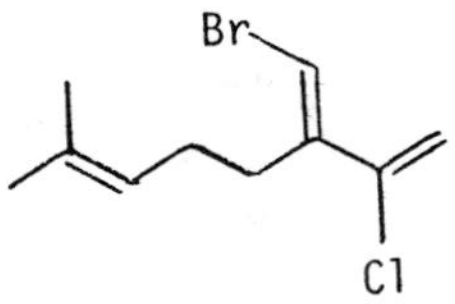 (E)-9-Bromo-2-chloromyrcene	Red alga: *Desmia hornemanni*			126
90. $C_{10}H_{14}BrCl$	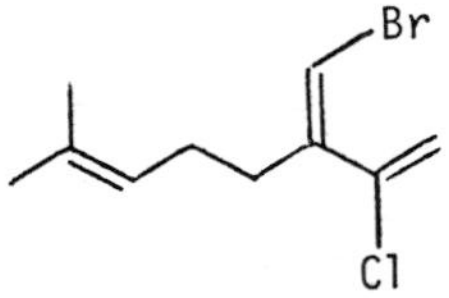 (Z)-9-Bromo-2-chloromyrcene	Red alga: *Desmia hornemanni*			126
91. $C_{10}H_{14}N_5O_7P$	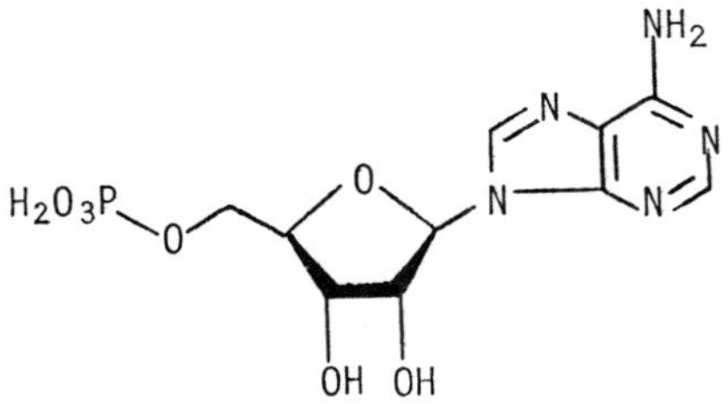 5′-Adenylic acid	Clam: *Tapes japonica* Fish	+	+	117 175 202
92. $C_{10}H_{15}Br$	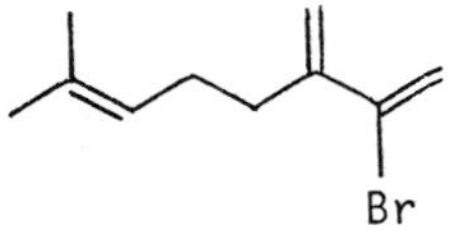 2-Bromomyrcene	Red alga: *Desmia hornemanni*			126

Table 27 (continued)

Molecular formula II/—	Structural formula	Name	Source of compound	Available by synthesis	Available nonmarine sources	Activity	References
C_{10}							
93. $C_{10}H_{15}Br$		(E)-9-Bromomyrcene	Red alga: *Desmia hornemanni*				126
94. $C_{10}H_{15}Br$		(Z)-9-Bromomyrcene	Red alga: *Desmia hornemanni*				126
95. $C_{10}H_{15}Cl$		2-Chloromyrcene	Red alga: *Desmia hornemanni*				126
96. $C_{10}H_{15}N_5O_{10}P_2$		Adenosine diphosphate	Clam: *Tapes japonica* Fish	+	+		117 175 202
97. $C_{10}H_{16}$		Myrcene	Red alga: *Desmia hornemanni*	+	+		126 202

Formula	Structure	Name	Source			Ref.
98. $C_{10}H_{16}N_4O_3$	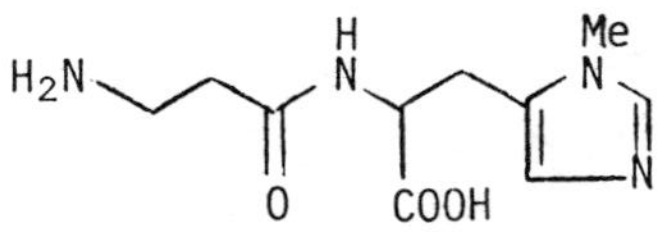	Anserine	Fish	+	+	202 254
99. $C_{10}H_{16}N_5O_{13}P_3$	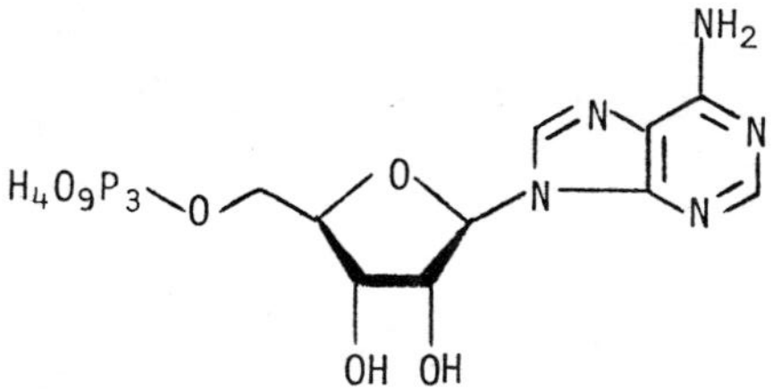	Adenosine triphosphate	Fish	+	+	175 202
C_{11}						
100. $C_{11}H_{11}Br_3O_4$		Acetoxyfimbrolide G	Red alga: *Delisea fimbriata*			162
101. $C_{11}H_{12}BrIO_4$	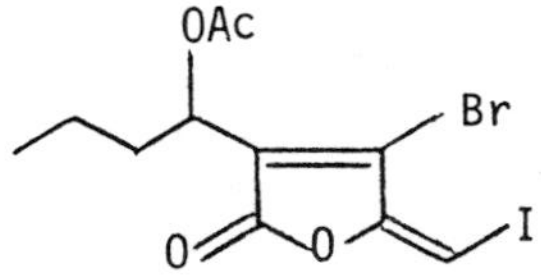	Acetoxyfimbrolide C	Red alga: *Delisea fimbriata*			162
102. $C_{11}H_{12}Br_2O_4$	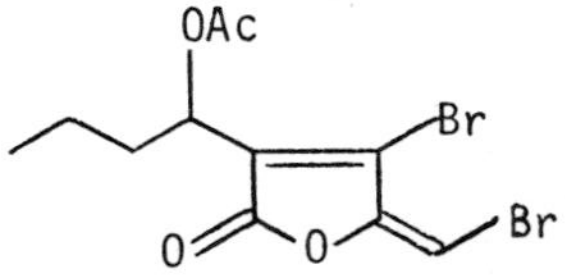	Acetoxyfimbrolide A	Red alga: *Delisea fimbriata*			162

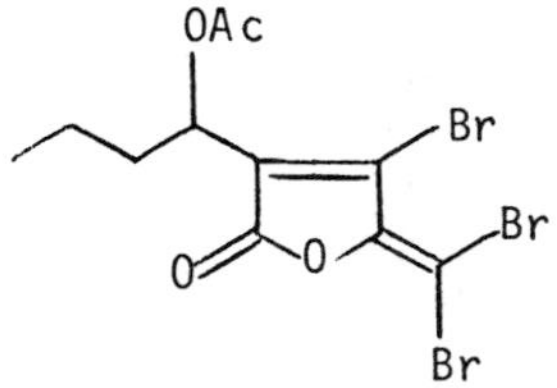

Table 27 (continued)

Molecular formula II/—	Structural formula	Name	Source of compound	Available by synthesis	Available nonmarine sources	Activity	References
C_{11}							
103. $C_{11}H_{12}Br_2O_4$	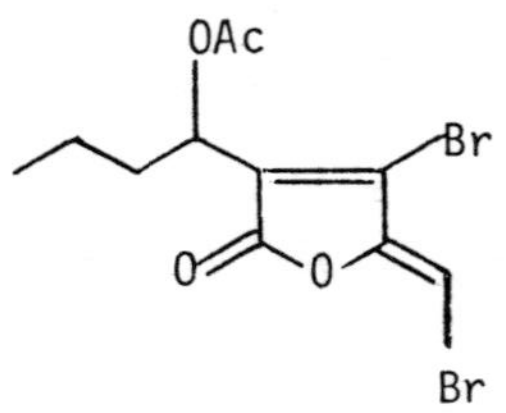	Acetoxyfimbrolide B	Red alga: *Delisea fimbriata*				162
104. $C_{11}H_{12}N_6$	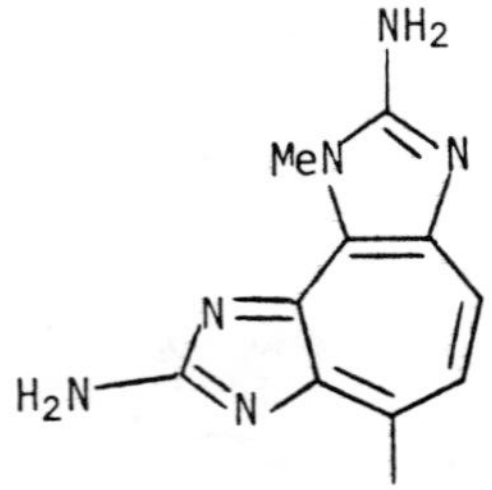	3-Norpseudozoantho-xanthin	Zoanthid: *Epizoanthus arena-ceus*				42
105. $C_{11}H_{13}N_5O$	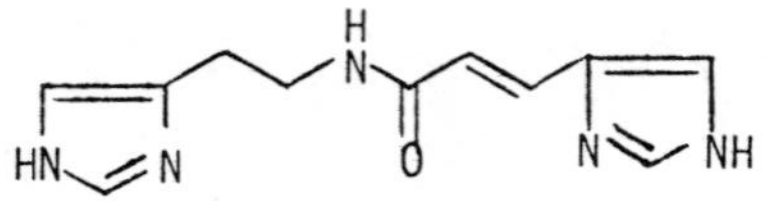	*N*-Urocanylhistamine	Mollusc: *Drupa concatenata*				236
106. $C_{11}H_{15}N_5O$	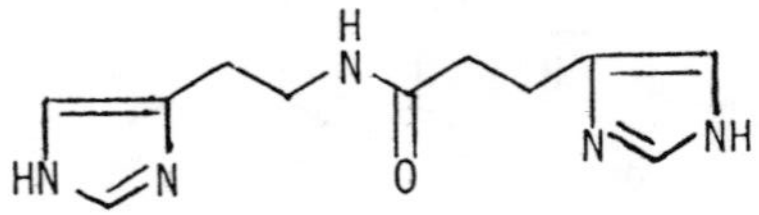	*N*(4-Imidazole-pro-pionyl)histamine	Mollusc: *Drupa concatenata*	+			235

No.	Structure	Name	Source		Ref.
107. $C_{11}H_{16}$	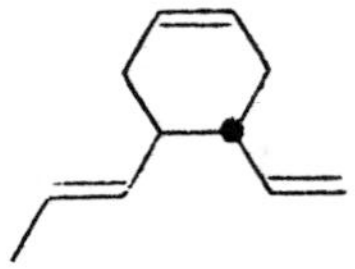	Aucantene	Brown alga: *Cutleria multifida*		134
108. $C_{11}H_{16}$	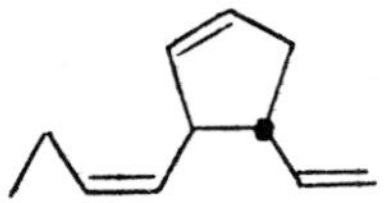	Multifidene	Brown alga: *Cutleria multifida*		134 219
109. $C_{11}H_{16}O$	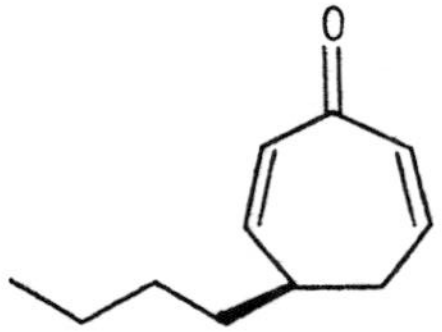	(+)-*R*-4-Butylcyclohepta-2,6-dienone	Brown algae: *Dictyopteris australis*, *D. plagiogramma*	+	213
110. $C_{11}H_{16}O$	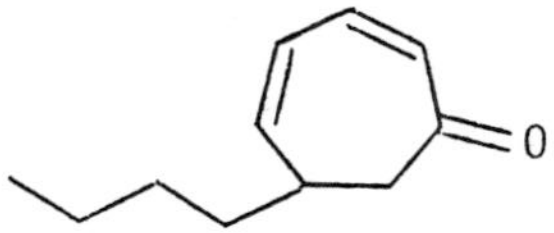	(+)-6-Butylcyclohepta-2,4-dienone	Brown algae: *Dictyopteris australis*, *D. plagiogramma*		213
111. $C_{11}H_{18}$	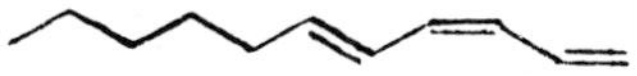	*cis, trans*-1,3,5-Undecatriene	Brown algae: *Dictyopteris australis*, *D. plagiogramma*		214
112. $C_{11}H_{18}$	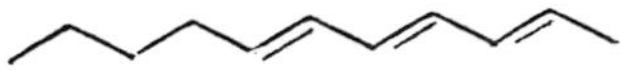	*trans, trans, trans*-2,4,6-Undecatriene	Brown algae: *Dictyopteris australis*, *D. plagiogramma*		214

Table 27 (continued)

Molecular formula II/—	Structural formula	Name	Source of compound	Available by synthesis	Available nonmarine sources	Activity	References
C_{11}							
113. $C_{11}H_{18}ClNO$	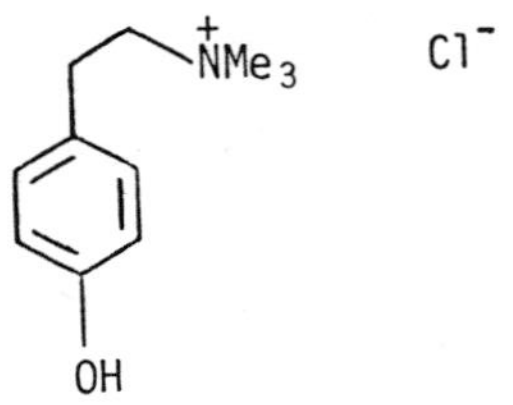	Candicine chloride	Mollusc: *Turbo argyrostoma*	+	+	Stimulating and paralyzing nicotinic and curariform actions	279
114. $C_{11}H_{19}NO_9$	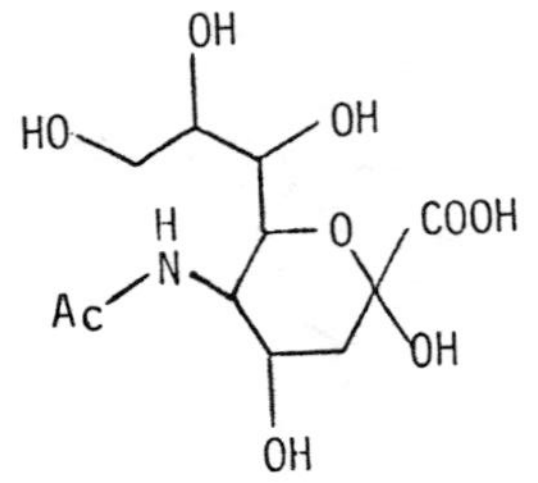	*N*-Acetylneuraminic acid	Sea urchins: *Anthocidaris crassispina* *Hemicentrotus pulcherrimus*	+	+		69 124 202
115. $C_{11}H_{19}NO_{10}$	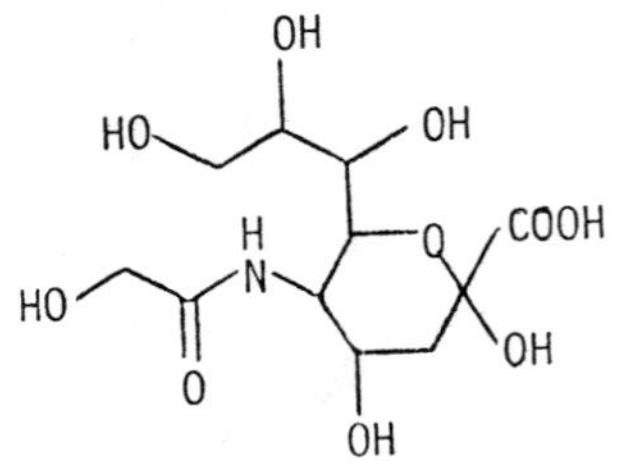	*N*-Glycolylneuraminic acid	Sea urchins: *Anthocidaris crassispina* *Hemicentrotus pulcherrimus* *Pseudocentrotus depressus*		+		124 202

	Compound	Name	Source		Ref.
C_{12}					
116. $C_{12}H_{10}O_7$	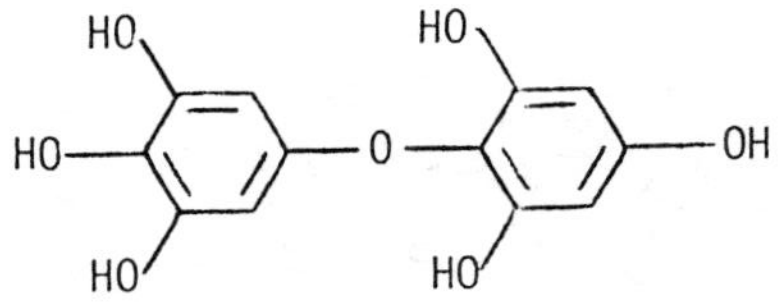	Bifuhalol	Brown alga: *Bifurcaria bifurcata*		96
117. $C_{12}H_{14}N_6$	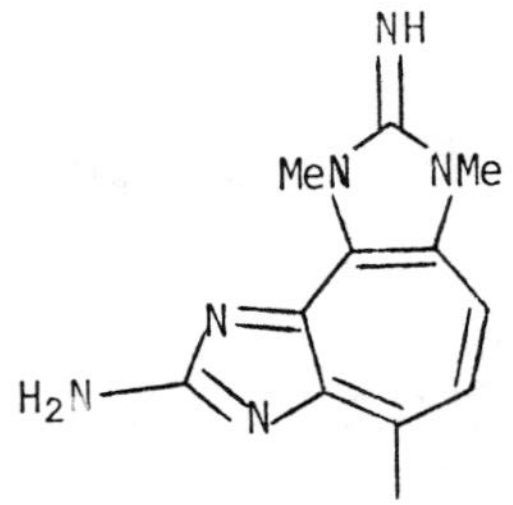	Pseudozoanthoxanthin	Zoanthid: *Epizoanthus arenaceus*		42
118. $C_{12}H_{21}N_3O_3$	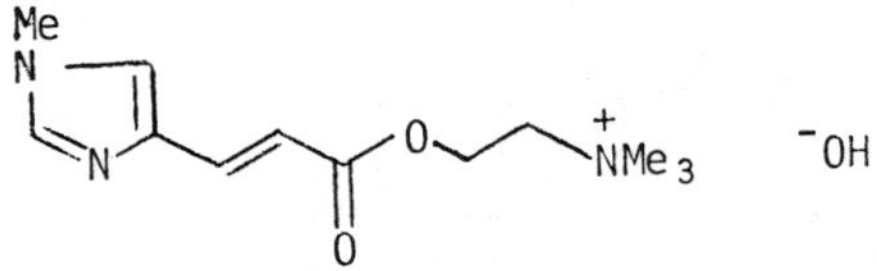	*N*-Methylmurexine	Mollusc: *Nucella emarginata*		15
C_{13}					
119. $C_{13}H_8Br_4O_2$	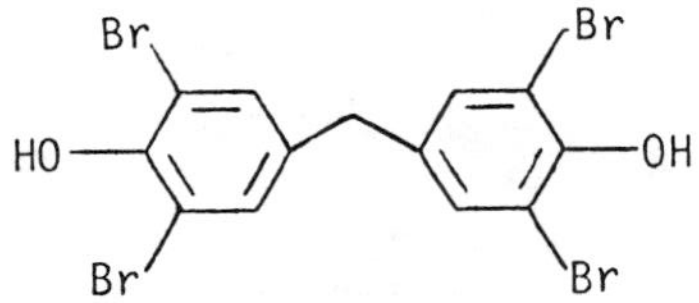	Bis(3,5-dibromo-4-hydroxyphenyl)methane	Sea worm: *Thelepus setosus*	+	119

Table 27 (continued)

Molecular formula II/—	Structural formula	Name	Source of compound	Available by synthesis	Available nonmarine sources	Activity	References
C_{13}							
120. $C_{13}H_{12}Br_2N_2O_5$	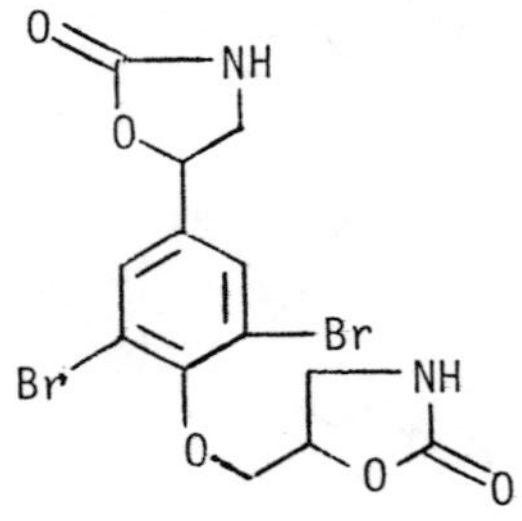	LL-PAA216	Sponge: *Verongia lacunosa*				29
121. $C_{13}H_{16}N_6$	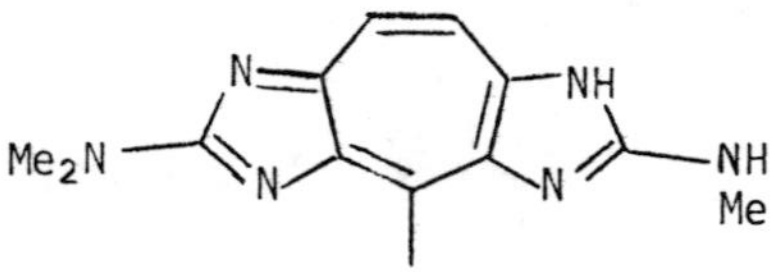	Epizoanthoxanthin A	Zoanthid: *Epizoanthus arenaceus*				41 42
122. $C_{13}H_{28}$		Tridecane	Brown algae: *Fucus distichus* *F. vesiculosus*				283
C_{14}							
123. $C_{14}H_9Br_3O_3$	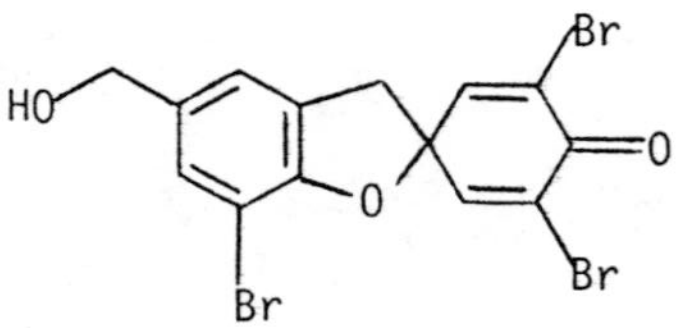	Thelepin	Sea worm: *Thelepus setosus*				119

No.	Structure	Name	Source		Ref.
124. $C_{14}H_{11}Br_3O_3$	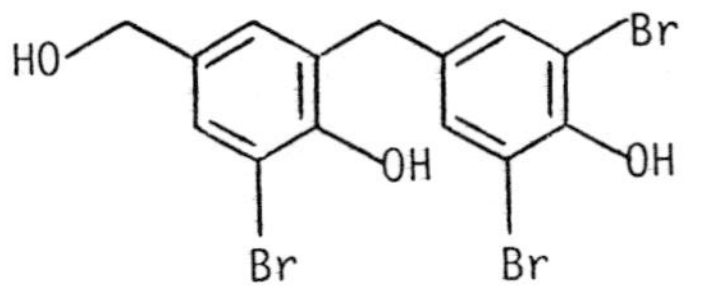	2,4′-Dihydroxy-5-hydroxymethyl-3,3′,5′-tribromodiphenyl methane	Sea worm: *Thelepus setosus*	+	119
125. $C_{14}H_{18}N_6$	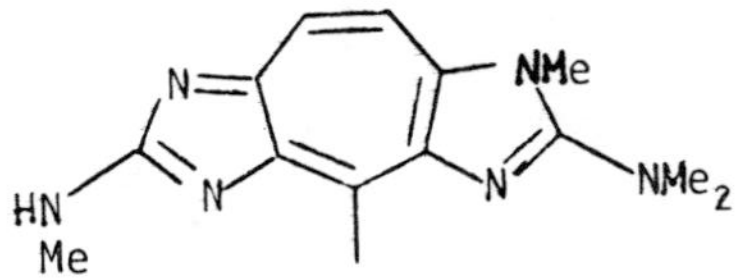	Epizoanthoxanthin B	Zoanthid: *Epizoanthus arenaceus*		42
126. $C_{14}H_{18}N_6$	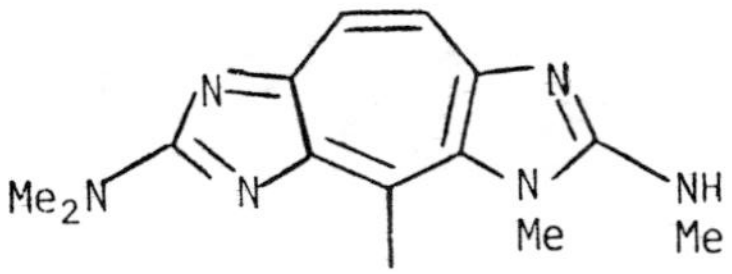	Parazoanthoxanthin E	Zoanthid: *Parazoanthus axinellae*		41
127. $C_{14}H_{18}N_6$	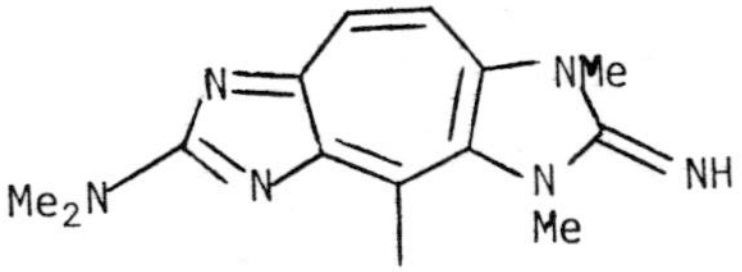	Parazoanthoxanthin F	Zoanthid: *Parazoanthus axinellae*		41
128. $C_{14}H_{23}NO_9$	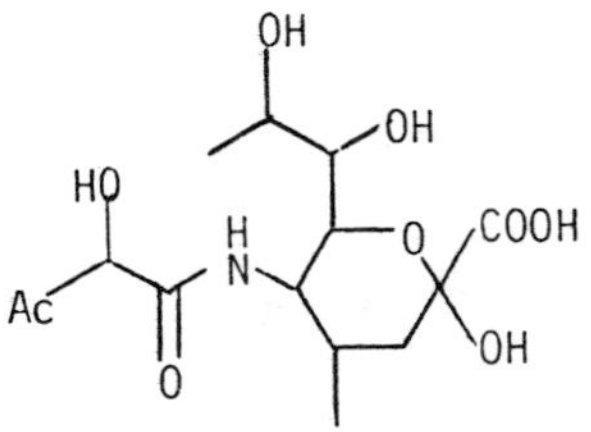	*N*-Acetoglycolyl-4-methyl-4,9-dideoxyneuraminic acid	Sea urchin: *Pseudocentrotus depressus*		124
129. $C_{14}H_{30}$	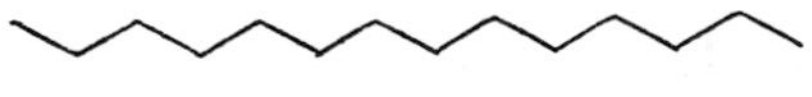	Tetradecane	Brown alga: *Fucus distichus*		283

Table 27 (continued)

Molecular formula II/—	Structural formula	Name	Source of compound	Available by synthesis	Available nonmarine sources	Activity	References
130. $C_{14}H_{30}O$	 OH	Myristyl alcohol Tetradecanol	Coelenterates	+			194 202
C_{15} 131. $C_{15}H_{18}BrClO_2$	Cl, Br, O, O	Rhodophytin	Red alga: *Laurencia* sp.				80
132. $C_{15}H_{19}BrO$	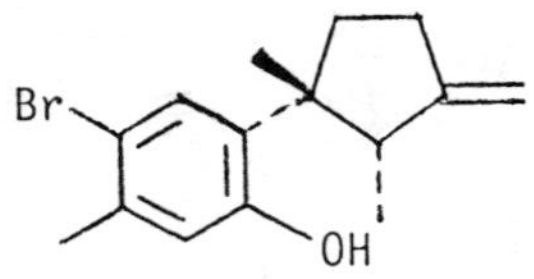 Br, OH	Allo-laurinterol	Red alga: *Laurencia filiformis* f. *dendritica*				162
133. $C_{15}H_{19}BrO$	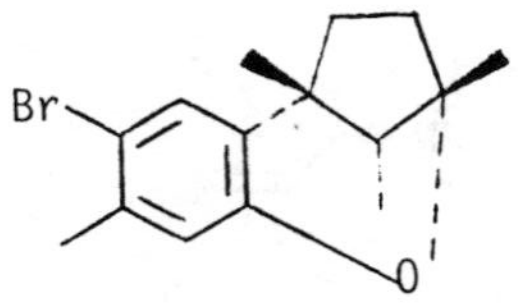 Br, O	Filiformin	Red alga: *Laurencia filiformis* f. *dendritica*				162

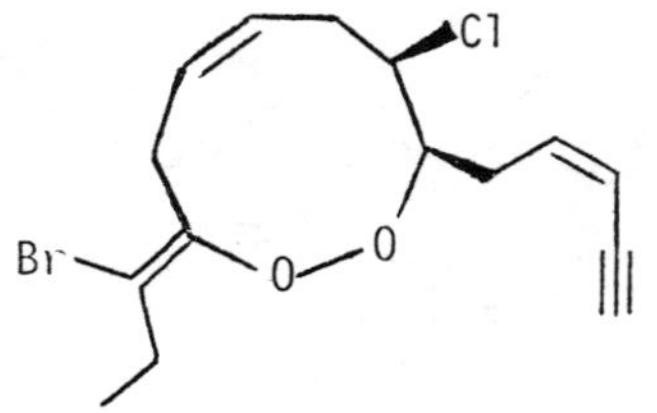

	Compound	Source	Ref.
134. $C_{15}H_{19}BrO_2$	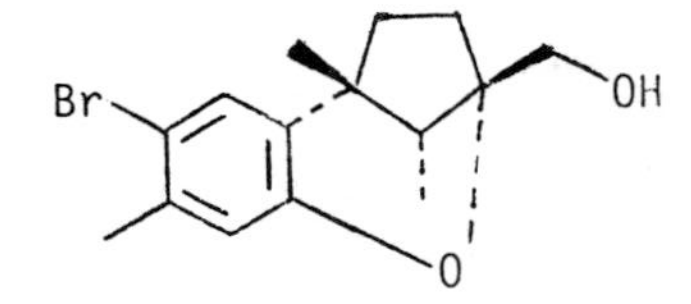 Filiforminol	Red alga: *Laurencia filiformis* f. *dentritica*	162
135. $C_{15}H_{19}Br_2ClO$	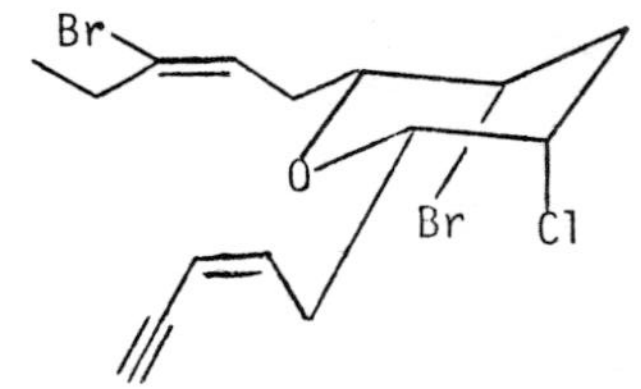 Dactylyne	Sea hare: *Aplysia dactylomela*	185 244
136. $C_{15}H_{20}Br_2O_2$	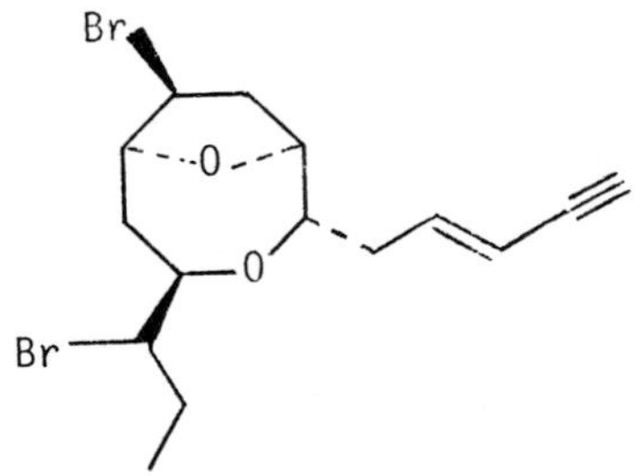 T-Isolaureatin 3-*trans*-Isolaureatin	Red alga: *Laurencia nipponica*	177
137. $C_{15}H_{20}Br_2O_2$	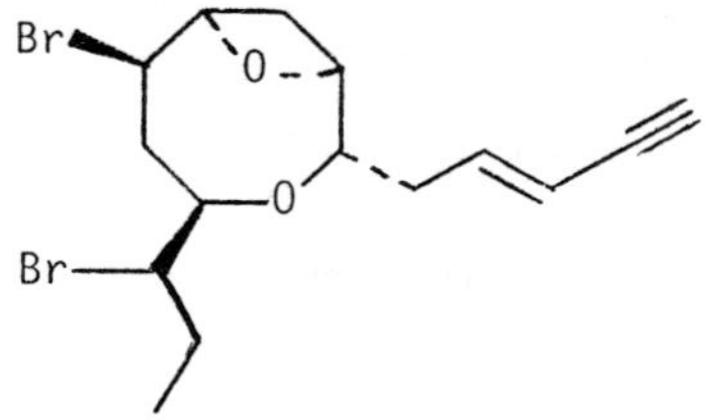 T-Laureatin 3-*trans*-Laureatin	Red alga: *Laurencia nipponica*	177

Table 27 (continued)

Molecular formula II/—	Structural formula	Name	Source of compound	Available by synthesis	Available nonmarine sources	Activity	References
C_{15}							
138. $C_{15}H_{21}BrO_2$	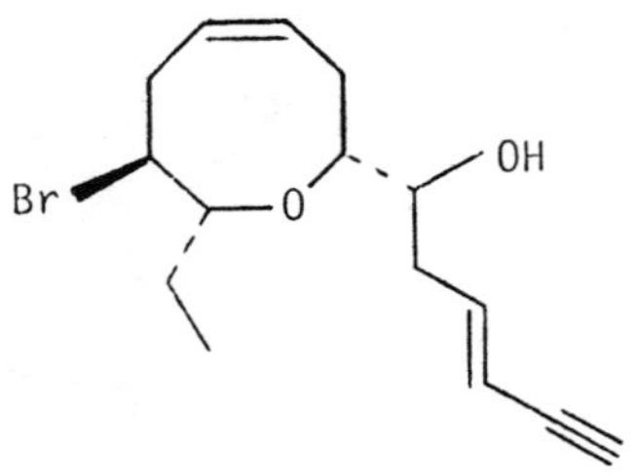	Deacetyllaurencin	Red alga: *Laurencia nipponica*				177
139. $C_{15}H_{21}Br_2ClO_2$	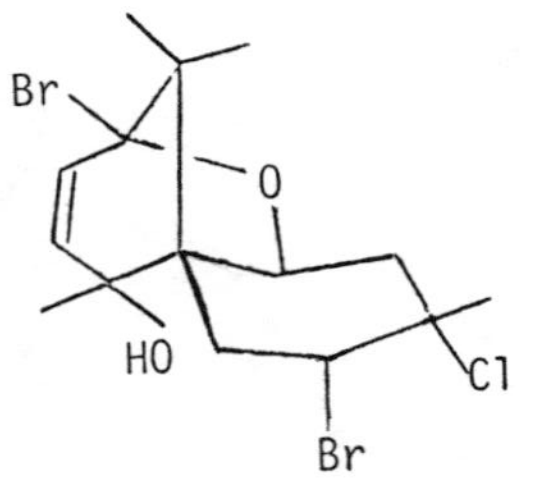	Pacifenol	Red algae: *Laurencia tasmanica*; *L. nidifica*; Sea hare: *Aplysia californica*				246, 251, 270
140. $C_{15}H_{21}Br_2ClO_3$	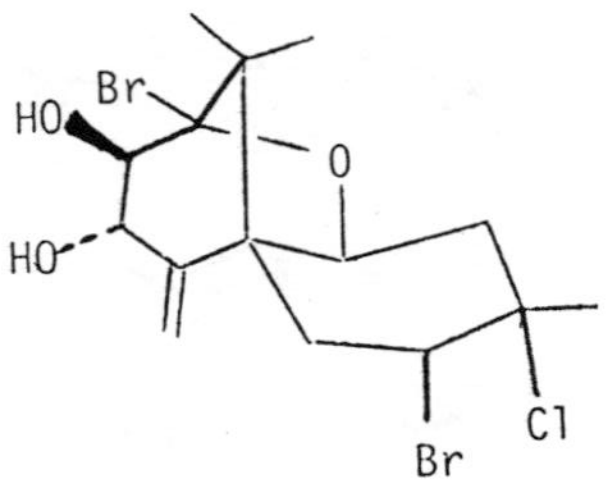	2,7-Dibromo-8- chloro-octahydro-8,10,10-trimethyl-5-methylene-6H-2,5a-methano-1-benzoxepin-3,4-diol	Sea hare: *Aplysia californica*				78

Formula	Name	Source	Ref.
141. $C_{15}H_{21}Br_2ClO_3$	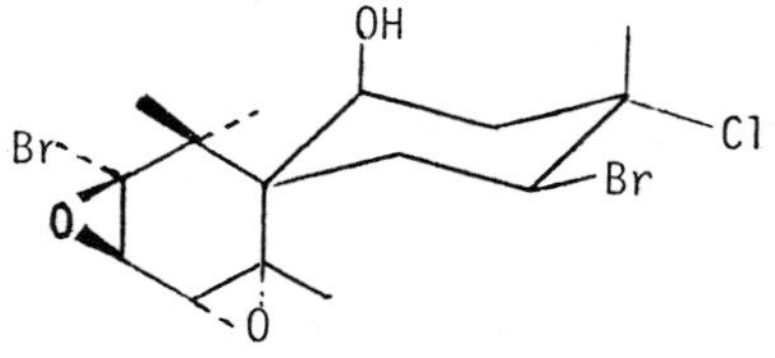 Prepacifenol epoxide	Red alga: *Laurencia johnstonii*; Sea hare: *Aplysia californica*	78
142. $C_{15}H_{22}$	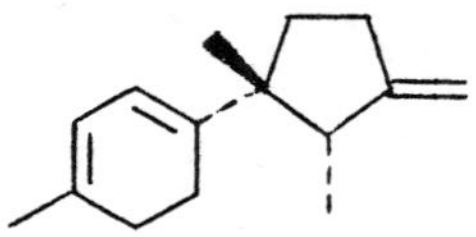 Dihydrolaurene	Red alga: *Laurencia filiformis* f. *dendritica*	162
143. $C_{15}H_{22}BrCl$	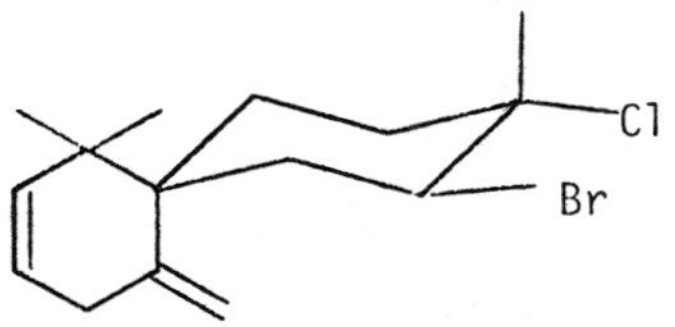 Nidifidiene	Red alga: *Laurencia nidifica*	270
144. $C_{15}H_{22}BrClO$	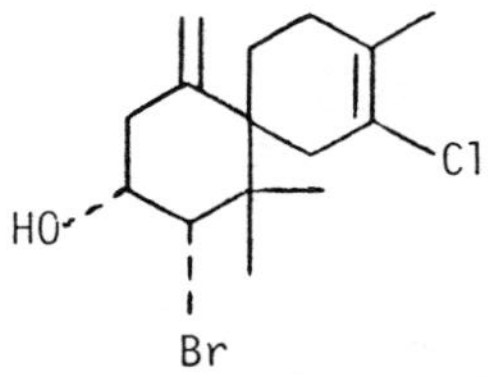 Elatol	Red alga: *Laurencia elata*	247
C_{15} 145. $C_{15}H_{22}O_2$	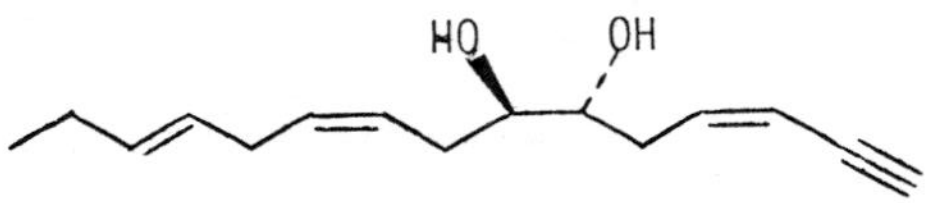 6*R*, 7*R* *cis*-Laurediol	Red alga: *Laurencia nipponica*	177

Table 27 (continued)

Molecular formula II/—	Structural formula	Name	Source of compound	Available by synthesis	Available nonmarine sources	Activity	References
C_{15}							
146. $C_{15}H_{22}O_2$	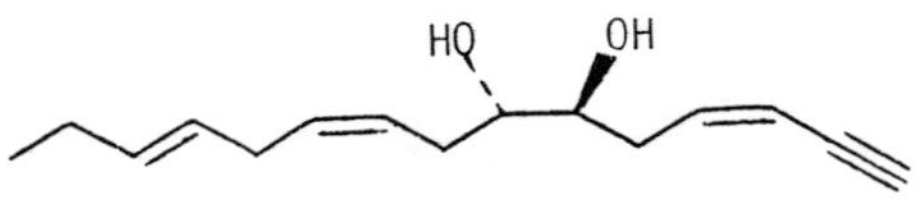	6*S*, 7*S* *cis*-Laurediol	Red alga: *Laurencia nipponica*				177
147. $C_{15}H_{22}O_2$	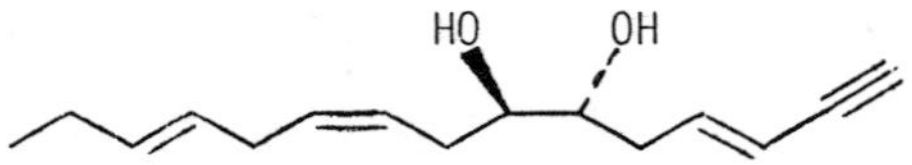	6*R*, 7*R* *trans*-Laurediol	Red alga: *Laurencia nipponica*				177
148. $C_{15}H_{22}O_2$	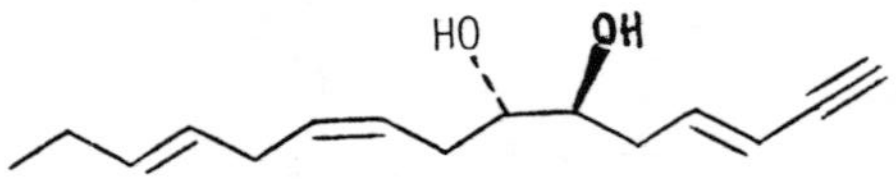	6*S*, 7*S* *trans*-Laurediol	Red alga: *Laurencia nipponica*				177
149. $C_{15}H_{23}BrO$	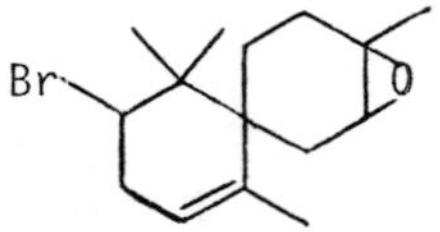	4-Bromo-α-chamigren-8,9-epoxide	Red alga: *Laurencia glanduli-fera*				255
150. $C_{15}H_{23}BrO$	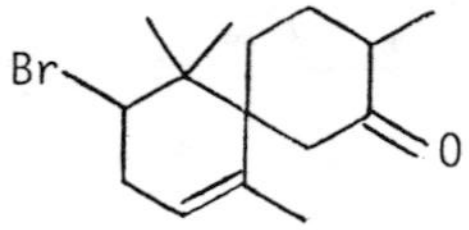	4-Bromo-α-chamigren-8-one	Red alga: *Laurencia glanduli-fera*				255

Formula	Name	Source	Ref.
151. $C_{15}H_{23}BrO$	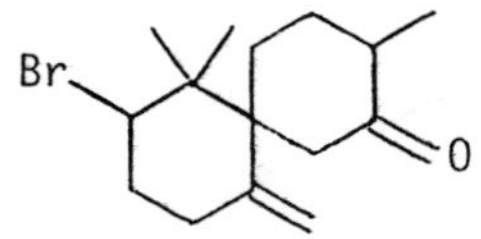 4-Bromo-β-chamigren-8-one	Red alga: *Laurencia glanduli-fera*	255
152. $C_{15}H_{23}Br_2Cl$	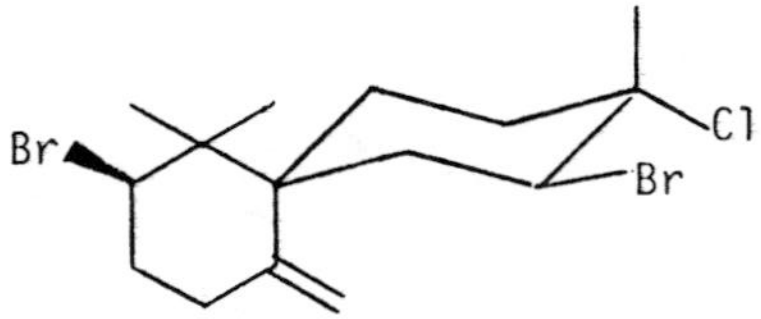 Nidificene	Red alga: *Laurencia nidifica*	270
153. $C_{15}H_{24}$	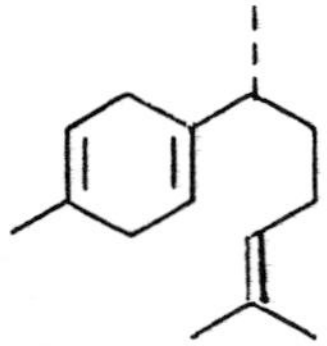 (−)-β-Curcumene	Gorgonians: *Muricea elongata* *Plexaurella nutans*	138
154. $C_{15}H_{24}BrClO$	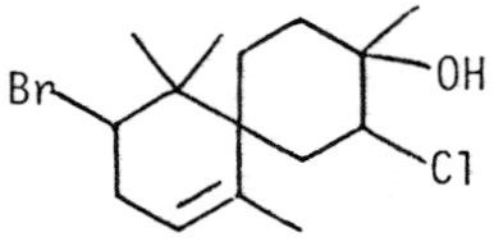 Glanduliferol	Red alga: *Laurencia glanduli-fera*	256
155. $C_{15}H_{24}O$	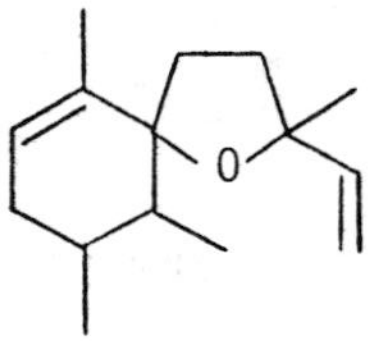 Dactyloxene-B	Sea hare: *Aplysia dactylomela*	243

Table 27 (continued)

Molecular formula II/—	Structural formula	Name	Source of compound	Available by synthesis	Available nonmarine sources	Activity	References
C_{15}							
156. $C_{15}H_{24}O_3$	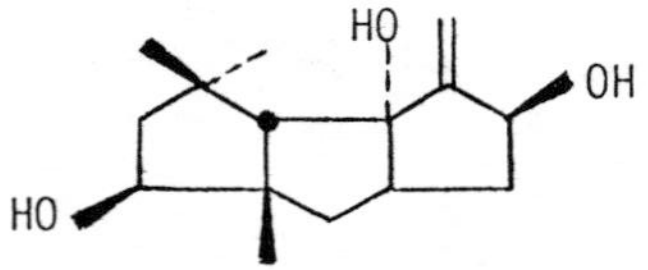 HO, OH, HO	Δ9(12)-Capnellene-3β, 8β,10α-triol	Soft coral: *Capnella imbricata*				141
157. $C_{15}H_{25}Br_2ClO_2$	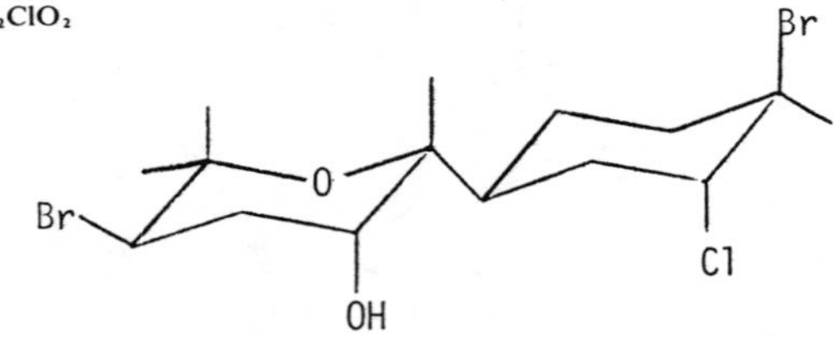 Br, O, Br, OH, Cl	Isocaespitol	Red alga: *Laurencia caespitosa*				99 101
158. $C_{15}H_{26}O$	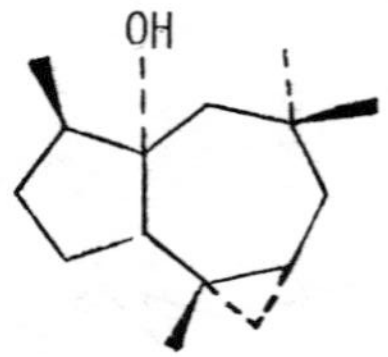 OH	Africanol	Soft coral: *Lemnalia africana*				266
159. $C_{15}H_{26}O$	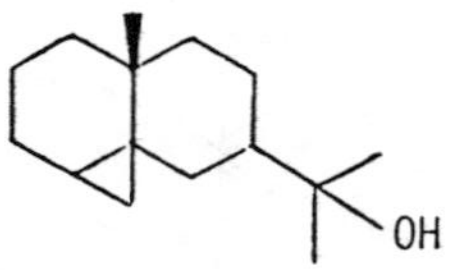 OH	Cycloeudesmol	Red alga: *Chondria oppositiclada*			Antibiotic	81

No.	Structure	Name	Source		Activity	Ref.
160. $C_{15}H_{32}$		Pentadecane	Algae			283
161. $C_{15}H_{32}O$	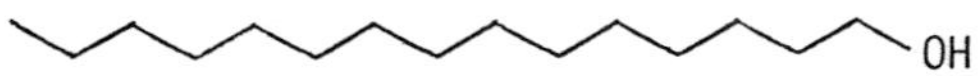 OH	Pentadecanol	Coelenterates			194
162. $C_{15}H_{32}O_3$	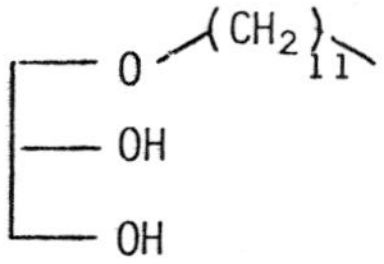 O–$(CH_2)_{11}$; OH; OH	1-*O*-Dodecylglycerol	Cod liver oil Herring: *Clupea harengus* Mussel: *Mytilus edulis*	+		114
C_{16} 163. $C_{16}H_{22}O_2$	OH; H; 2; OH	2-Diprenyl-1,4-benzo-quinol	Tunicates: *Aplidium crateri-ferum* *Aplidium* sp.			83 162
164. $C_{16}H_{25}N$	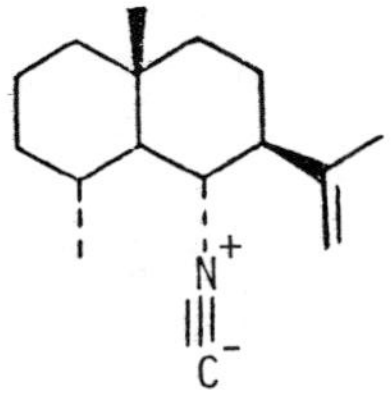 N^+; C^-	Acanthellin-1	Sponge: *Acanthella acuta*		Antibacterial	206
165. $C_{16}H_{25}N$	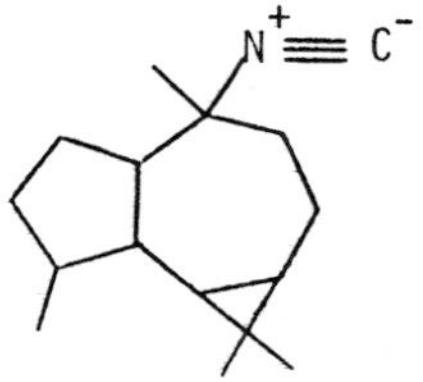 $N^+ \equiv C^-$	Axisonitrile-2	Sponge: *Axinella cannabina*			77

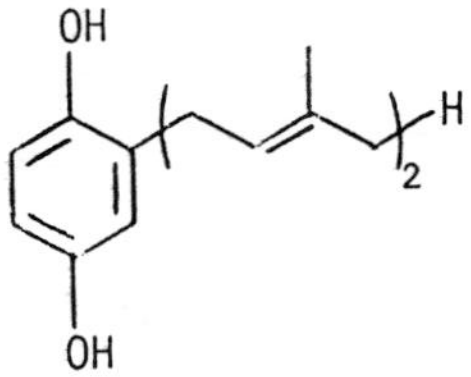

Table 27 (continued)

Molecular formula II/—	Structural formula	Name	Source of compound	Available by synthesis	Available nonmarine sources	Activity	References
C_{16}							
166. $C_{16}H_{25}N$	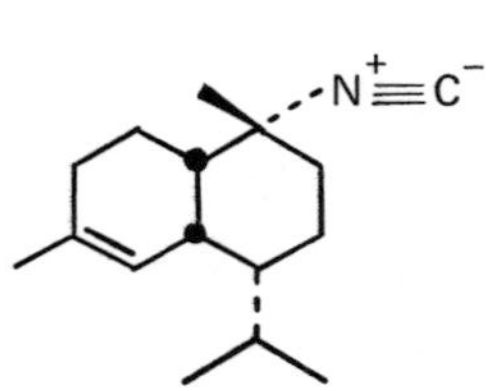	1,2,3,4,4a,7,8,8a-Octahydro-4-isopropyl-1,6-dimethyl-1-naphthyl isocyanide	Sponge: *Halichondria* sp.				36 37
167. $C_{16}H_{25}NS$	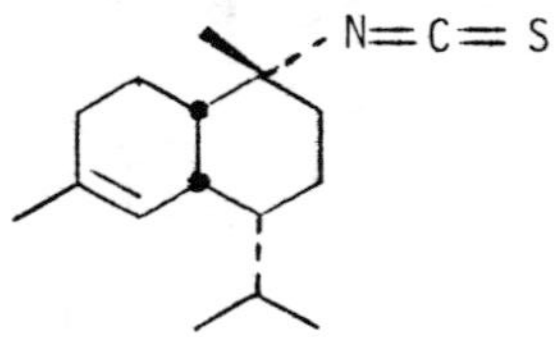	1,2,3,4,4a,7,8,8a-Octahydro-4-isopropyl-1,6-dimethyl-1-naphthyl isothiocyanate	Sponge: *Halichondria* sp.				36 37
168. $C_{16}H_{26}O_2$	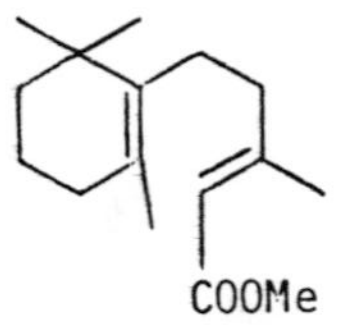	Methyl-*trans*-monocyclofarnesate	Sponge: *Halichondria panicea*				45
169. $C_{16}H_{27}NO$	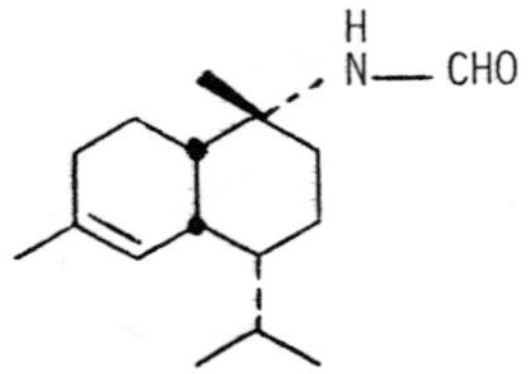	*N*-(1,2,3,4,4a,7,8,8a-Octahydro-4-isopropyl-1,6-dimethyl-1-naphthyl)formamide	Sponge: *Halichondria* sp.				36 37

No. / Formula	Name	Source			Ref.
170. $C_{16}H_{34}$	 Hexadecane	Algae: *Ascophyllum nodosum*, *Fucus vesiculosus*, *Rhodomela confervoides*			283
171. $C_{16}H_{34}O$	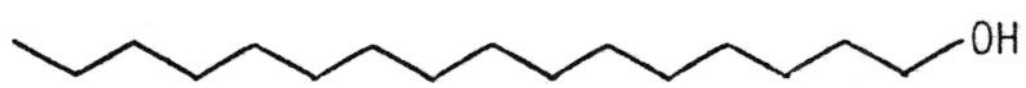 Cetyl alcohol Hexadecanol	Algae Coelenterates	+	+	194 202
C_{17} 172. $C_{17}H_{15}N_3O_4S$	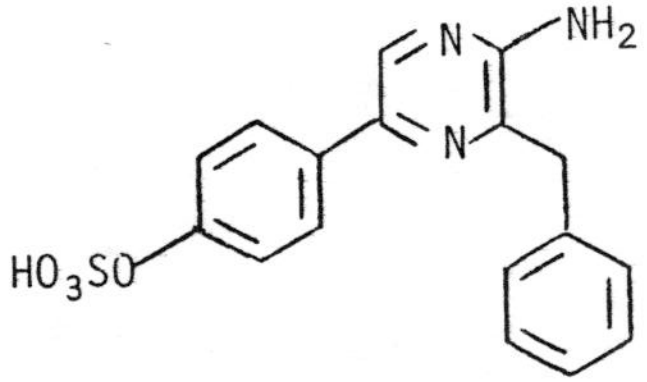 AF-350 monosulfate	Squid: *Watasenia scintillans*	+		109
173. $C_{17}H_{20}N_4O_6$	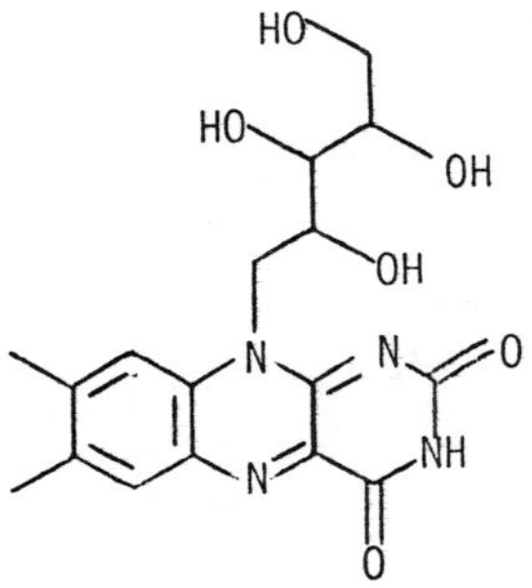 Riboflavine	Copepods Fish	+	+	60 202 210

Table 27 (continued)

Molecular formula II/—	Structural formula	Name	Source of compound	Available by synthesis	Available nonmarine sources	Activity	References
C_{17}							
174. $C_{17}H_{27}Br_2ClO_3$	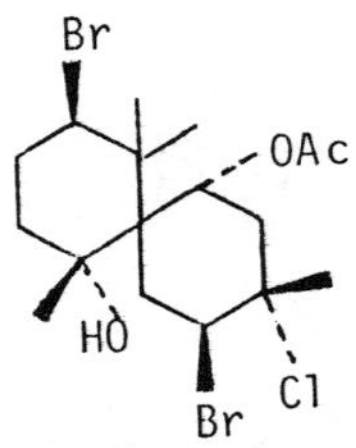	Acetoxyintricatol	Red alga: *Laurencia intricata*				187
175. $C_{17}H_{34}O_2$	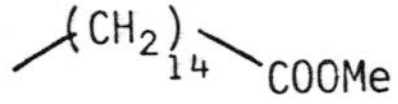	Methylpalmitate	Red alga: *Desmia hornemanni*				126
176. $C_{17}H_{34}O_4$	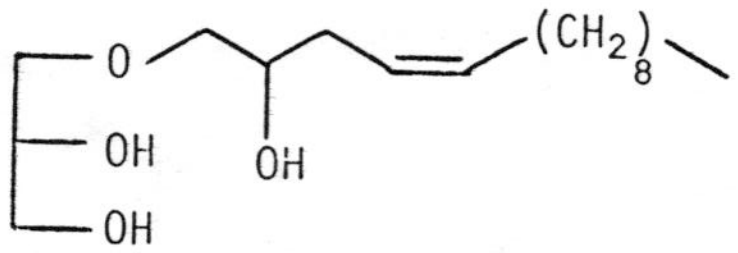	1-*O*-(2-Hydroxy-4-tetradecenyl)glycerol	Shark: *Somniosus microcephalus*				113 115
177. $C_{17}H_{36}$		Heptadecane	Algae				85,283
178. $C_{17}H_{36}O$	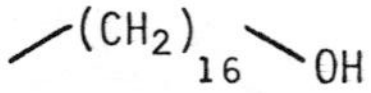	Heptadecanol	Coelenterates				194
179. $C_{17}H_{36}O_3$	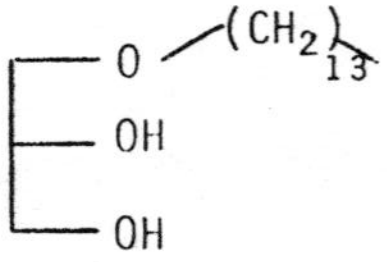	1-*O*-Tetradecylglycerol	Cod liver oil Crayfish Herrings Mackerel Mussel Shrimp		+		114

No.	Compound	Source			References
180. $C_{17}H_{36}O_4$	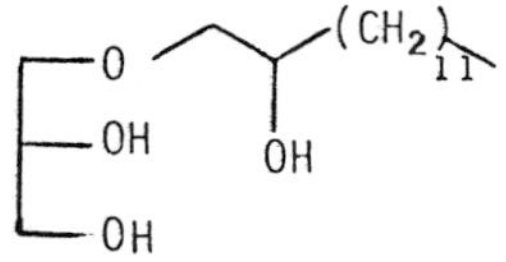 1-*O*-(2-Hydroxytetra-decyl)glycerol	Shark: *Somniosus microcephalus*			115
C_{18} 181. $C_{18}H_{14}O_{10}$	Trifuhalol	Brown alga: *Halidrys siliquosa*	+		95
182. $C_{18}H_{22}O_2$	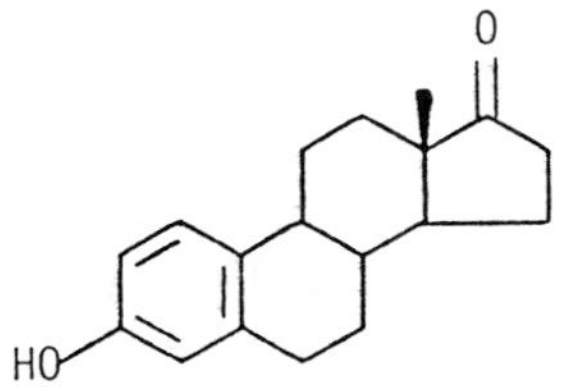 Estrone	Dogfish: *Squalus suckleyi* Mollusc: *Pecten hericius*	+	+	32 202 273
183. $C_{18}H_{24}O_2$ 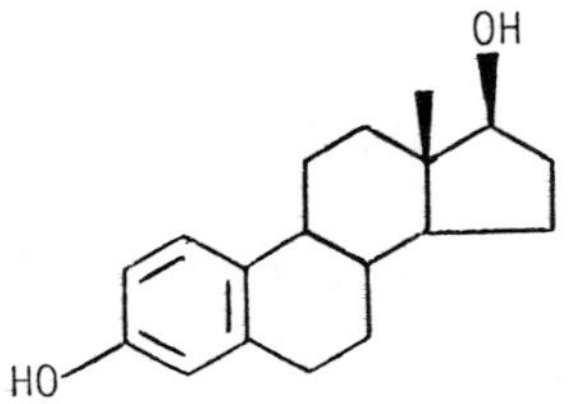	β-Estradiol	Dogfish: *Squalus suckleyi* Mollusc: *Pecten hericius* Sea urchin: *Strongylocentrotus franciscanus* Starfish: *Pisaster ochraceous*	+	+	31 32 202 273
184. $C_{18}H_{36}O$	 Oleyl alcohol *cis*-9-Octadecen-1-ol	Coelenterates	+		194 202

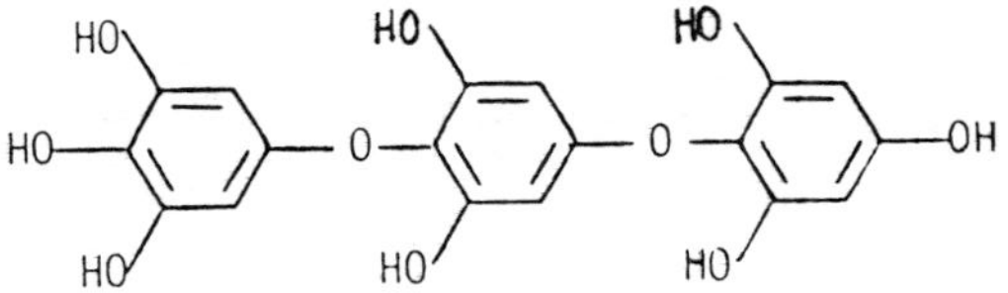

Table 27 (continued)

Molecular formula II/—	Structural formula	Name	Source of compound	Available by synthesis	Available nonmarine sources	Activity	References
C_{18}							
185. $C_{18}H_{38}$	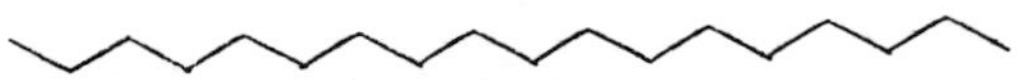	Octadecane	Algae				283
186. $C_{18}H_{38}O$	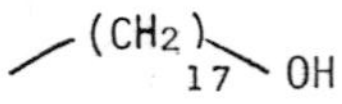 $(CH_2)_{17}$ OH	Stearyl alcohol Octadecanol	Algae Coelenterates	+			194 202
187. $C_{18}H_{38}O_3$	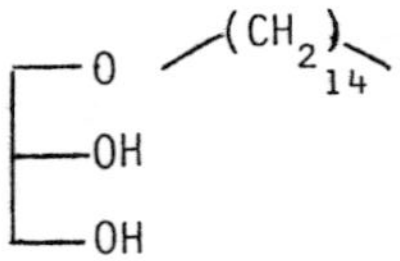 O, OH, OH; $(CH_2)_{14}$	1-*O*-Pentadecylglycerol	Cod liver oil Crayfish Herrings Mackerel Mussel Shrimp		+		114
188. $C_{18}H_{38}O_4$	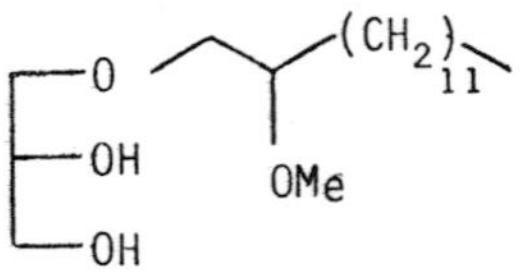 O, OH, OH; $(CH_2)_{11}$; OMe	1-*O*-(2-Methoxytetradecyl)glycerol	Cod liver oil Crayfish: *Nephrops norvegicus* Herring: *Clupea harengus*	+			114
C_{19}							
189. $C_{19}H_{30}$		*cis, cis, cis, cis, cis*-3,6,9,12,15-Nonadecapentaene	Brown algae: *Porphyra leucostica* *Pylaiella littoralis* *Scytosiphon lomentaria*				283
190. $C_{19}H_{32}$		*cis, cis,cis, cis*-4,7,10,13-Nonadecatetraene	Algae: *Porphyra leucostica* *Scytosiphon lomentaria*				283

No. / Formula	Structure	Name	Source			Ref.
191. $C_{19}H_{38}O_4$	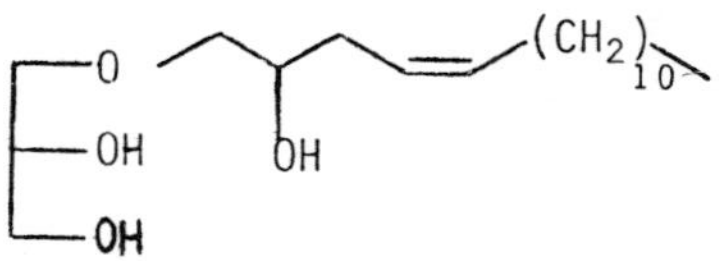	1-*O*-(2-Hydroxy-4-hexadecenyl)glycerol	Shark: *Somniosus microcephalus*	+		113 115
192. $C_{19}H_{40}$		Nonadecane	Algae			283
193. $C_{19}H_{40}O$	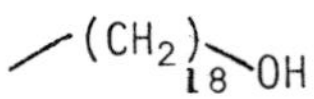	Nonadecanol	Coelenterates			194
194. $C_{19}H_{40}O_3$	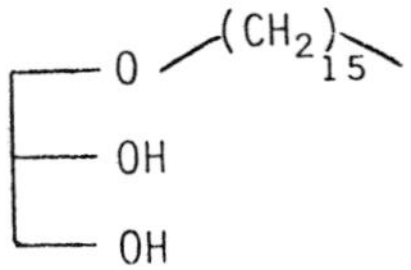	Chimyl alcohol 1-*O*-Hexadecylglycerol	Algae Cod liver oil Coelenterates Crayfish Herrings Mackerel Mussel Shrimp	+	+	114 194 202
195. $C_{19}H_{40}O_4$	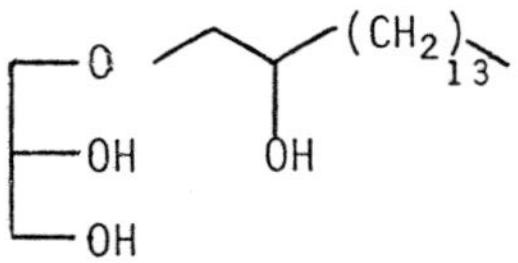	1-*O*-(2-Hydroxyhexadecyl)glycerol	Shark: *Somniosus microcephalus*	+		115
196. $C_{19}H_{40}O_4$	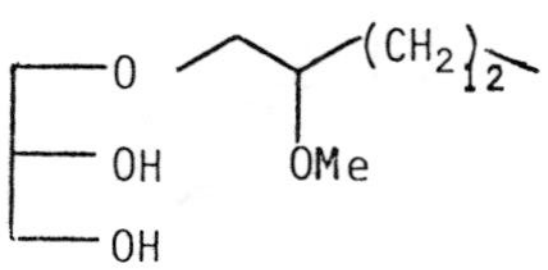	1-*O*-(2-Methoxypentadecyl)glycerol	Cod liver oil Crayfish Herrings Mackerel	+		114

Table 27 (continued)

Molecular formula II/—	Structural formula	Name	Source of compound	Available by synthesis	Available nonmarine sources	Activity	References
C_{20}							
197. $C_{20}H_{13}N_3O_3$	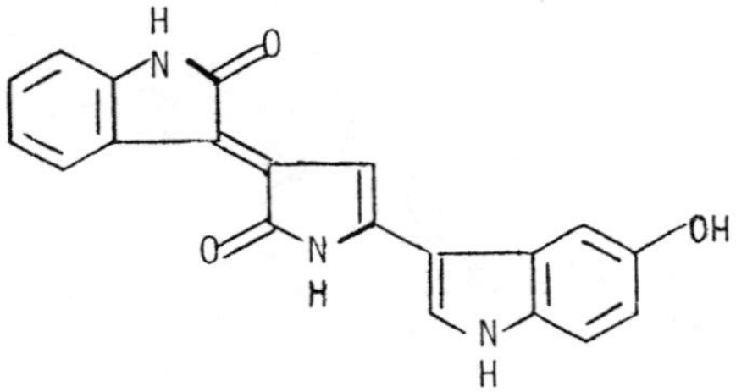	Violacein	Bacterium: *Chromobacterium* sp.	+			6 202
198. $C_{20}H_{17}N_3O_2$	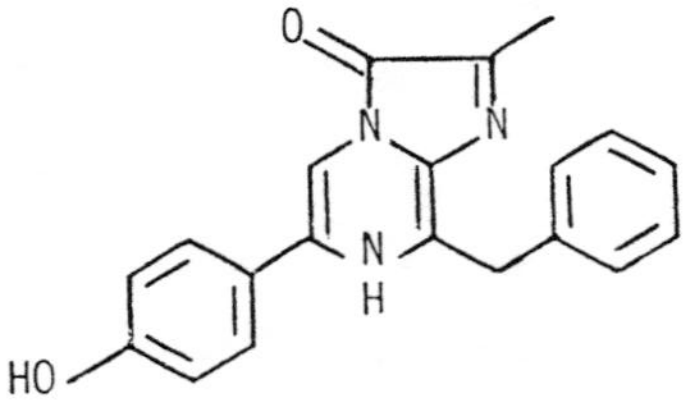	Renilla luciferin	Soft coral: *Renilla reniformis*	+			53,54 55,56 57,104 121,122 123,184
199. $C_{20}H_{24}O_4$	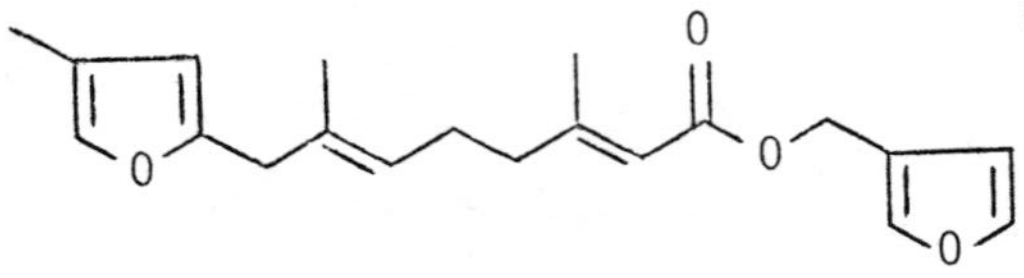	Pleraplysillin-2	Sponge: *Pleraplysilla spinifera*				48

No.	Structure	Name	Source	Activity	Ref.
200. $C_{20}H_{28}O_3$	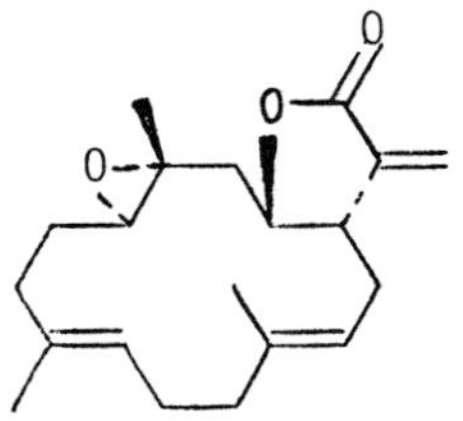	Lobophytolide	Soft coral: *Lobophytum crista-galli*		267
201. $C_{20}H_{28}O_3$	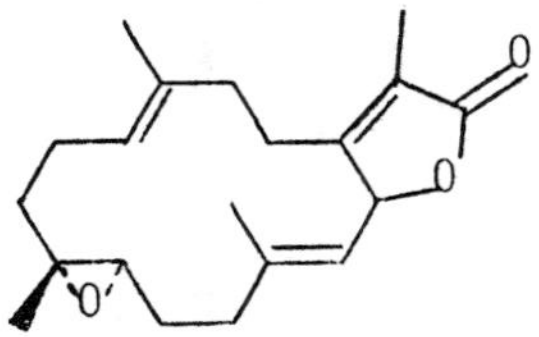	2,3,6,7,10a,13,14,14a-Octahydro-1a,5,8,12-tetramethyloxireno[9,10]cyclotetradeca[1,2-b]furan-9(1aH)-one	Soft coral: *Sarcophyton glaucum*		150
202. $C_{20}H_{28}O_3$	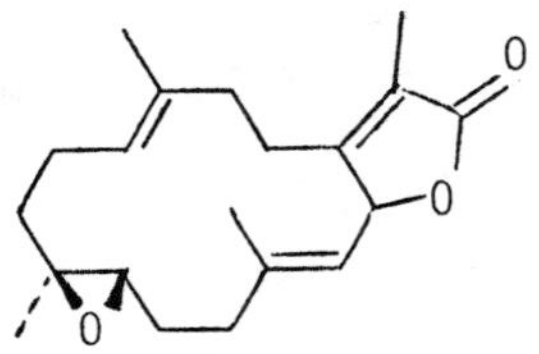	2,3,6,7,10a,13,14,14a-Octahydro-1a,5,8,12-tetramethyloxireno[9,10]cyclotetradeca[1,2-b]furan-9(1aH)-one	Soft coral: *Sarcophyton glaucum*		150
203. $C_{20}H_{28}O_3$	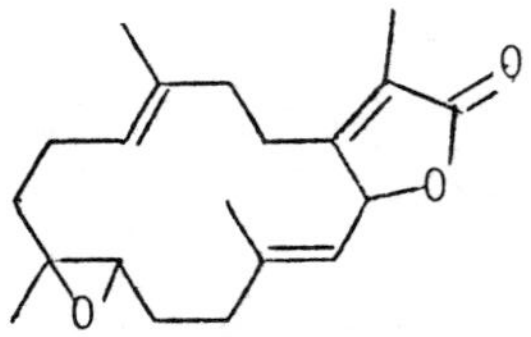	Sarcophine	Soft coral: *Sarcophyton glaucum* *S. trocheliophorum*	Antiacetyl-choline Cholinesterase inhibitor	17 162 222

Table 27 (continued)

Molecular formula II/—	Structural formula	Name	Source of compound	Available by synthesis	Available nonmarine sources	Activity	References
C_{20} 204. $C_{20}H_{28}O_4$	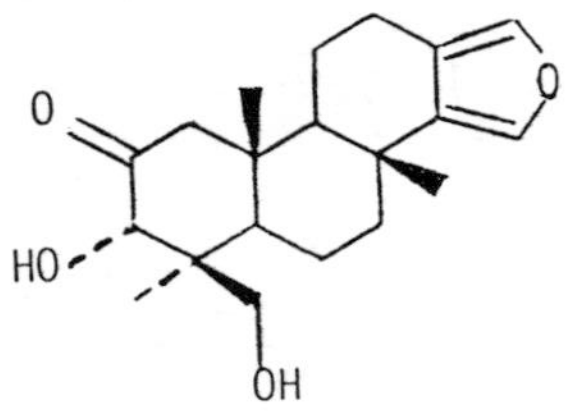	Spongiadiol	Sponges: *Spongia* sp.				162
205. $C_{20}H_{28}O_4$	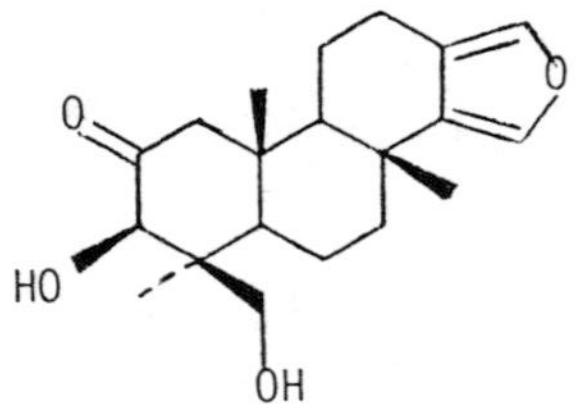	*epi*-Spongiadiol	Sponges: *Spongia* sp.				162
206. $C_{20}H_{28}O_5$	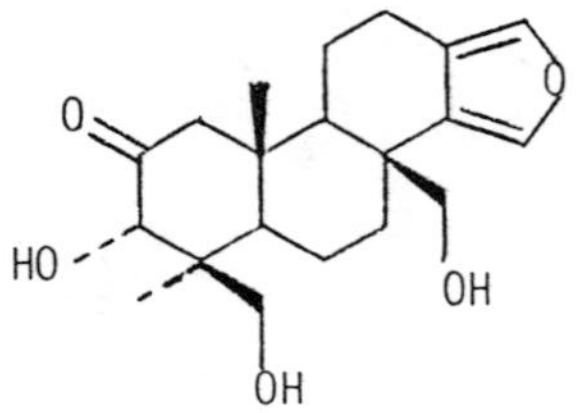	Spongiatriol	Sponges: *Spongia* sp.				162

No.	Structure	Name	Source			Ref.
207. $C_{20}H_{28}O_5$	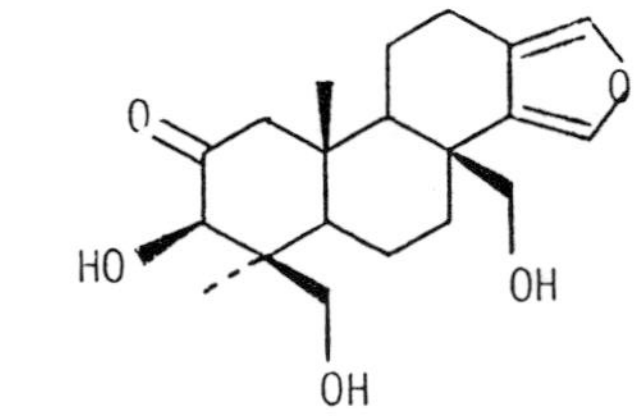	*epi*-Spongiatriol	Sponges: *Spongia* sp.			162
208. $C_{20}H_{30}O$	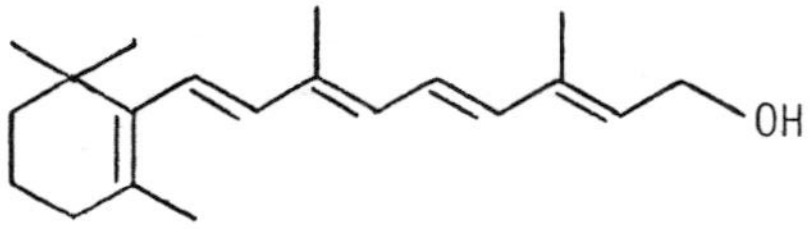	Retinol Vitamin A	Crustaceans Fish	+	+	14,62 86,87 202
209. $C_{20}H_{30}O_2$	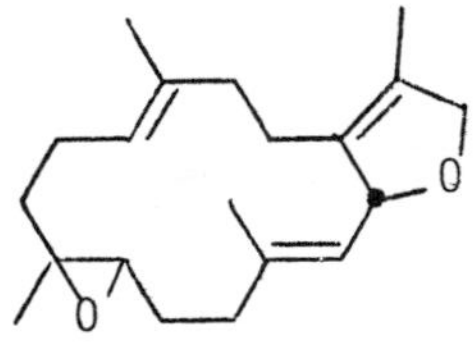	1a,2,3,6,7,9,10a,13,14,14a-Decahydro-1a,5,8,12-tetramethyloxireno-[9,10]cyclotetradeca-[1,2-b]furan	Soft coral: *Sarcophyton glaucum*			150
210. $C_{20}H_{30}O_2$	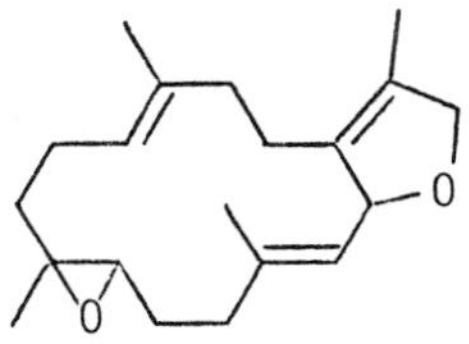	1a,2,3,6,7,9,10a,13,14,14a-Decahydro-1a,5,8,12-tetramethyloxireno-[9,10]cyclotetradeca-[1,2-b]furan	Soft coral: *Sarcophyton glaucum*			150
211. $C_{20}H_{30}O_2$	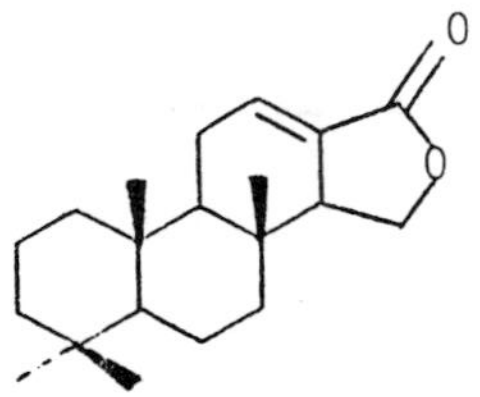	Isoagatholactone	Sponge: *Spongia officinalis*			50

Table 27 (continued)

Molecular formula II/—	Structural formula	Name	Source of compound	Available by synthesis	Available nonmarine sources	Activity	References
C_{20}							
212. $C_{20}H_{30}O_4$	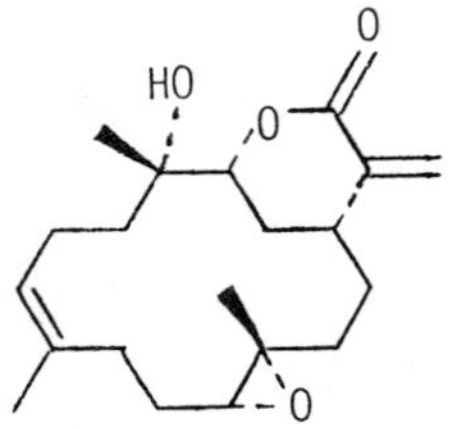	Flexibilide	Soft coral: *Sinularia flexibilis*				161
213. $C_{20}H_{32}O$	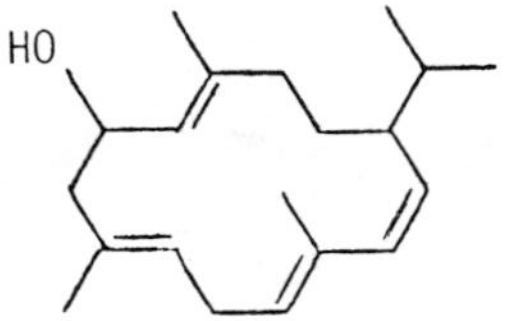	6-Isopropyl-3,9,13-trimethyl-2,7,9,12-cyclotetradecatetraen-1-ol	Soft coral: *Sarcophyton glaucum*				150
214. $C_{20}H_{32}O_2$	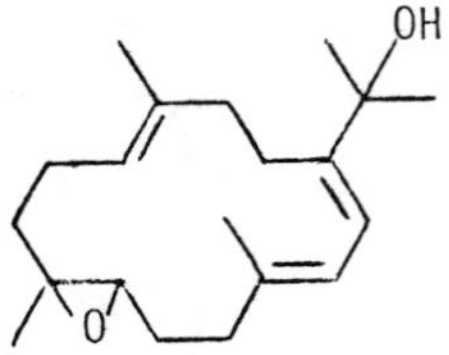	Dehydroepoxynephthenol	Soft coral: *Lobophytum* sp.				51

No.	Structure	Name	Source		Ref.
215. $C_{20}H_{32}O_4$	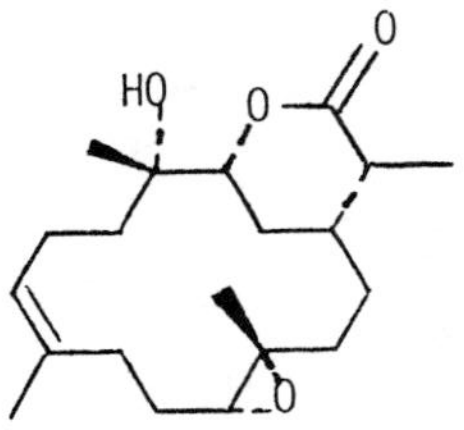	Dihydroflexibilide	Soft coral: *Sinularia flexibilis*		161
216. $C_{20}H_{34}O$	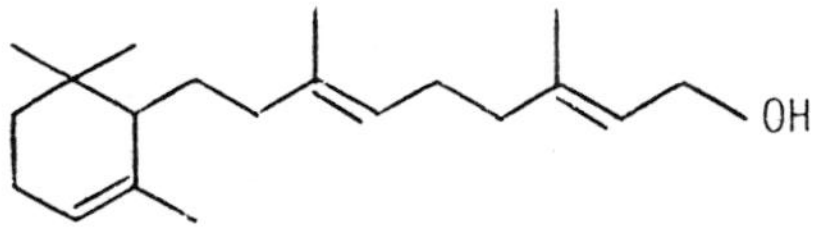	Caulerpol	Green alga: *Caulerpa brownii*		21
217. $C_{20}H_{34}O$	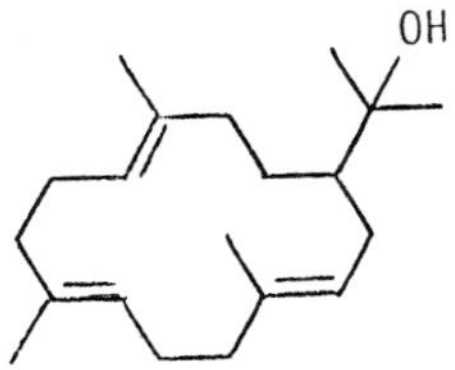	Nephthenol	Soft coral: *Nephthea* sp.		242
218. $C_{20}H_{34}O_5$	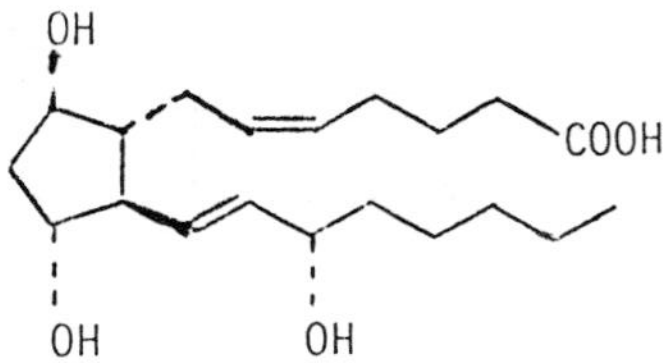	$PGF_{2\alpha}$	Tuna: *Thunnus thynnus*	+	202 224

Table 27 (continued)

Molecular formula II/—	Structural formula	Name	Source of compound	Available by synthesis	Available nonmarine sources	Activity	References
C_{20}							
219. $C_{20}H_{36}O_5$	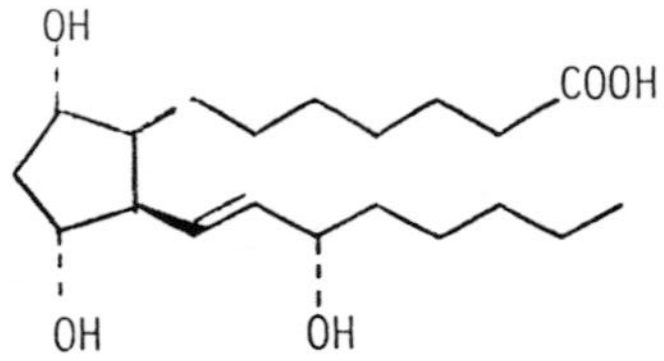	$PGF_{1\alpha}$	Salmon: *Oncorhynchus keta*		+		202 224
220. $C_{20}H_{40}O$	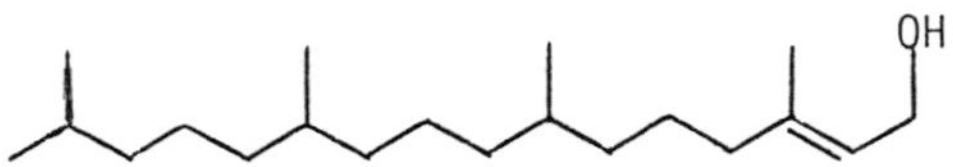	Phytol	Brown alga: *Hizikia fusiformis*	+		Lipase-activa-tor	174 202
221. $C_{20}H_{40}O_4$	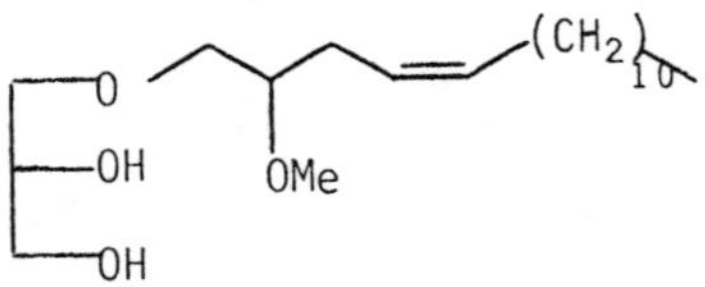	1-*O*-(2-Methoxy-4-hex-adecenyl)glycerol	Shark: *Somniosus microce-phalus*	+			113
222. $C_{20}H_{42}$	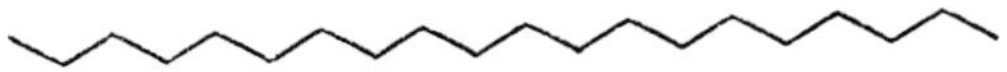	Eicosane	Algae				283
223. $C_{20}H_{42}O$	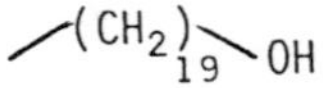	Eicosanol	Coelenterates				194
224. $C_{20}H_{42}O_3$	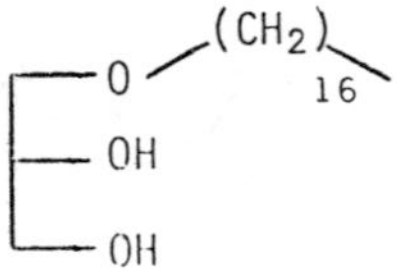	1-*O*-Heptadecylglycerol	Cod liver oil Coelenterates Crayfish Herrings Mackerel Mussel Shrimp		+		114 194

No. / Formula	Name	Source			Activity	Ref.
225. $C_{20}H_{42}O_4$	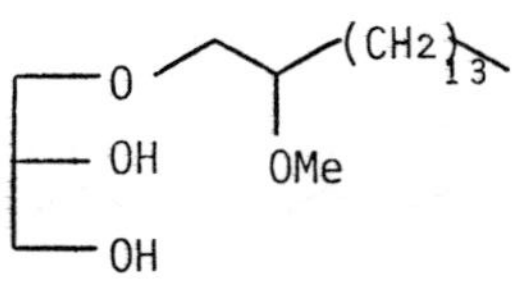 1-*O*-(2-Methoxyhexadecyl)glycerol	Cod liver oil Crayfish Herrings Mackerel Mussel Shark Shrimp	+	+	Antitumor	26 113 114
C_{21}						
226. $C_{21}H_{26}O_3$	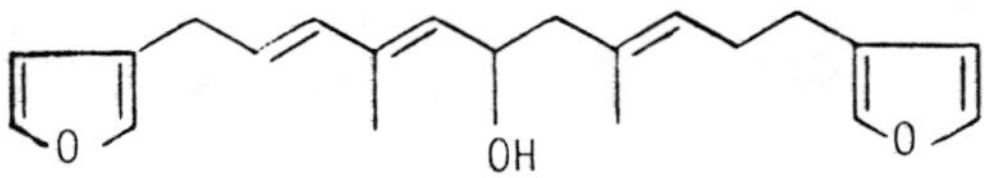 Tetradehydrofurospongin-1	Sponges: *Spongia* sp.				162
227. $C_{21}H_{28}O_2$	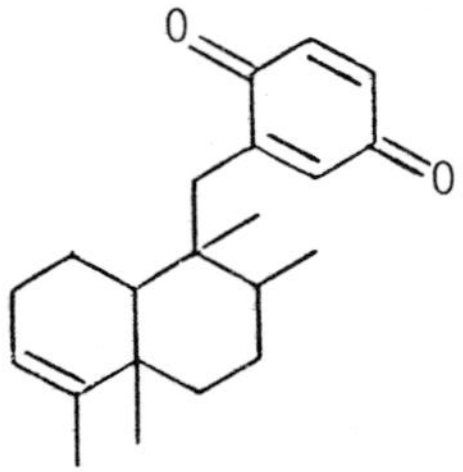 Avarone	Sponge: *Dysidea avara*				207
228. $C_{21}H_{28}O_3$	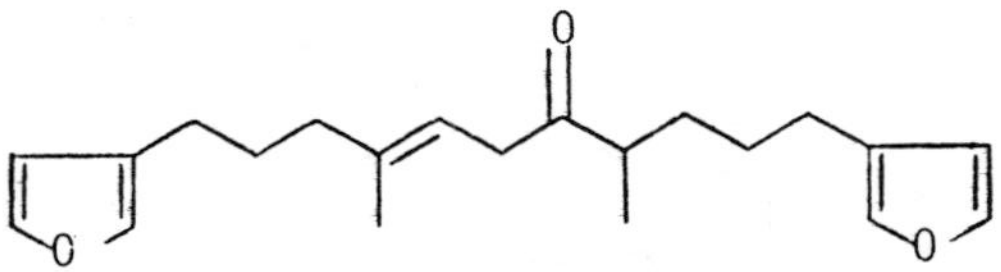 Furospongenone	Sponge: *Spongia* sp.				162
229. $C_{21}H_{30}O_2$	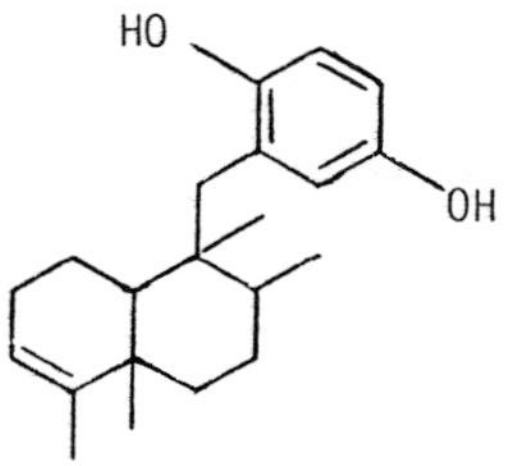 Avarol	Sponge: *Dysidea avara* *Dysidea* sp.				162 207

Table 27 (continued)

Molecular formula II/—	Structural formula	Name	Source of compound	Available by synthesis	Available nonmarine sources	Activity	References
C_{21}							
230. $C_{21}H_{30}O_2$	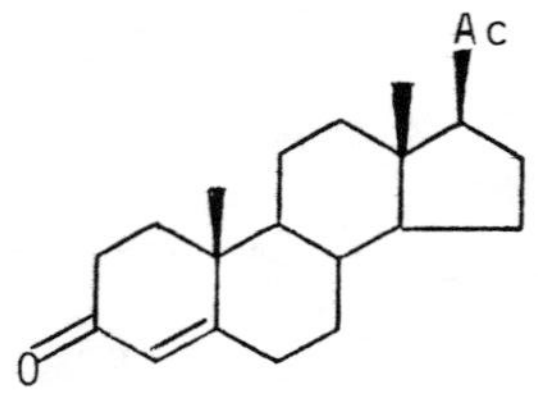	Progesterone	Dogfish: *Squalus suckleyi* Mollusc: *Pecten hericius* Sea urchin: *Strongylocentrotus franciscanus* Starfish: *Pisaster ochraceous*	+	+		31 32 202 273
231. $C_{21}H_{30}O_2$	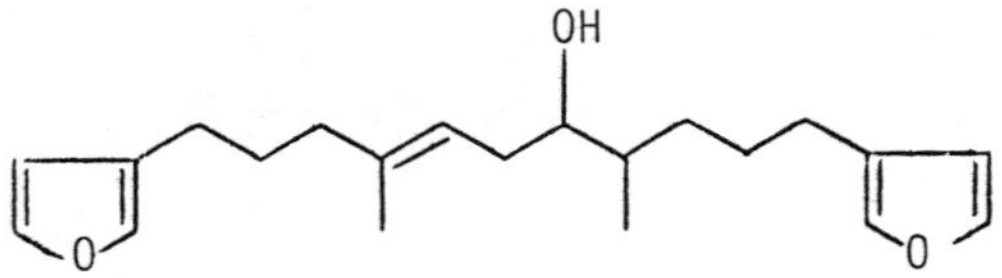	Furospongenol	Sponges: *Spongia* sp.				162
232. $C_{21}H_{30}O_5$	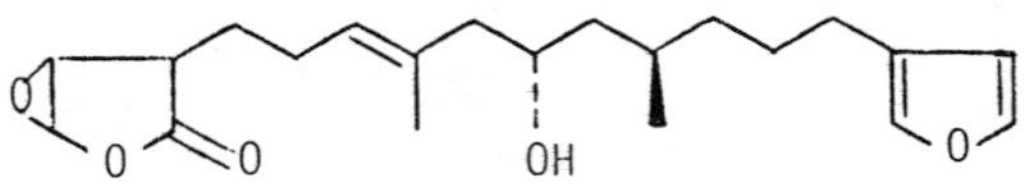	4-[11-(3-Furyl)-6-hydroxy-4,8-dimethyl-3-undecenyl]-2,6-dioxabicyclo[3.1.0] hexan-3-one	Sponge: *Spongia officinalis*				47
233. $C_{21}H_{30}O_5$	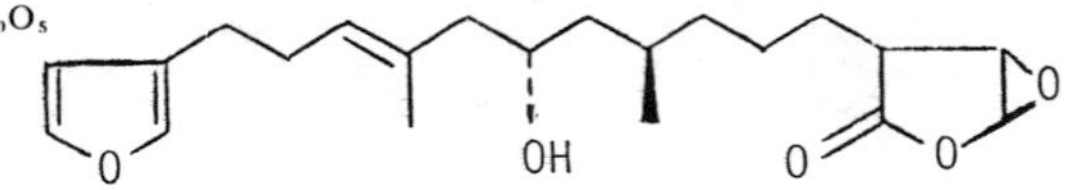	4-[11-(3-Furyl)-6-hydroxy-4,8-dimethyl-8-undecenyl]-2,6-dioxabicyclo[3.1.0] hexan-3-one	Sponge: *Spongia officinalis*				47
234. $C_{21}H_{30}O_5$	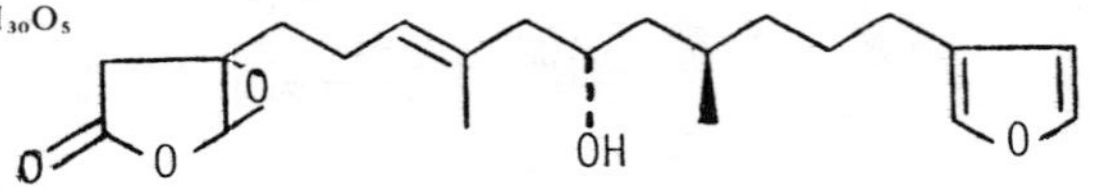	5-[11-(3-Furyl)-6-hydroxy-4,8-dimethyl-3-undecenyl]-2,6-dioxabicyclo[3.1.0] hexan-3-one	Sponge: *Spongia officinalis*				47

235. $C_{21}H_{30}O_5$	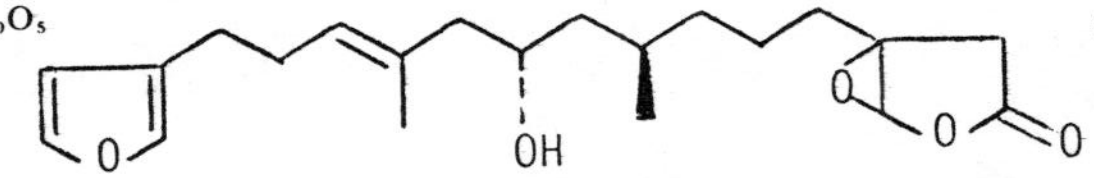 5-[11-3-Furyl)-6-hydroxy-4,8-dimethyl-8-undecenyl]-2,6-dioxabicyclo[3.1.0] hexan-3-one	Sponge: *Spongia officinalis*	47
236. $C_{21}H_{30}O_5$	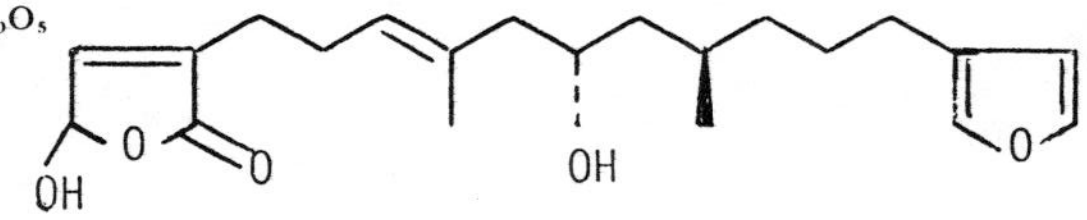 3-[11-(3-Furyl)-6-hydroxy-4,8-dimethyl-3-undecenyl]-5-hydroxy-2(5H)-furanone	Sponge: *Spongia officinalis*	47
237. $C_{21}H_{30}O_5$	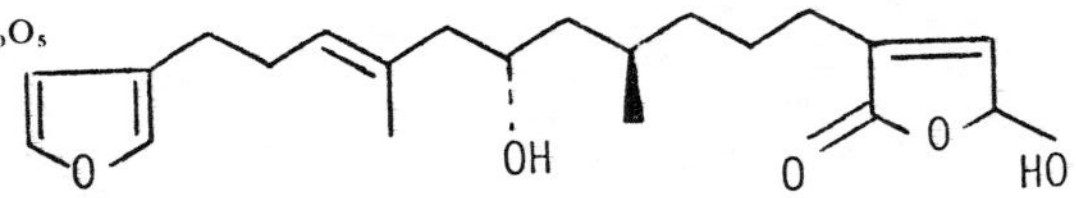 3-[11-(3-Furyl)-6-hydroxy-4,8-dimethyl-8-undecenyl]-5-hydroxy-2(5H)-furanone	Sponge: *Spongia officinalis*	47
238. $C_{21}H_{30}O_5$	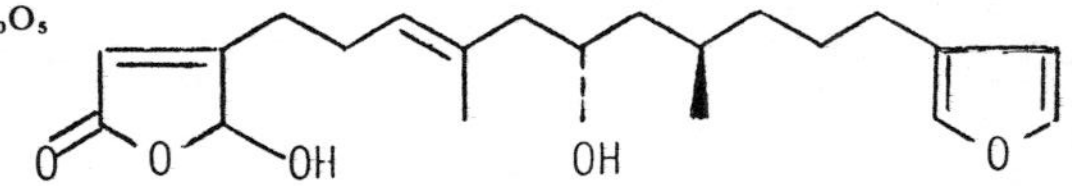 4[11-(3-Furyl)-6-hydroxy-4,8-dimethyl-3-undecenyl]-5-hydroxy-2(5H)-furanone	Sponge: *Spongia officinalis*	47
239. $C_{21}H_{30}O_5$	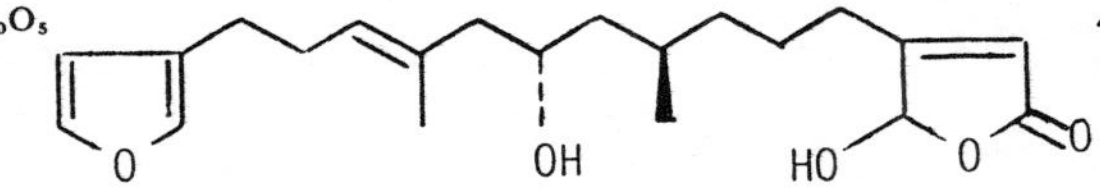 4-[11-(3-Furyl)-6-hydroxy-4,8-dimethyl-8-undecenyl]-5-hydroxy-2(5H)-furanone	Sponge: *Spongia officinalis*	47
240. $C_{21}H_{32}O_2$	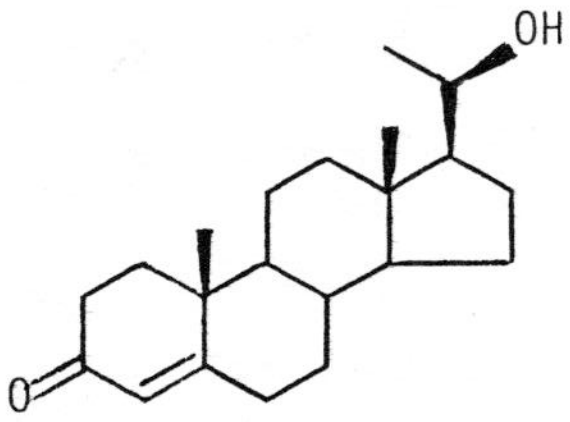 Δ^4-3-Keto-pregnen-20β-ol	Mollusc: *Pecten hericius* Sea urchin: *Strongylocentrotus franciscanus*	32

Table 27 (continued)

Molecular formula II/—	Structural formula	Name	Source of compound	Available by synthesis	Available nonmarine sources	Activity	References
C_{21}							
241. $C_{21}H_{32}O_4$	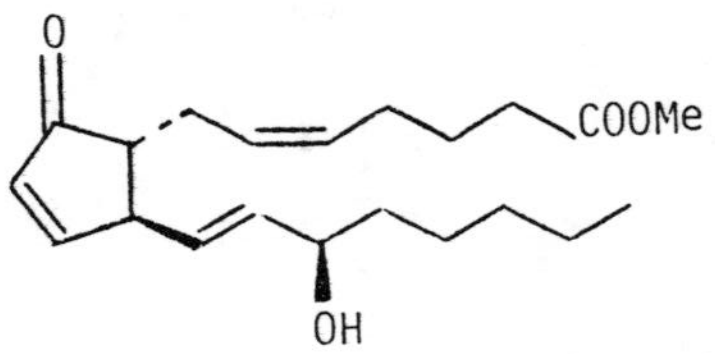 O; COOMe; OH	15 *R*-PGA₂ methyl ester	Gorgonian: *Plexaura homomalla*				10
242. $C_{21}H_{33}N$	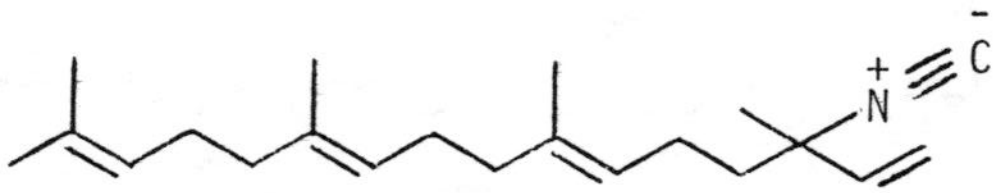 $\overset{+}{N}\equiv\overset{-}{C}$	1,5,9,13-Tetramethyl-1-vinyl-4,8,12-tetradecatrienyl isocyanide	Sponge: *Halichondria* sp.				36
243. $C_{21}H_{33}NS$	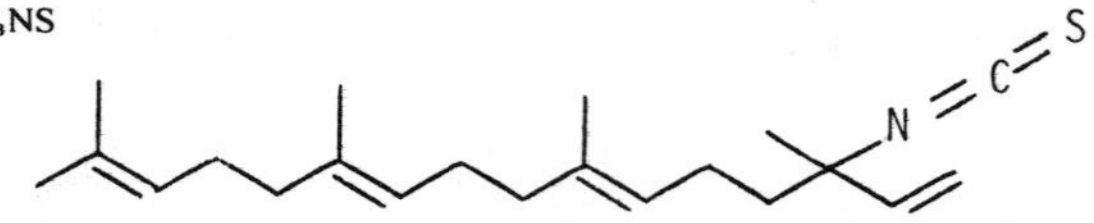 N=C=S	1,5,9,13-Tetramethyl-1-vinyl-4,8,12-tetradecatrienyl isothiocyanate	Sponge: *Halichondria* sp.				36
244. $C_{21}H_{34}$		*cis,cis,cis,cis,cis*-3,6,9,12,15-Heneicosapentaene	Green algae: *Monostroma* sp. *Spongomorpha arcta*				283
245. $C_{21}H_{35}NO$	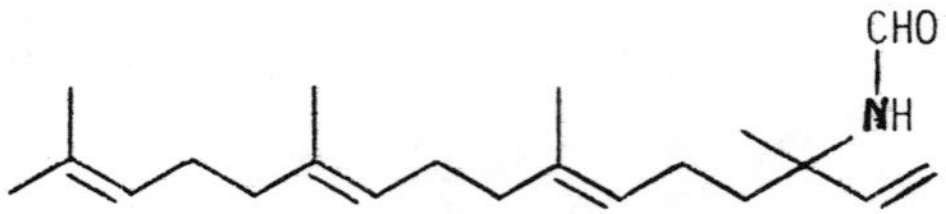 CHO; NH	*N*-(1,5,9,13-Tetramethyl-1-vinyl-4,8,12-tetradecatrienyl)-formamide	Sponge: *Halichondria* sp.				36

No.	Compound	Source		References
246. $C_{21}H_{42}O_4$	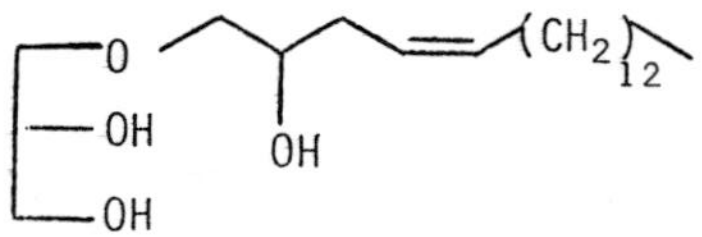 1-*O*-(2-Hydroxy-4-octadecenyl)glycerol	Shark: *Somniosus microcephalus*		113 115
247. $C_{21}H_{44}$	 Heneicosane	Algae: *Fucus distichus* *F. vesiculosus* *Laminaria saccharina* *Prasiola stipitata* *Rhodomela confervoides*		283
248. $C_{21}H_{44}O_4$	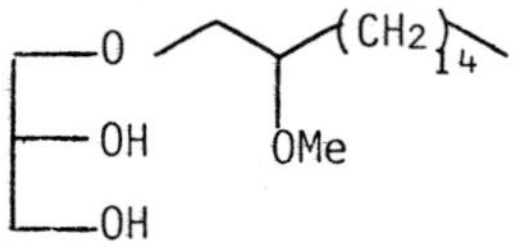 1-*O*-(2-Methoxyheptadecyl)glycerol	Cod liver oil Crayfish Herrings Mackerel Mussel Shrimp	+	114
C_{22}				
249. $C_{22}H_{27}N_7O$	Cypridina luciferin	Crustacean: *Cypridina hilgendorfii*	+	18,55 104,105 106,107 108,149 166,184

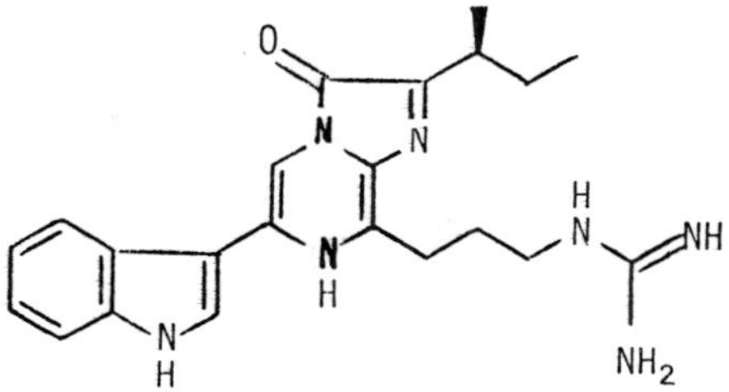

Table 27 (continued)

Molecular formula II/—	Structural formula	Name	Source of compound	Available by synthesis	Available nonmarine sources	Activity	References
C_{21}							
250. $C_{22}H_{28}O_4$	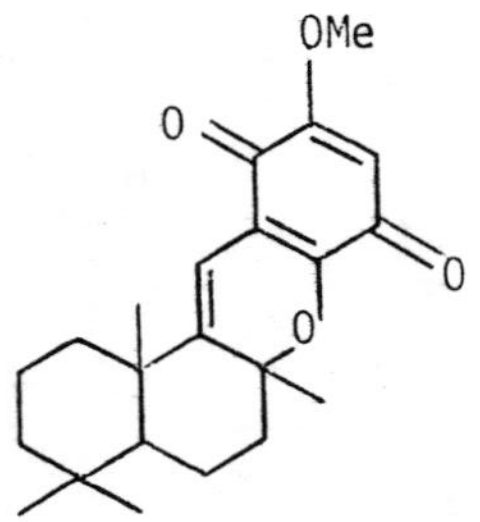	Didehydrocyclospong-iaquinone-1	Sponge				162
251. $C_{22}H_{30}O_4$	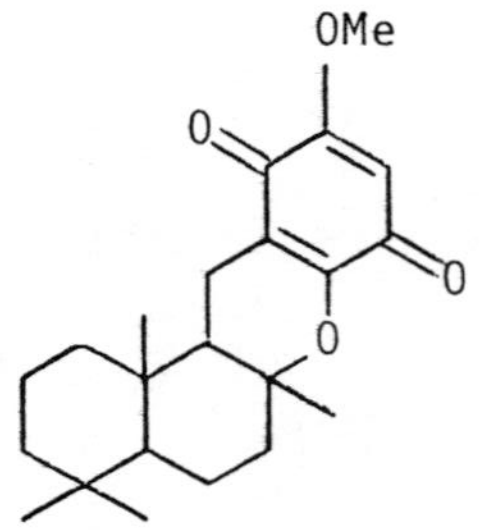	Cyclospongiaquinone-1	Sponge				162

No.	Formula	Name	Source	Ref.
252.	$C_{22}H_{30}O_4$	Cyclospongiaquinone-2	Sponge	162
253.	$C_{22}H_{30}O_4$	Isospongiaquinone	Sponge	162
254.	$C_{22}H_{30}O_4$	Spongiaquinone	Sponge	162

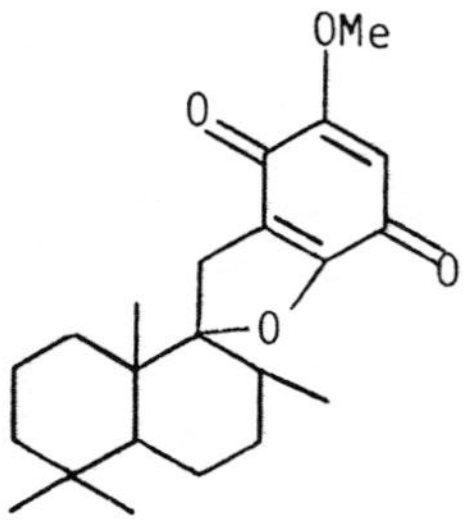

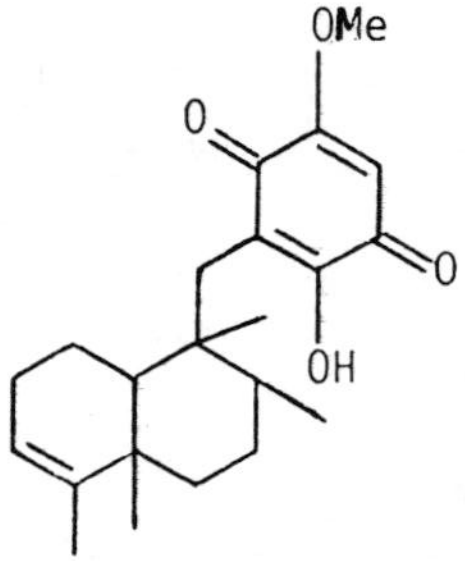

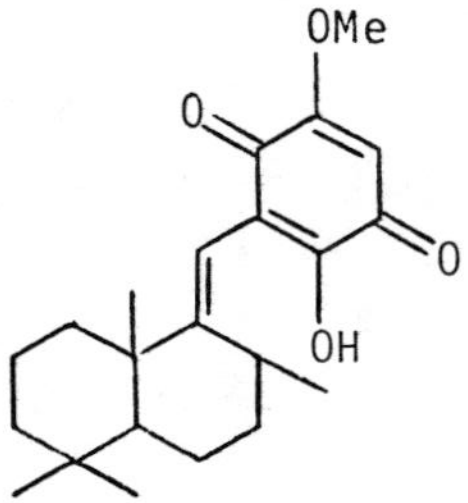

Table 27 (continued)

Molecular formula II/—	Structural formula	Name	Source of compound	Available by synthesis	Available nonmarine sources	Activity	References
C_{22}							
255. $C_{22}H_{36}O_2$	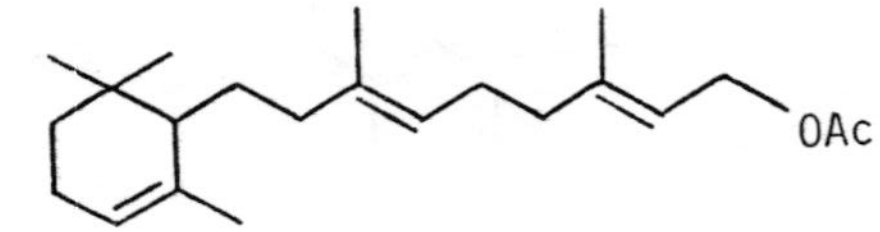	Caulerpol acetate	Green alga: *Caulerpa brownii*				21
256. $C_{22}H_{36}O_3$	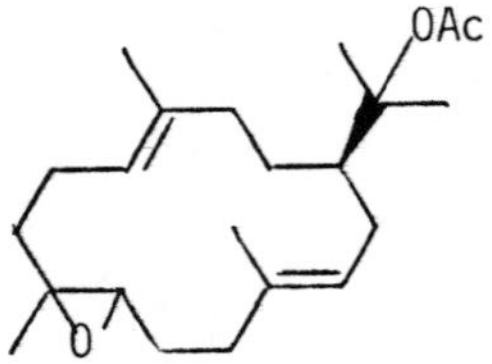	Epoxynephthenol acetate	Soft coral: *Nephthea* sp.				242
257. $C_{22}H_{44}O_4$	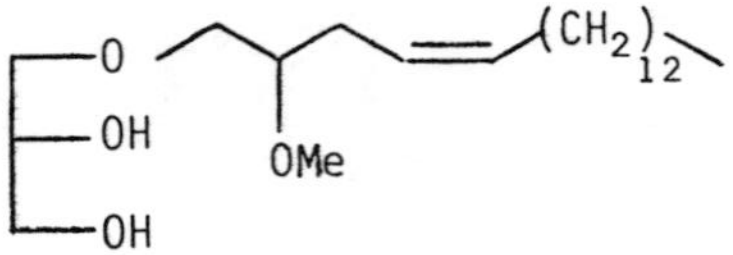	1-*O*-(2-Methoxy-4-octadecenyl)glycerol	Shark: *Somniosus microcephalus*	+			113
258. $C_{22}H_{46}$	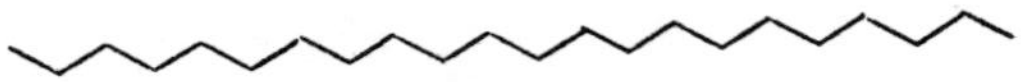	Docosane	Algae: *Fucus vesiculosus* *Laminaria saccharina* *Prasiola stipitata* *Rhodomela confervoides*				283

259. $C_{22}H_{46}O_3$	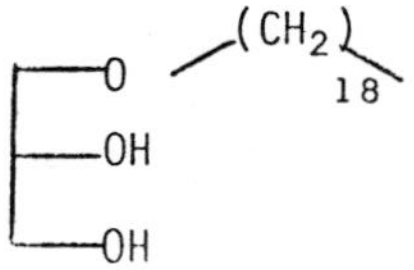 O, OH, OH, $(CH_2)_{18}$	1-*O*-Nonadecylglycerol	Cod liver oil Coelenterates Crayfish Herring Mackerel Mussel Shrimp	+	114 194
260. $C_{22}H_{46}O_4$	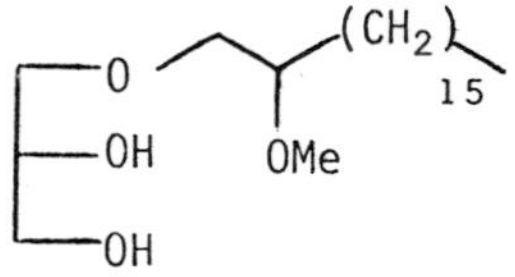 O, OH, OH, OMe, $(CH_2)_{15}$	1-*O*-(2-Methoxyoctade-cyl)glycerol	Cod liver oil Crayfish Herrings Mackerel Mussel Shrimp	+	114
C_{23}					
261. $C_{23}H_{32}O_2$	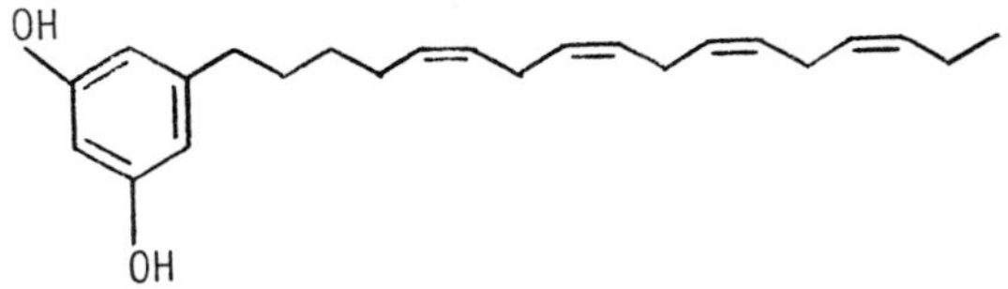 OH, OH	1-(Heptadeca-5,8,11,14-tetraenyl)-3,5-dihydroxybenzene	Brown alga: *Cystophora torulosa*		162
262. $C_{23}H_{48}$		Tricosane	Algae: *Fucus distichus* *F. vesiculosus* *Laminaria saccharina* *Prasiola stipitata* *Rhodomela confervoides*		283
263. $C_{23}H_{48}O_3$	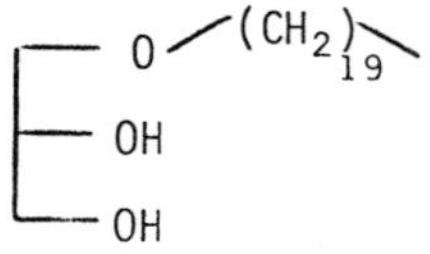 O, OH, OH, $(CH_2)_{19}$	1-*O*-Eicosylglycerol	Cod liver oil Coelenterates Crayfish Herrings Mackerel Mussel Shrimp	+	114 194

Table 27 (continued)

Molecular formula II/—	Structural formula	Name	Source of compound	Available by synthesis	Available nonmarine sources	Activity	References
C_{22}							
264. $C_{23}H_{48}O_3$	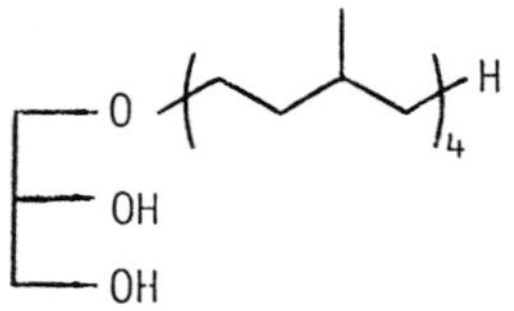	1-*O*-Phytanylglycerol	Cod liver oil	+			114
265. $C_{23}H_{48}O_4$	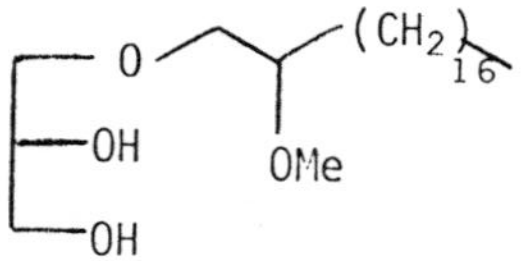	1-*O*-(2-Methoxynona-decyl)glycerol	Cod liver oil Crayfish Herrings Mackerel Mussel Shrimp		+		114
C_{24}							
266. $C_{24}H_{32}O_6$	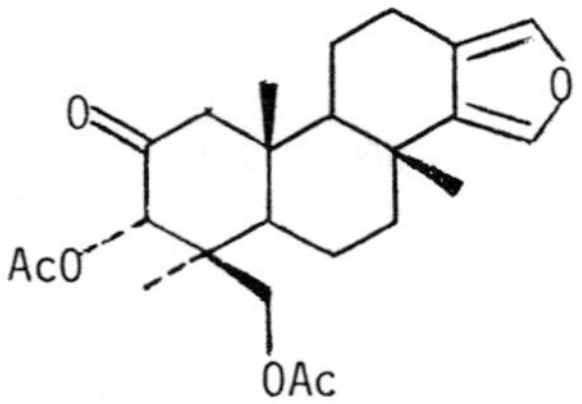	Spongiadiol diacetate	Sponges: *Spongia* sp.				162
267. $C_{24}H_{32}O_6$	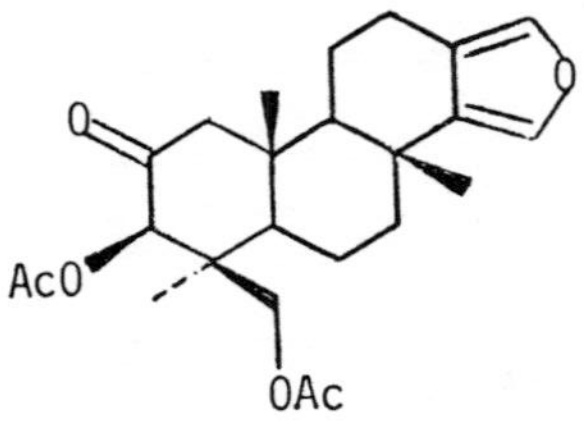	*epi*-Spongiadiol diace-tate	Sponges: *Spongia* sp.				162

No.	Structure	Name	Source		Ref.
268. $C_{24}H_{42}O_4$	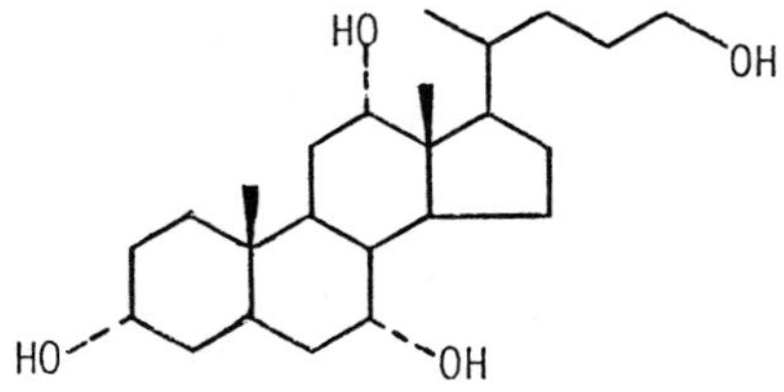 Occurs as 24-sulfate ester	Petromyzonol	Fish: *Petromyzon marinus*		258
269. $C_{24}H_{50}$	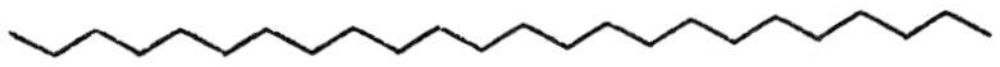	Tetracosane	Algae: *Fucus distichus*, *F. vesiculosus*, *Laminaria saccharina*, *Prasiola stipitata*, *Rhodomela confervoides*		283
270. $C_{24}H_{50}O_3$	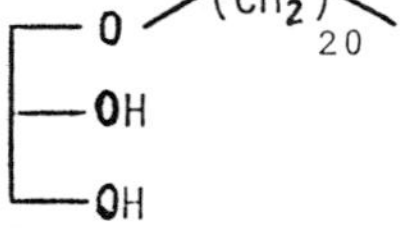	1-*O*-Heneicosylglycerol	Cod liver oil; Herring: *Clupea harengus*; Mussel: *Mytilus edulis*	+	114
271. $C_{24}H_{50}O_4$	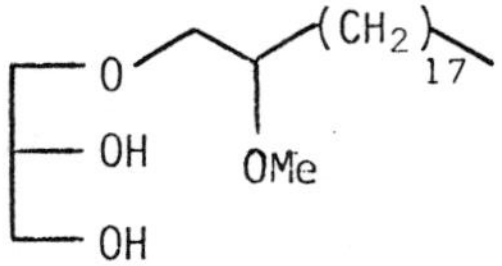	1-*O*-(2-Methoxyeicosyl) glycerol	Cod liver oil; Crayfish; Herrings; Mussel; Shrimp	+	114

Table 27 (continued)

Molecular formula II/—	Structural formula	Name	Source of compound	Available by synthesis	Available nonmarine sources	Activity	References
C_{25}							
272. $C_{25}H_{21}N_3O_9S_2$	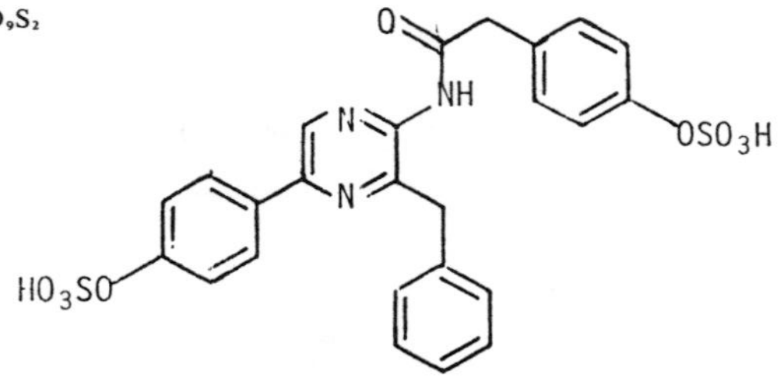	Watasenia oxyluciferin	Squid: *Watasenia scintillans*	+			109
273. $C_{25}H_{28}O_4$	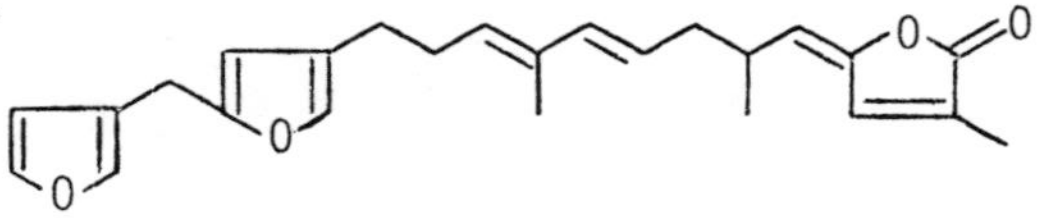	Ircinolide	Sponge: *Thorecta marginalis*				162
274. $C_{25}H_{28}O_5$	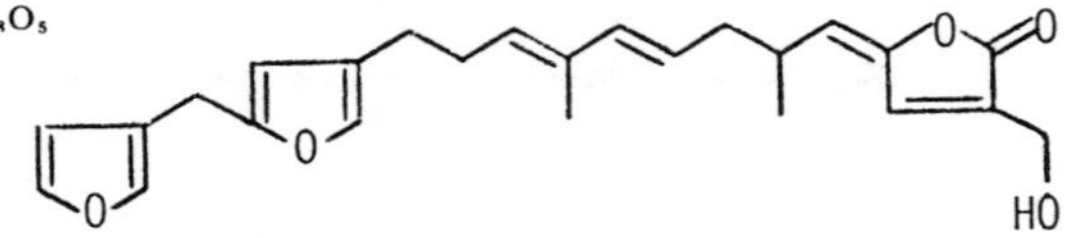	24-Hydroxyircinolide	Sponge: *Thorecta marginalis*				162
275. $C_{25}H_{32}O_4$	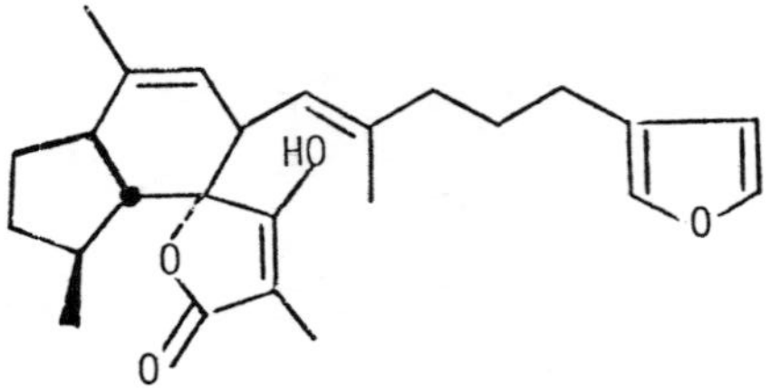	Ircinianin	Sponge: *Ircinia* sp.				72

No.	Formula	Structure	Name	Source		Ref.
276.	$C_{25}H_{34}O_4$	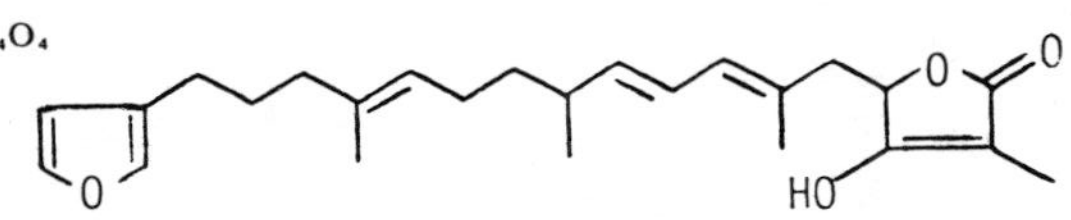	Halmiformin-1	Sponge: *Ircinia halmiformis*		162
277.	$C_{25}H_{36}O_3$	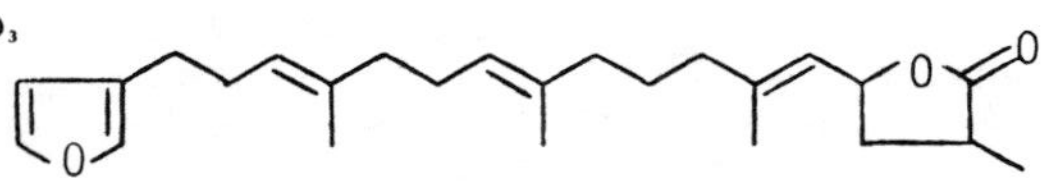	Fasciospongin	Sponge: *Fasciospongia* sp.		162
278.	$C_{25}H_{42}O$	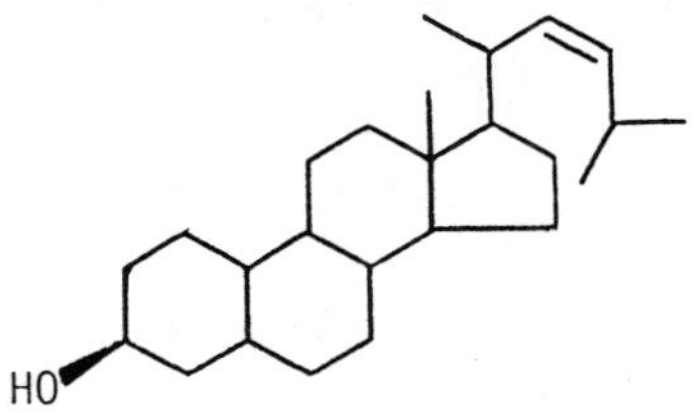	19,24-Bisnorcholest-*cis*-22-en-3β-ol	Sponge: *Axinella polypoides*		204
279.	$C_{25}H_{42}O$	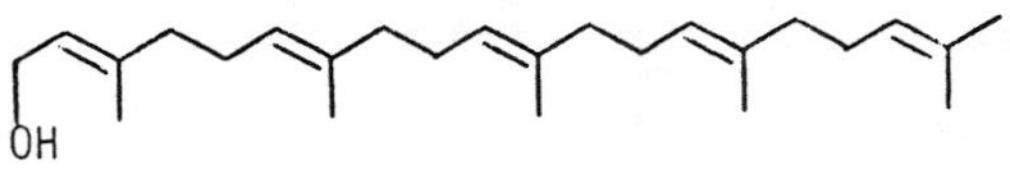	Geranylfarnesol	Sponge: *Fasciospongia fovea*		162
280.	$C_{25}H_{52}O_3$	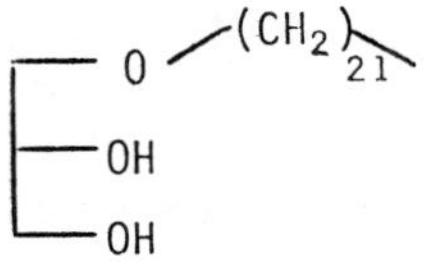 O–$(CH_2)_{21}$; OH; OH	1-*O*-Docosylglycerol	Cod liver oil Herring: *Clupea harengus* Mussel: *Mytilus edulis*	+	114
281.	$C_{25}H_{52}O_4$	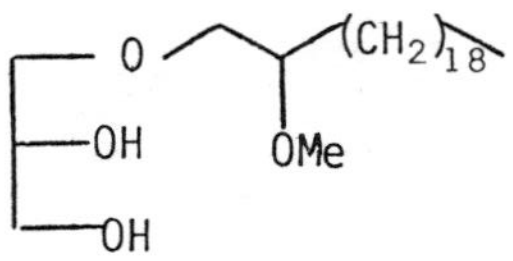 O–$(CH_2)_{18}$; OMe; OH; OH	1-*O*-(2-Methoxyheneicosyl)glycerol	Herring: *Clupea harengus*	+	114

Table 27 (continued)

Molecular formula II/—	Structural formula	Name	Source of compound	Available by synthesis	Available nonmarine sources	Activity	References
C_{26}							
282. $C_{26}H_{34}O_8$	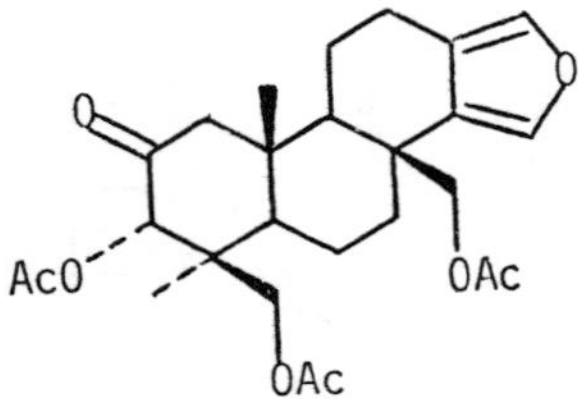	Spongiatriol triacetate	Sponges: *Spongia* sp.				162
283. $C_{26}H_{34}O_8$	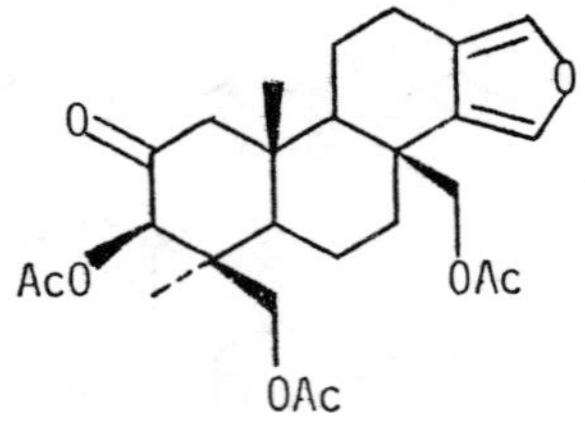	*epi*-Spongiatriol triacetate	Sponges: *Spongia* sp.				162
284. $C_{26}H_{42}NO_6S^-$	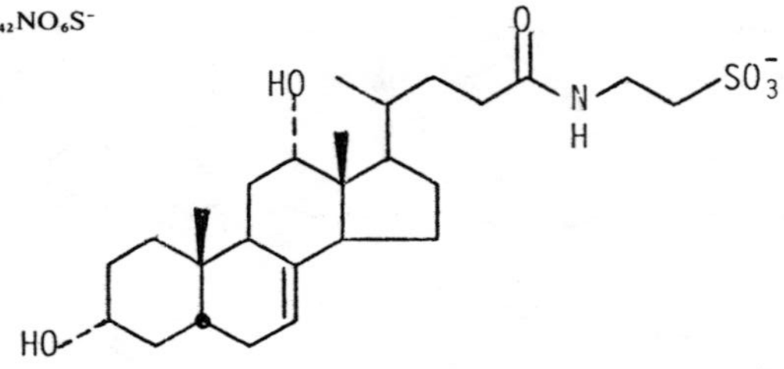	Tauro-3α,12α-dihydroxy-5β-chol-7-en-24-oate	Fish: *Myoxocephalus quadricornis*				258

Formula	Compound	Source	Ref.
285. $C_{26}H_{42}O$	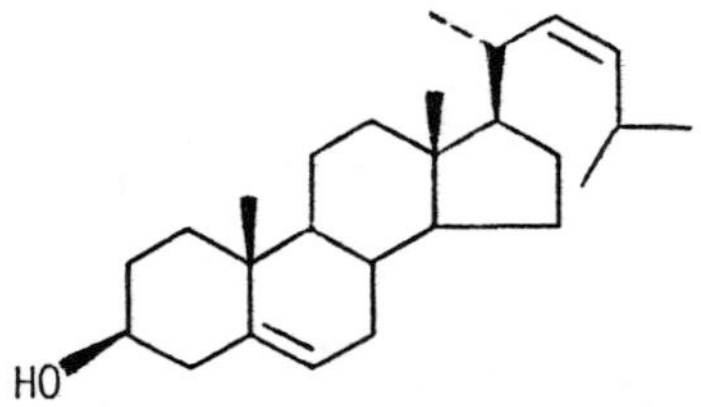 22-*cis*-24-Norcholesta-5,22-dien-3β-ol	Sponges: *Verongia aerophoba*, *V. archeri*, *V. fistularis*, *V. thiona*	68
286. $C_{26}H_{44}NO_6S^-$	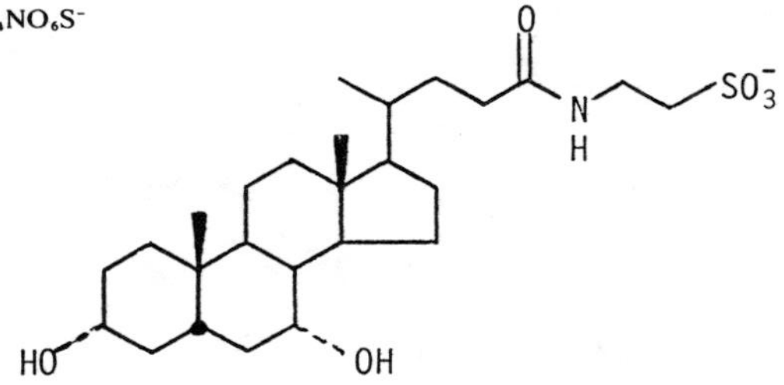 Taurochenodeoxycholate	Fish bile	258
287. $C_{26}H_{44}NO_6S^-$	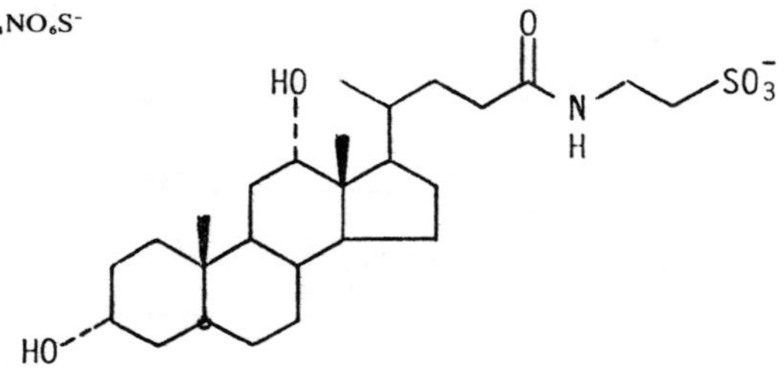 Taurodeoxycholate	Fish: *Fugu rubripes*, *Gadus callarias*, *Myoxocephalus quadricornis*	258

Table 27 (continued)

Molecular formula II/—	Structural formula	Name	Source of compound	Available by synthesis	Available nonmarine sources	Activity	References
C_{26}							
288. $C_{26}H_{44}NO_6S^-$	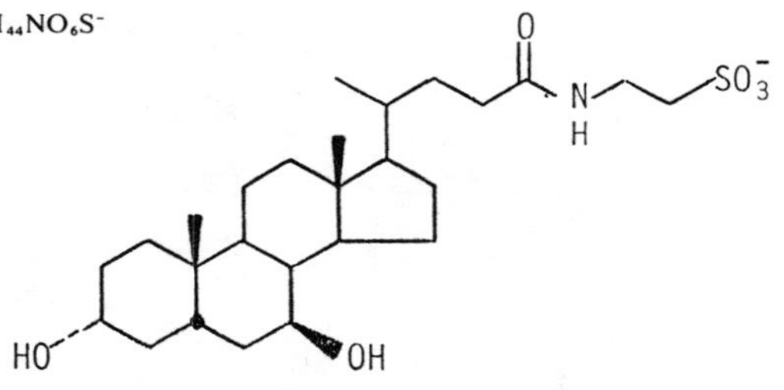	Tauro-3α,7β-dihydroxy-5β-cholan-24-oate	Fish: *Gadus callarias*				258
289. $C_{26}H_{44}NO_7S^-$	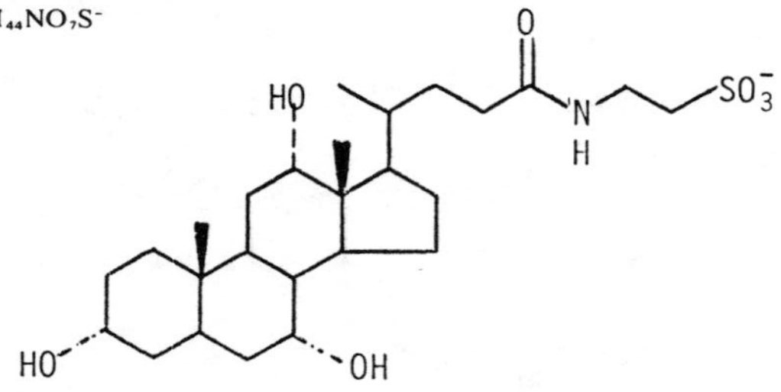	Tauroallocholate	Fish: *Fugu rubripes* *Tetrodon porphyreus*				258
290. $C_{26}H_{44}NO_7S^-$	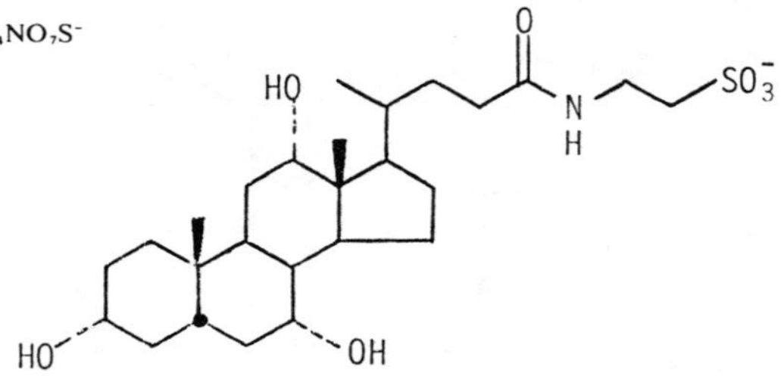	Taurocholate	Fish bile				258

No.	Formula	Name	Source	Ref.
291.	$C_{26}H_{44}NO_7S^-$	Taurohaemulcholate	Fish: *Parapristipoma trilineatum*, *Plectorhynchus cinctus*	258
292.	$C_{26}H_{44}NO_7S^-$	Tauro-3α,7β-12α-trihydroxy-5β-cholan-24-oate	Eel: *Conger myriaster*; Fish: *Gadus callarias*	258
293.	$C_{26}H_{44}O$	19-Norcholest-*trans*-22-en-3β-ol	Sponge: *Axinella polypoides*	204
294.	$C_{26}H_{46}O$	19-Norcholestan-3β-ol	Sponge: *Axinella polypoides*	204

Table 27 (continued)

Molecular formula II/—	Structural formula	Name	Source of compound	Available by synthesis	Available nonmarine sources	Activity	References
C_{26}							
295. $C_{26}H_{46}O_2$	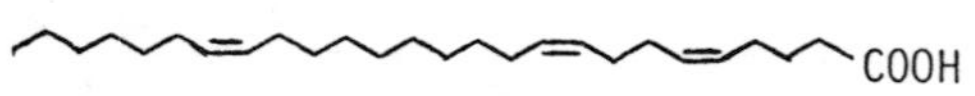	*cis, cis, cis*-5,9,19-Hexacosatrienoic acid	Sponge: *Microciona prolifera*				136
296. $C_{26}H_{48}O_2$		*cis, cis*-5,9-Hexacosadienoic acid	Sponge: *Microciona prolifera*				136
297. $C_{26}H_{54}O_4$	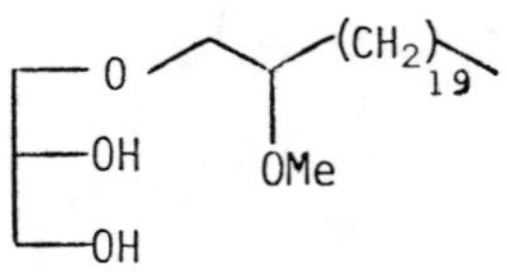	1-*O*-(2-Methoxydocosyl) glycerol	Herring: *Clupea harengus*		+		114
C_{27}							
298. $C_{27}H_{40}O_2$	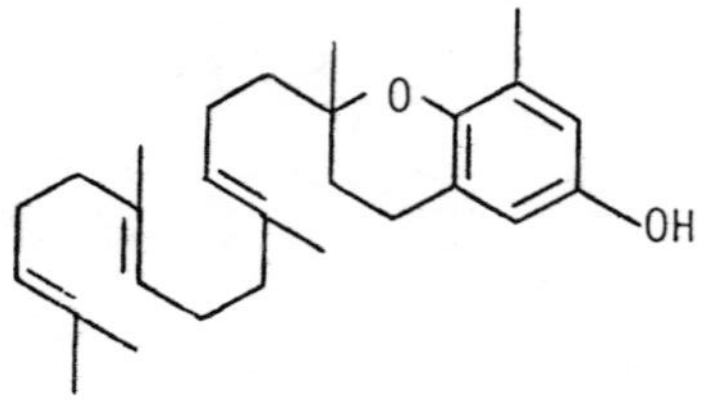	δ-Tocotrienol	Brown alga: *Sargassum tortile*	+			176
299. $C_{27}H_{40}O_3$	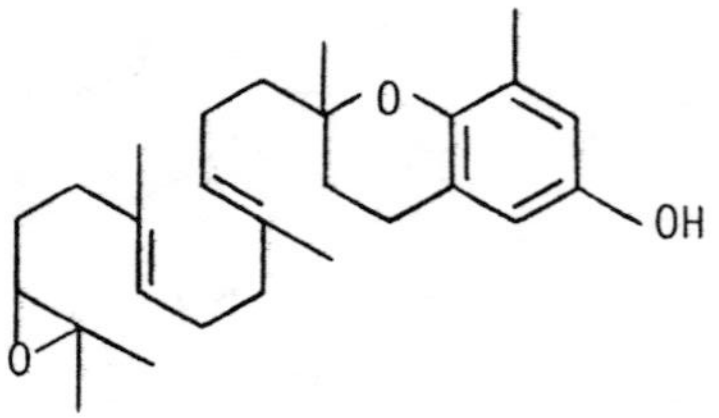	δ-Tocotrienol epoxide	Brown alga: *Sargassum tortile*	+			176

No. / Formula	Structure	Name	Source		Ref.
300. $C_{27}H_{40}O_4$	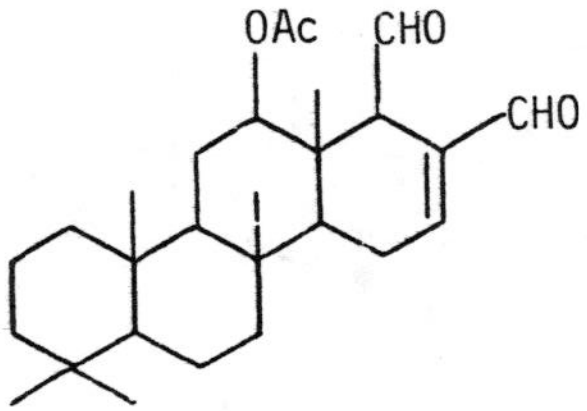 OAc, CHO, CHO	Scalaradial	Sponge: *Cacospongia mollior*		49
301. $C_{27}H_{42}O_3$	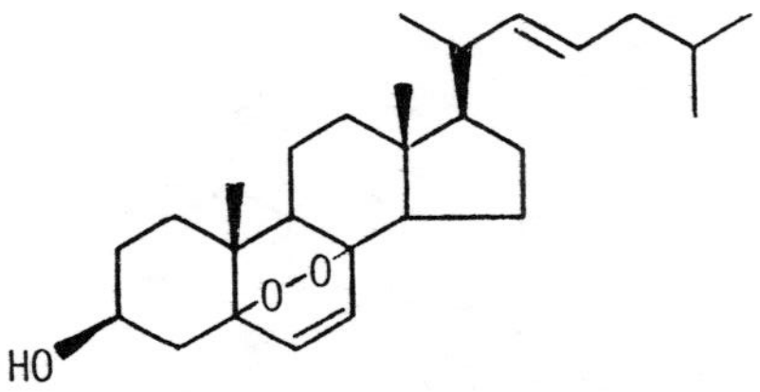 HO, O–O	5,8-Epidioxycholesta-6,22-dien-3β-ol	Sponge: *Axinella cannabina*	+	76
302. $C_{27}H_{44}O$	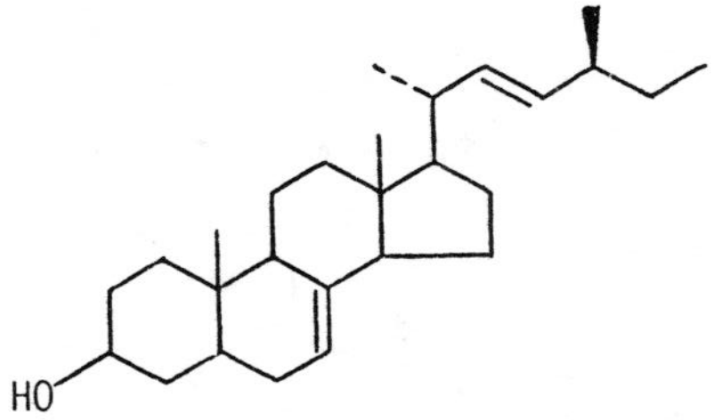 HO	Amuresterol	Starfish	+	171
303. $C_{27}H_{44}O$	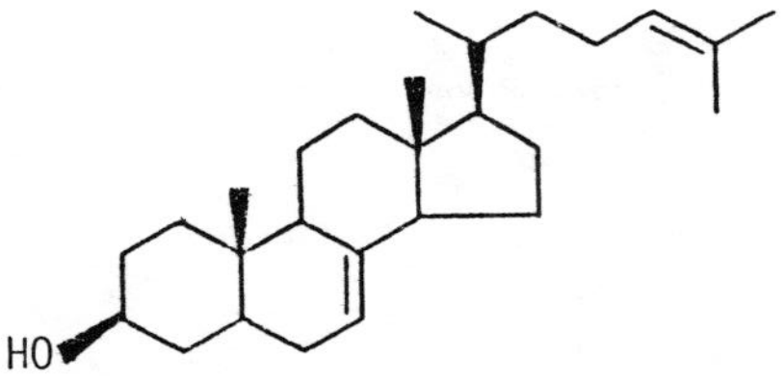 HO	5α-Cholesta-7,24-dien-3β-ol	Starfish: *Asterias rubens*		249

Table 27 (continued)

Molecular formula II/—	Structural formula	Name	Source of compound	Available by synthesis	Available nonmarine sources	Activity	References
C_{27}							
304. $C_{27}H_{44}O$	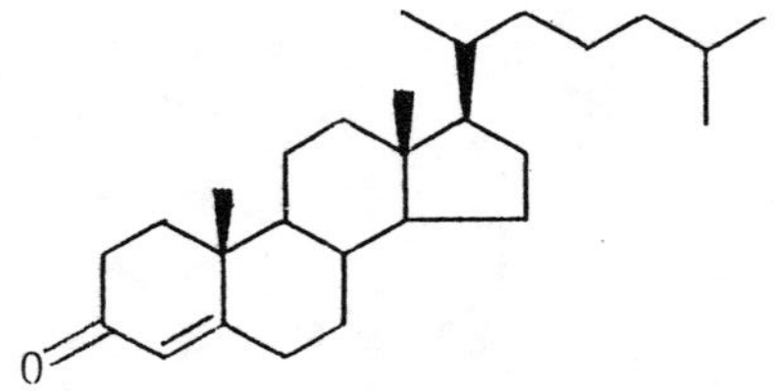	Cholest-4-en-3-one	Red algae: *Gracilaria textorii* *Meristotheca papulosa*				143 144
305. $C_{27}H_{44}O$	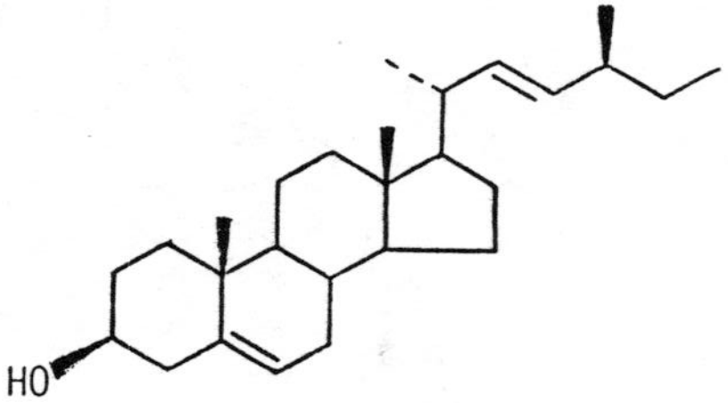	Occelasterol	Crustacean Molluscs Sea anemone Sea cucumber Sea urchin Sea worm Tunicate	+			172
306. $C_{27}H_{44}O_2$	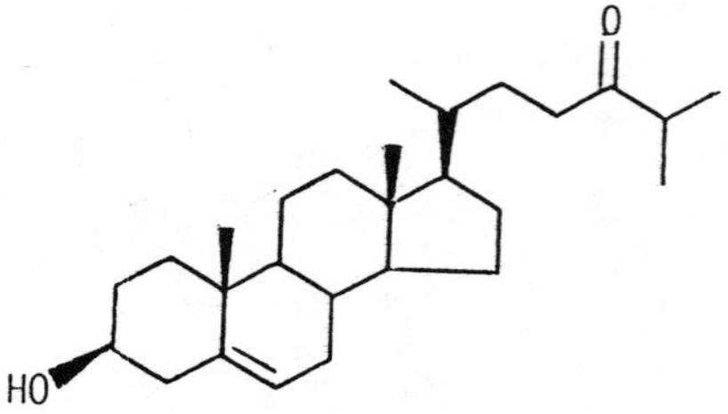	24-Ketocholesterol 24-Oxocholesterol	Brown algae: *Ascophyllum nodosum* *Laminaria saccharina* *Pelvetia canaliculata*				216 240

No.	Compound	Source			Ref.
307. $C_{27}H_{46}O$	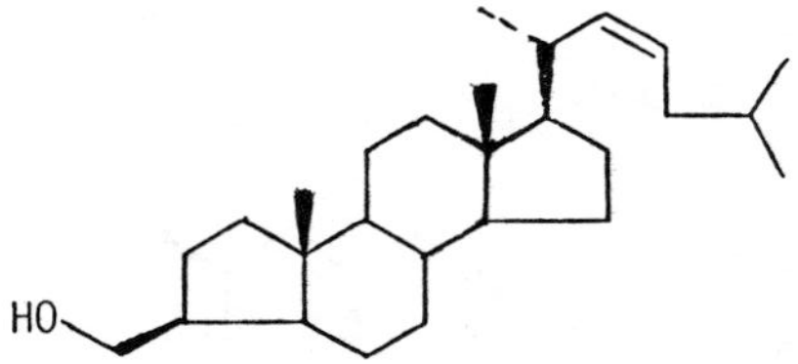 3β-Hydroxymethyl-A-nor-5α-*cis*-cholest-22-ene	Sponge: *Axinella verrucosa*			205
308. $C_{27}H_{46}O$	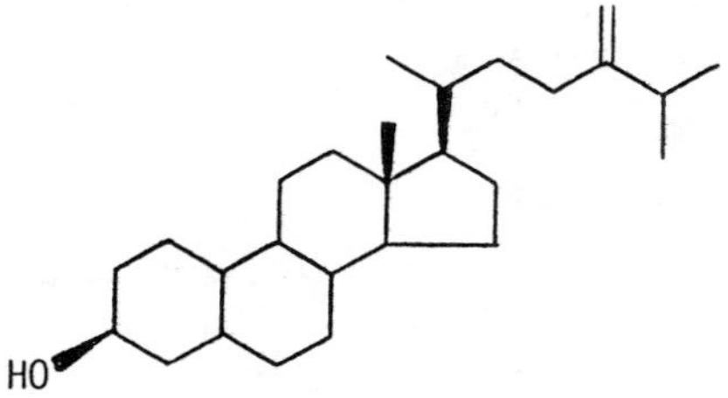 24-Methylene-19-nor-cholestan-3β-ol	Sponge: *Axinella polypoides*			204
309. $C_{27}H_{46}O$	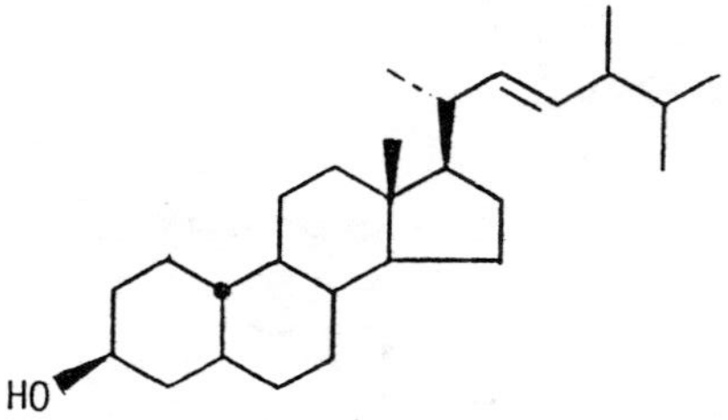 19-Nor-5α,10β-ergost-22-en-3β-ol	Sponge: *Axinella polypoides*			204
310. $C_{27}H_{46}O_2$	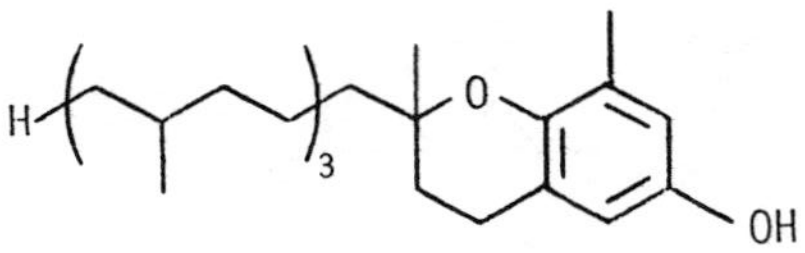 δ-Tocopherol	Brown algae: *Ascophyllum nodosum*, *Fucus serratus*, *F. spiralis*, *F. vesiculosus*, *Pelvetia canaliculata*	+	+	139, 202

Table 27 (continued)

Molecular formula II/—	Structural formula	Name	Source of compound	Available by synthesis	Available nonmarine sources	Activity	References
C_{27}							
311. $C_{27}H_{46}O_4S$	HO₃SO	Cholesteryl sulfate	Sea urchin: *Anthocidaris crassispina* Starfish: *Asterias rubens*	+	+		20 282
312. $C_{27}H_{48}O$	HO	3β-Hydroxymethyl-A-nor-5α-cholestane	Sponge: *Axinella verrucosa*				205
313. $C_{27}H_{48}O$	HO	24-Methyl-19-norcholestan-3β-ol	Sponge: *Axinella polypoides*				204

Compound	Name	Source	Ref.
314. $C_{27}H_{48}O_4$	3β,6α,15α,24ξ-Tetrahydroxy-5α-cholestane	Starfish: *Asterias amurensis*	142
315. $C_{27}H_{48}O_5$ Occurs as 26 or 27-sulfate ester	5α-Cyprinol	Fish: *Latimeria chalumnae*	258
316. $C_{27}H_{48}O_5$ Occurs as 26 or 27-sulfate ester	5β-Cyprinol	Eels: *Anguilla japonica* *Conger myriaster* *Muraenesox cinereus*	258

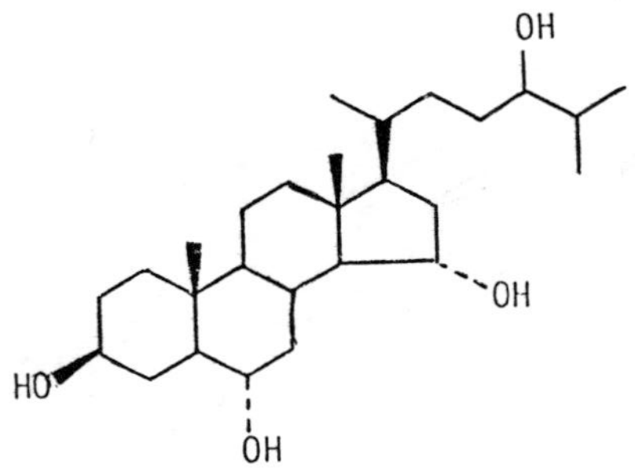

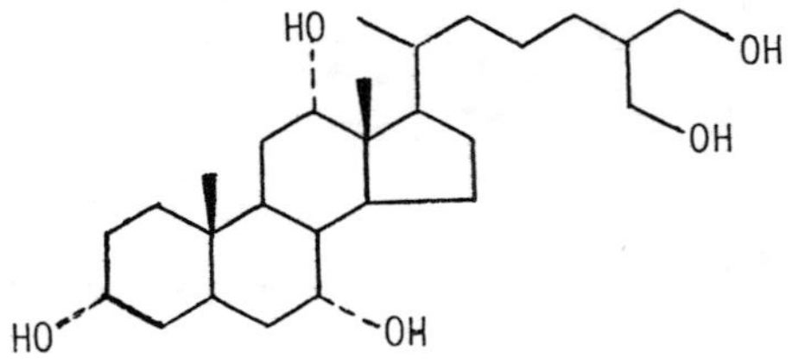

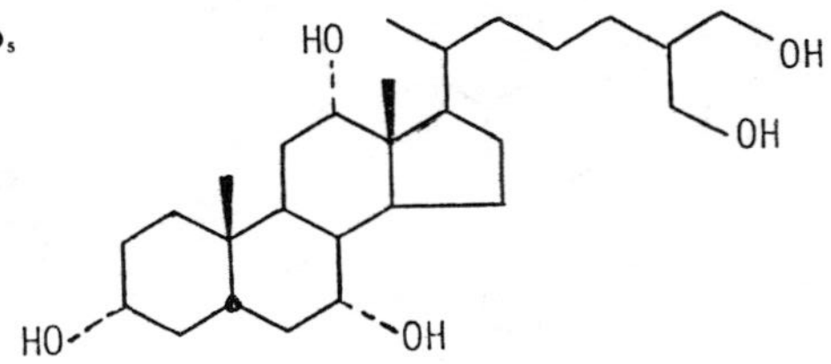

Table 27 (continued)

Molecular formula II/—	Structural formula	Name	Source of compound	Available by synthesis	Available nonmarine sources	Activity	References
C_{27}							
317. $C_{27}H_{48}O_5$	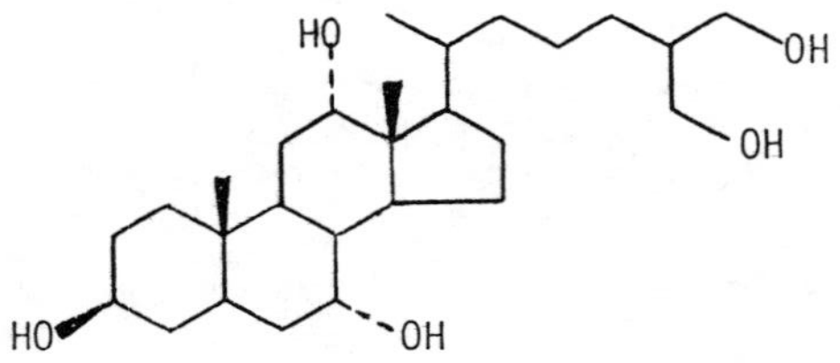 HO, OH, OH, HO, OH Occurs as 26 or 27-sulfate ester	Latimerol	Fish: *Latimeria chalumnae*				258
C_{28}							
318. $C_{28}H_{42}O_2$	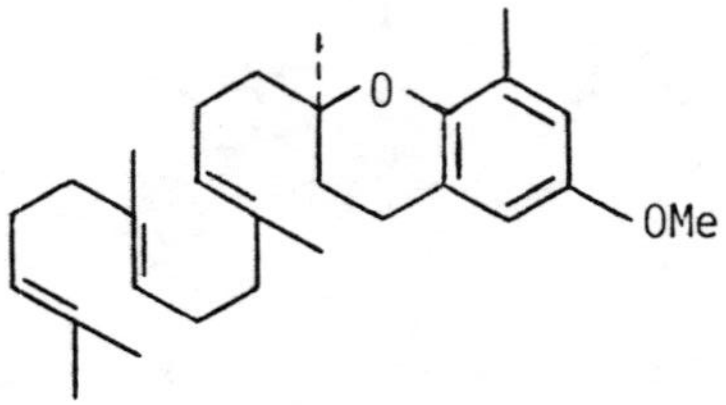 O, OMe	δ-Tocotrienol methyl ether	Brown alga: *Cystophora torulosa*				162
319. $C_{28}H_{42}O_4$	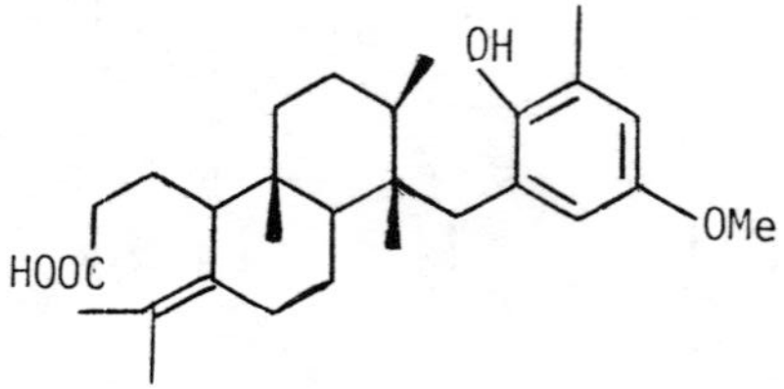 OH, OMe, HOOC	Atomaric acid	Brown alga: *Taonia atomaria*				100

No.	Formula	Structure	Name	Source			Ref.
320.	$C_{28}H_{42}O_4$	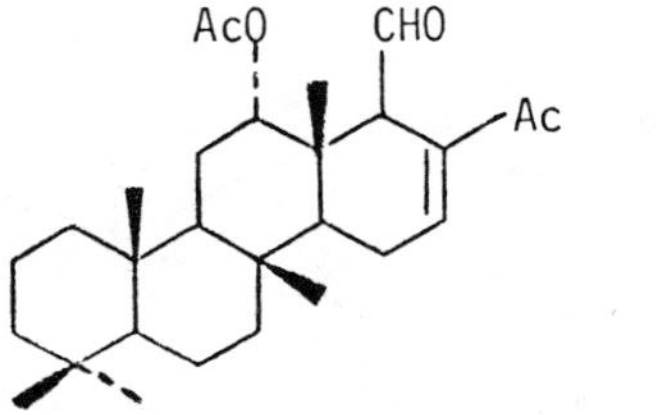 AcO, CHO, Ac	*epi*-Dendalone acetate	Sponge: *Phyllospongia dendyi*			162
321.	$C_{28}H_{44}O$	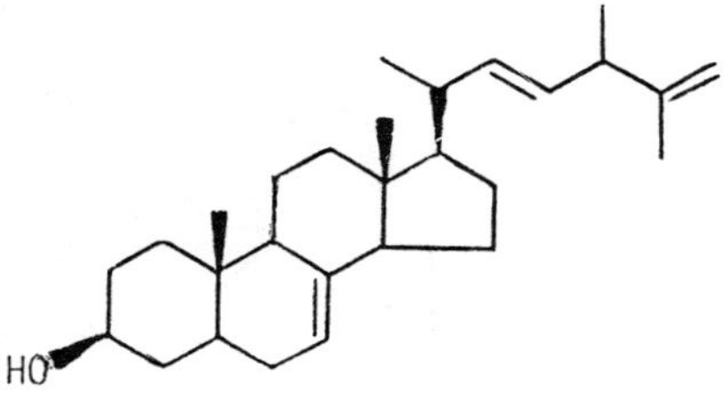 HO	24-Methylcholesta-7,22,25-trien-3β-ol	Starfish: *Leiaster leachii*		+	261
322.	$C_{28}H_{44}O_3$	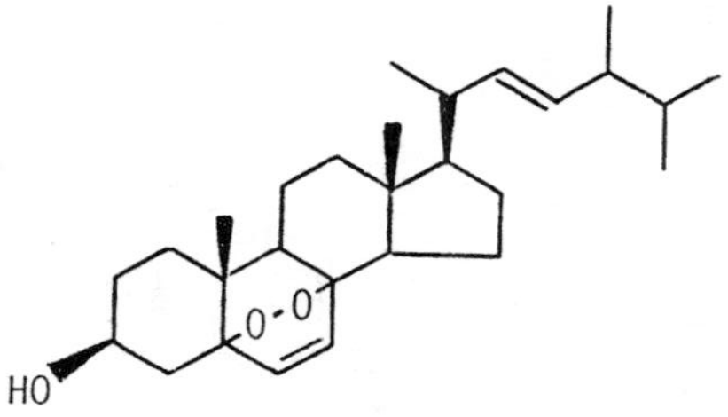 HO, O-O	Ergosterol peroxide	Sponge: *Axinella cannabina*	+	+	13 76
323.	$C_{28}H_{46}O$	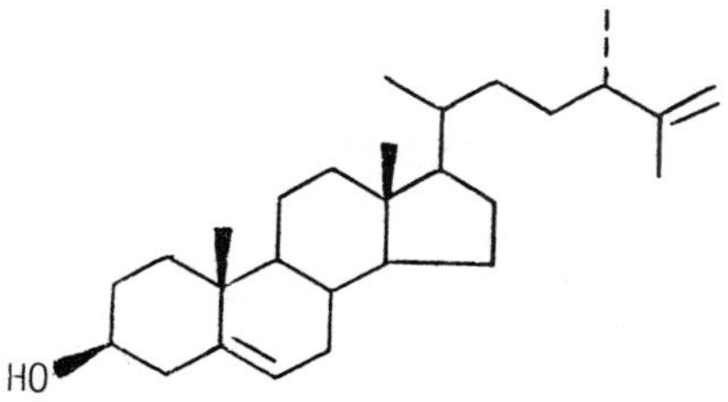 HO	Codisterol	Green alga: *Codium fragile*			237

Table 27 (continued)

Molecular formula II/—	Structural formula	Name	Source of compound	Available by synthesis	Available nonmarine sources	Activity	References
C_{28}							
324. $C_{28}H_{46}O$	HO	4α-Methyl-5α-cholesta-7,24-dien-3β-ol	Starfish: *Asterias rubens*				249
325. $C_{28}H_{46}O$	HO	24-Methylcholesta-5,22-dien-3β-ol	Brown alga: *Sargassum fluitans*				250
326. $C_{28}H_{46}O$	HO	24α-Methylcholesta-5,22-dien-3β-ol	Brittle star: *Ophiocomina nigra*; Diatom: *Phaeodactylum tricornutum*				238

No.	Name	Source	Ref.
327. $C_{28}H_{46}O$	24-Methylcholesta-7,22-dien-3β-ol	Starfish: *Leiaster leachii*	261
328. $C_{28}H_{48}O$	24-Ethyl-19-norcholest-*trans*-22-en-3β-ol	Sponge: *Axinella polypoides*	204
329. $C_{28}H_{48}O$	3β-Hydroxymethyl-24-methyl-A-nor-5α-*cis*-cholest-22-ene	Sponge: *Axinella verrucosa*	205
330. $C_{28}H_{48}O$	4α-Methyl-5α-cholest-7-en-3β-ol	Starfish: *Asterias rubens*	249

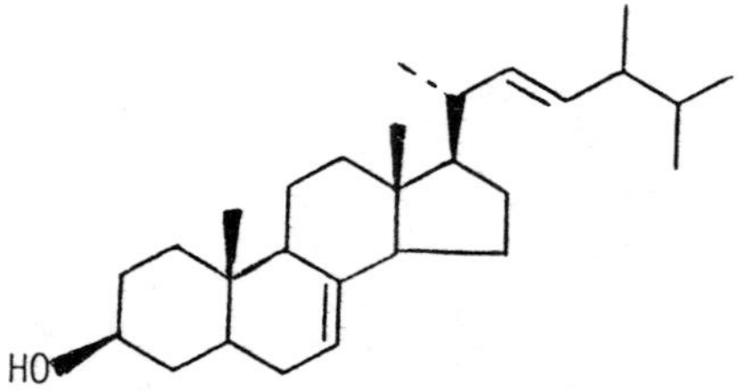

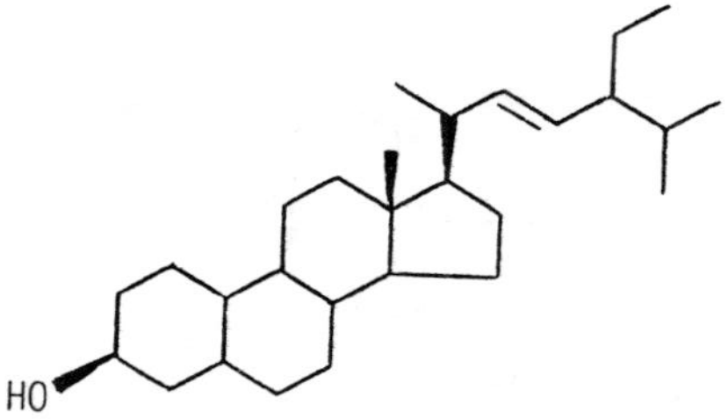

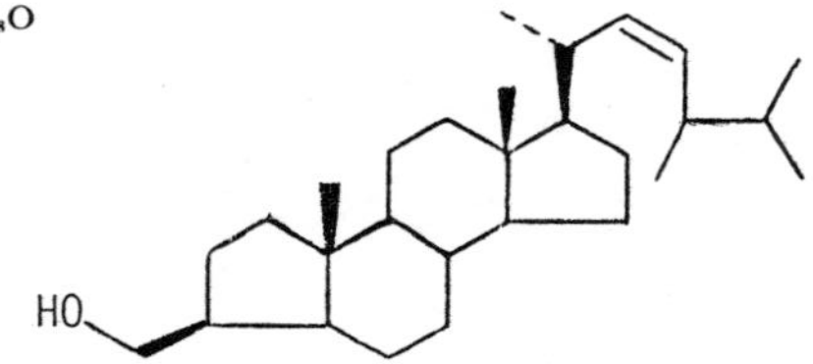

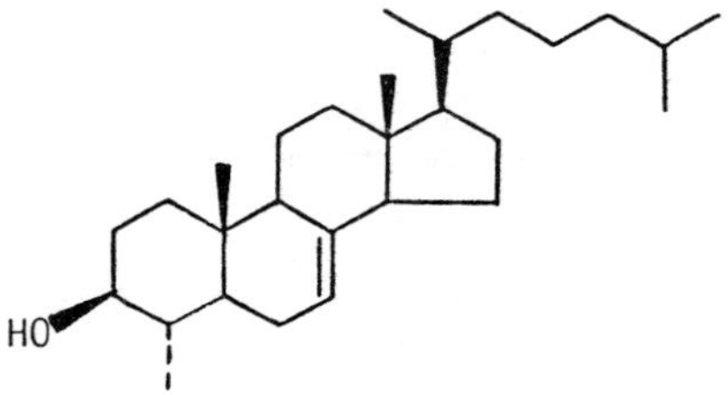

Table 27 (continued)

Molecular formula II/—	Structural formula	Name	Source of compound	Available by synthesis	Available nonmarine sources	Activity	References
C_{28}							
331. $C_{28}H_{48}O$	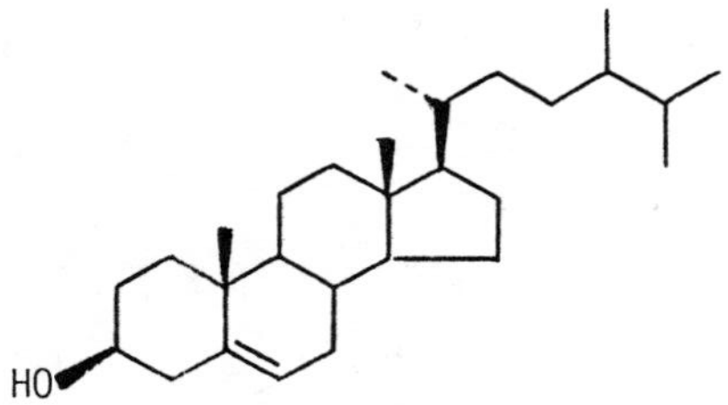	24-Methyl-5α-cholest-5-en-3β-ol	Brown alga Gorgonians Jellyfish Sea anemone Sea urchins Sea worm				22 169 172 250 275 276
332. $C_{28}H_{48}O$	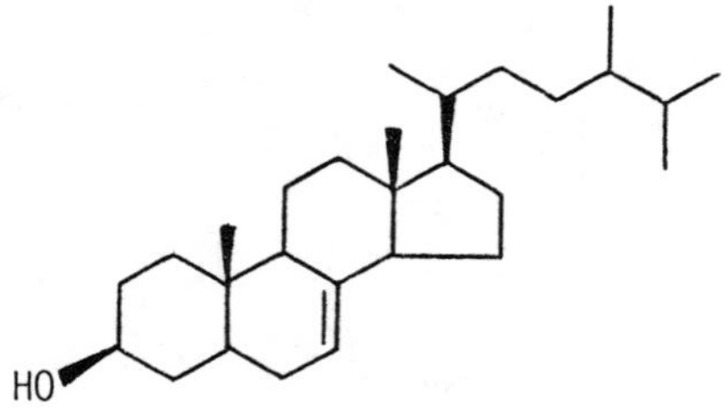	24-Methyl-5α-cholest-7-en-3β-ol	Starfish: *Asterias rubens* *Coscinasterias acutispina* *Leiaster leachii*				146 249
333. $C_{28}H_{48}O_2$	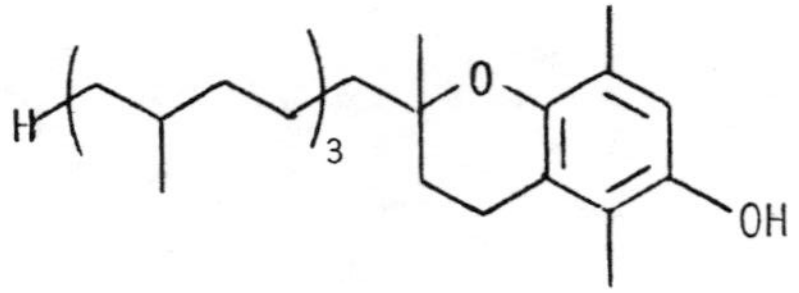	β-Tocopherol	Brown algae: *Ascophyllum nodosum* *Fucus serratus* *F. spiralis* *F. vesiculosus* *Pelvetia canaliculata*		+		139 202

No.	Name	Source	Activity	Ref.
334. $C_{28}H_{48}O_2$	γ-Tocopherol	Brown algae: *Ascophyllum nodosum*, *Fucus serratus*, *F. spiralis*, *F. vesiculosus*, *Pelvetia canaliculata*	+	139 202
335. $C_{28}H_{50}O$	24-Ethyl-19-norcholestan-3β-ol	Sponge: *Axinella polypoides*		204
336. $C_{28}H_{50}O$	3β-Hydroxymethyl-24-methyl-A-nor-5α-cholestane	Sponge: *Axinella verrucosa*		205
C_{29}				
337. $C_{29}H_{44}O_2$	Methyl-δ-tocotrienol methyl ether	Brown alga: *Cystophora torulosa*		162

Table 27 (continued)

Molecular formula II/—	Structural formula	Name	Source of compound	Available by synthesis	Available nonmarine sources	Activity	References
C_{29}							
338. $C_{29}H_{44}O_6$	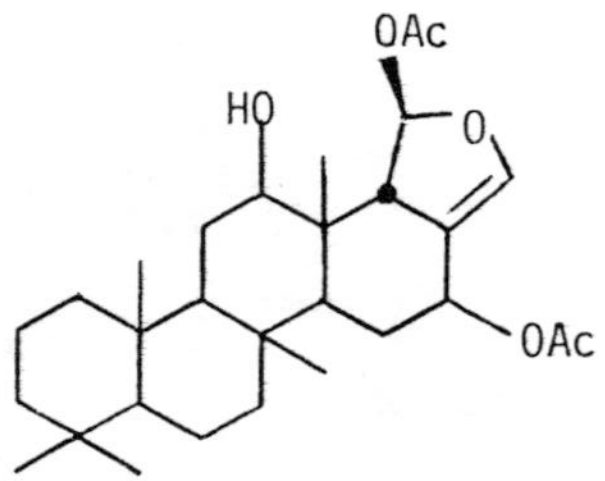	Heteronemin	Sponge: *Heteronema erecta*				162
339. $C_{29}H_{49}O$	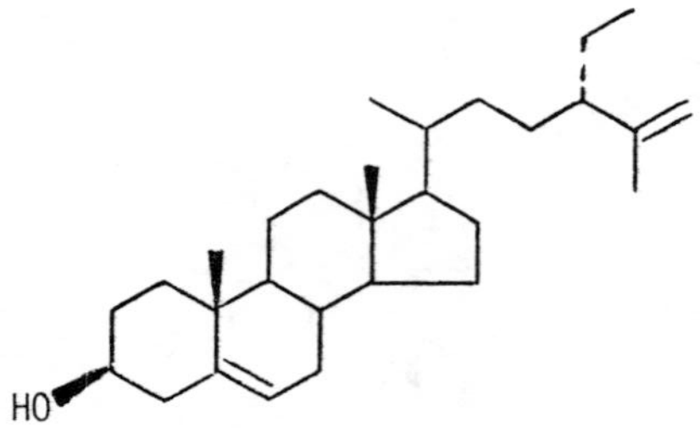	Clerosterol	Green alga: *Codium fragile*		+		237
340. $C_{29}H_{48}O$	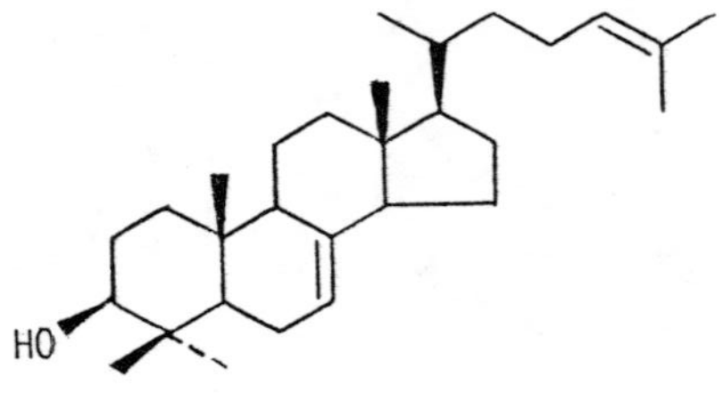	4,4-Dimethyl-5α-cholesta-7,24-dien-3β-ol	Starfish: *Asterias rubens*				249

No.	Formula	Name	Source	Ref.
341.	$C_{29}H_{48}O$	4,4-Dimethyl-5α-cholesta-8(9),24-dien-3β-ol	Starfish: *Asterias rubens*	249
342.	$C_{29}H_{48}O$	23,24-Dimethylcholesta-5,22-dien-3β-ol	Soft coral: *Sarcophyton elegans*	147
343.	$C_{29}H_{48}O$	24-Ethylcholesta-5,22-dien-3β-ol	Brown alga: *Sargassum fluitans* Sea worm: *Pseudopotamilla occelata*	169 172 250

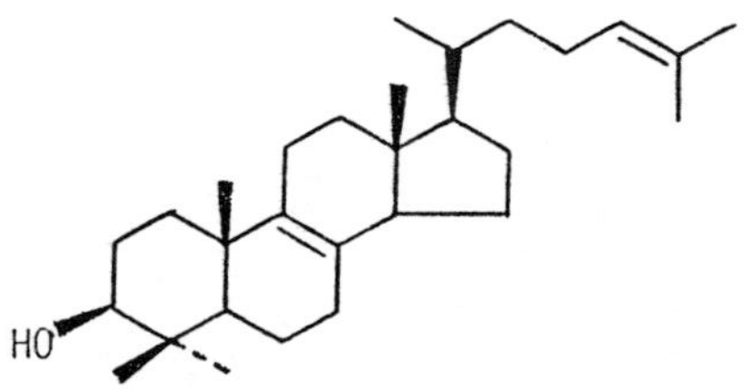

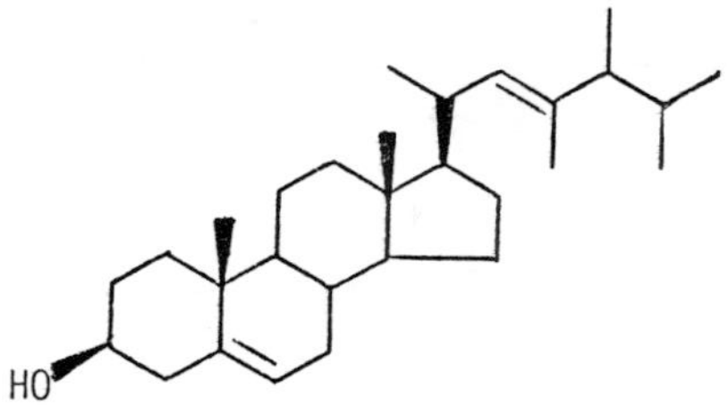

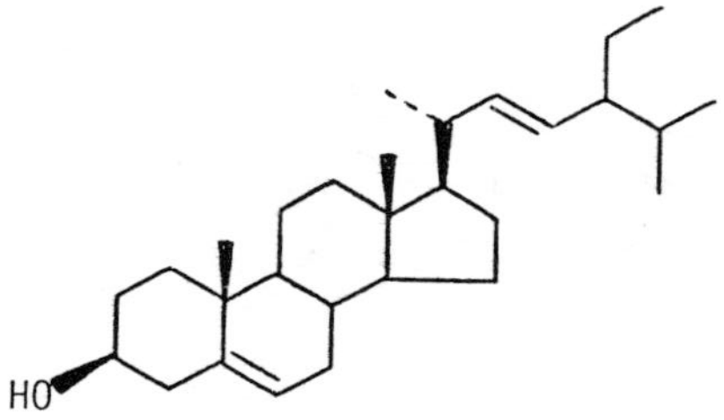

Table 27 (continued)

Molecular formula II/—	Structural formula	Name	Source of compound	Available by synthesis	Available nonmarine sources	Activity	References
C_{29}							
344. $C_{29}H_{48}O$	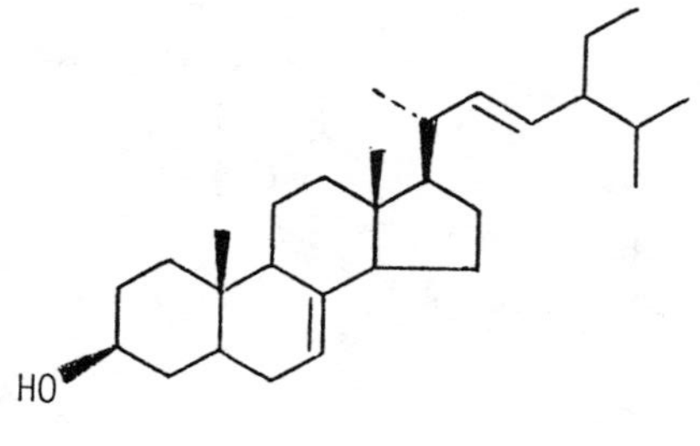	24-Ethylcholesta-7,22-dien-3β-ol	Starfish: *Asterias rubens*, *Coscinasterias acutispina*, *Leiaster leachii*				146, 249, 261
345. $C_{29}H_{48}O$	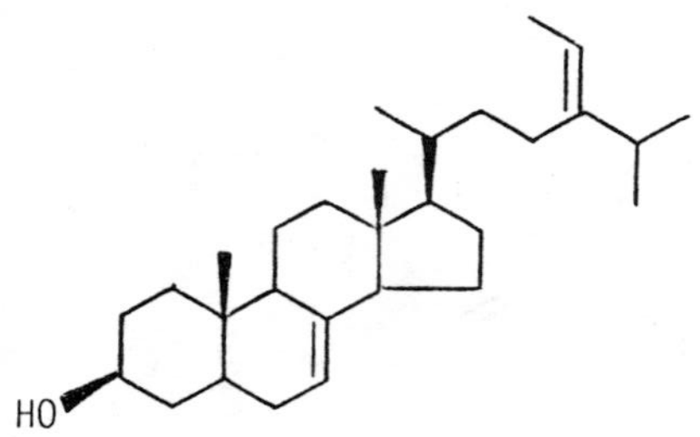	(E)-24-Ethylidene-5α-cholest-7-en-3β-ol	Starfish: *Asterias rubens*, *Leiaster leachii*				146, 249

Compound	Name	Source	Ref.
346. $C_{29}H_{48}O$	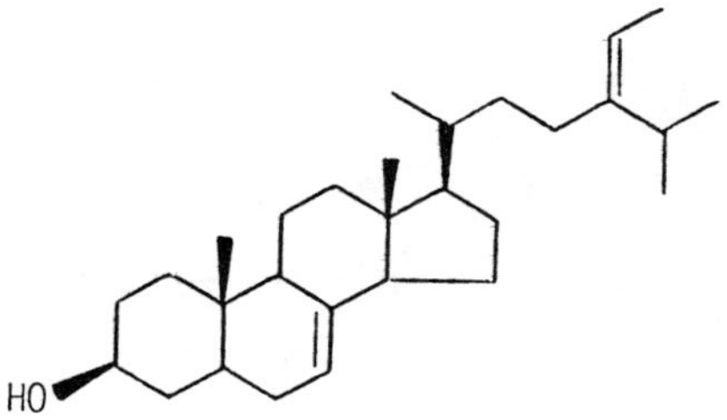 (Z)-24-Ethylidene-5α-cholest-7-en-3β-ol	Starfish: *Asterias rubens*	249
347. $C_{29}H_{50}O$	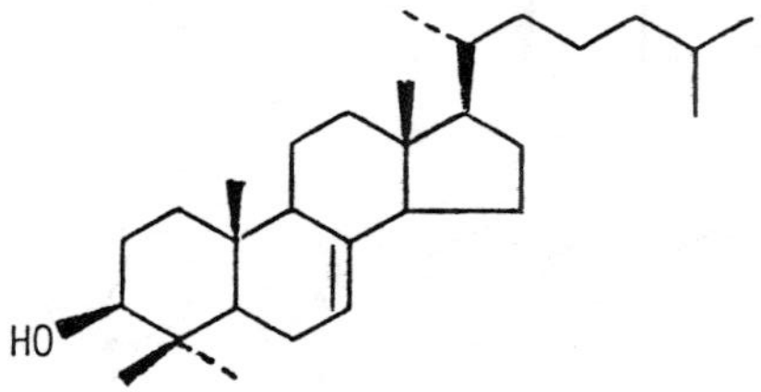 4,4-Dimethyl-5α-cholest-7-en-3β-ol	Starfish: *Asterias rubens*	249
348. $C_{29}H_{50}O$	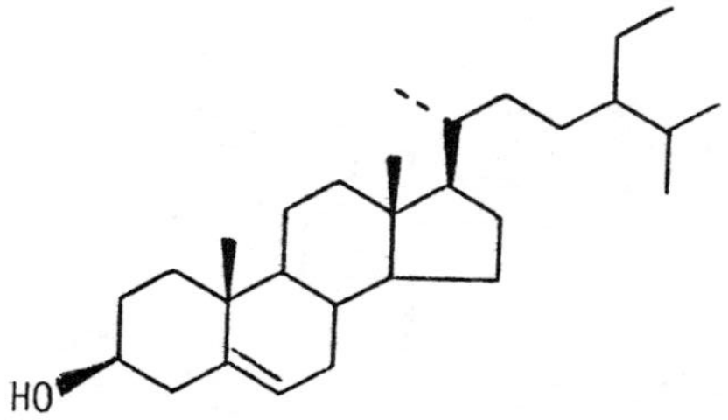 24-Ethylcholest-5-en-3β-ol	Brown alga Jellyfish Sea anemone Seaurchins Sea worm	169 172 250 275 276
349. $C_{29}H_{50}O$	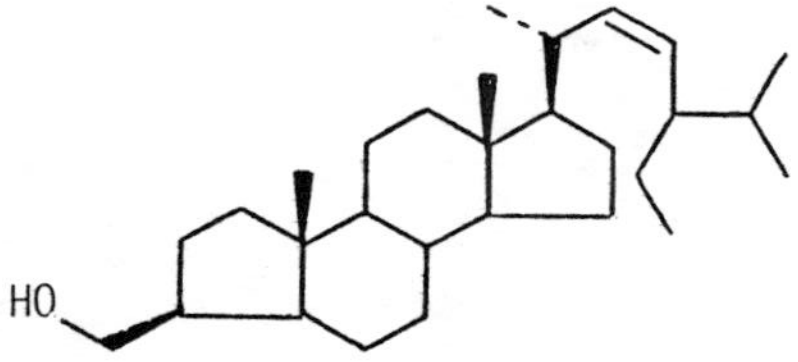 24-Ethyl-3β-hydroxymethyl-A-nor-5α-*cis*-cholest-22-ene	Sponge: *Axinella verrucosa*	205

Table 27 (continued)

Molecular formula II/—	Structural formula	Name	Source of compound	Available by synthesis	Available nonmarine sources	Activity	References
C_{29}							
350. $C_{29}H_{50}O$	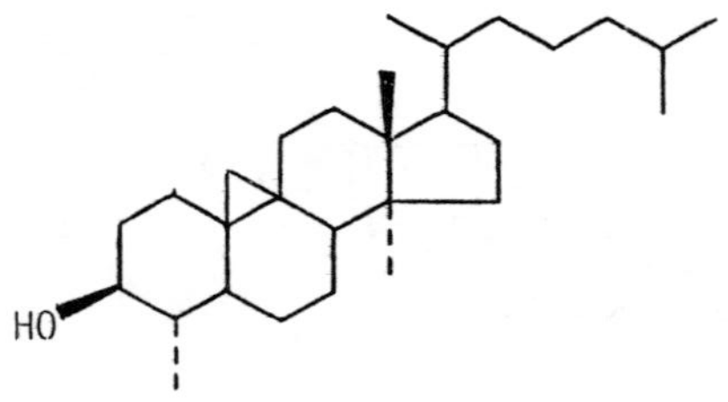	31-Norcycloartanol	Red alga: *Rhodymenia palmata* Starfish: *Asterias rubens*				84 249
351. $C_{29}H_{50}O_2$	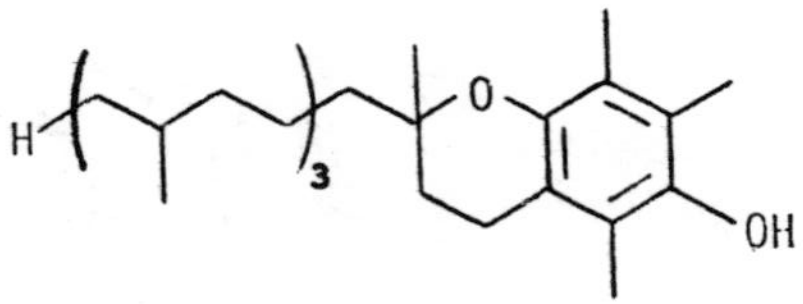	α-Tocopherol Vitamin E	Algae	+	+		139
352. $C_{29}H_{52}O$	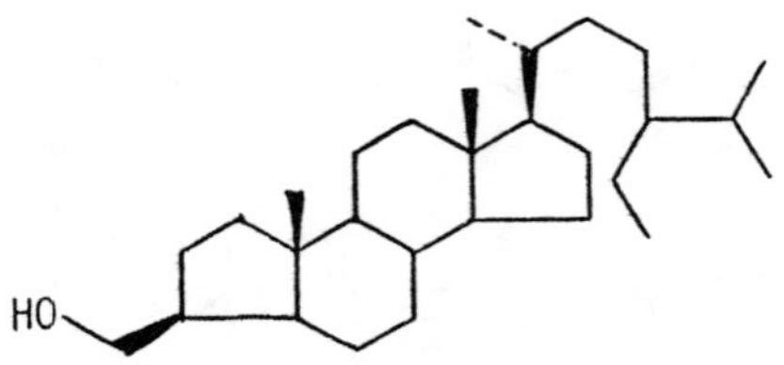	24-Ethyl-3β-hydroxymethyl-A-nor-5α-cholestane	Sponge: *Axinella verrucosa*				205

C_{30}

No. / Formula	Name	Source	Ref.
353. $C_{30}H_{46}O_5$	Dendalone 3-hydroxy-butyrate	Sponge: *Phyllospongia dendyi*	162
354. $C_{30}H_{46}O_5$ (replaces No. I/431.)	Stichopogenin A_4	Sea cucumber: *Stichopus japonicus*	168
355. $C_{30}H_{48}O_7$	Hippurin-1	Gorgonian: *Isis hippuris*	162

Table 27 (continued)

Molecular formula II/—	Structural formula	Name	Source of compound	Available by synthesis	Available nonmarine sources	Activity	References
C_{30}							
356. $C_{30}H_{50}O$	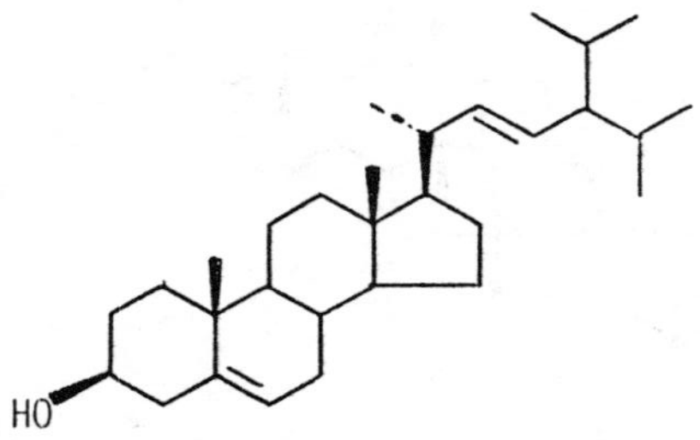 HO	24-Isopropylcholesta-5,22-dien-3β-ol	Sponge: *Pseudaxinyssa* sp.				120
357. $C_{30}H_{52}O$	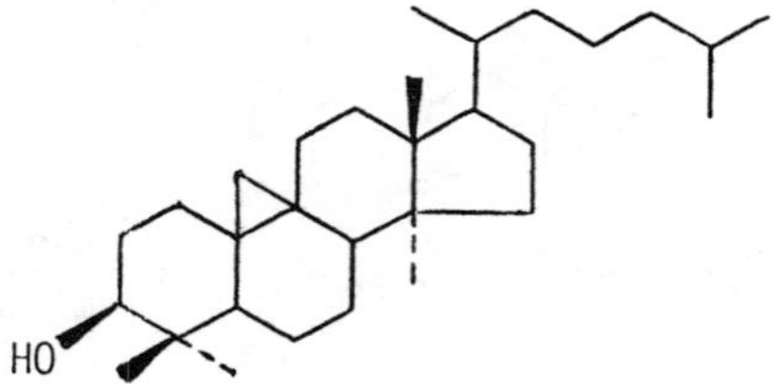 HO	Cycloartanol	Red alga: *Rhodymenia palmata* Starfish: *Asterias rubens*				84 249
358. $C_{30}H_{52}O$	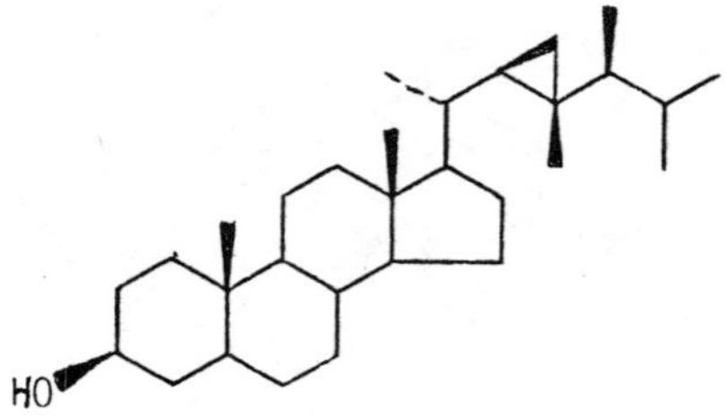 HO	Gorgostanol	Starfish: *Acanthaster planci*				148

C_{30}

No.	Formula	Name	Source		Ref.
359.	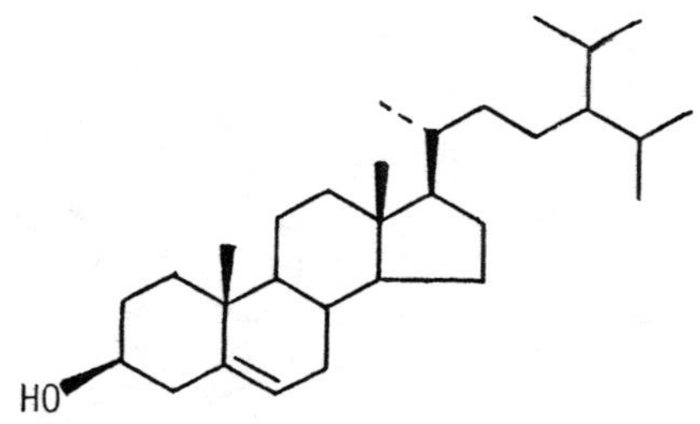 $C_{30}H_{52}O$	24-Isopropylcholest-5-en-3β-ol	Sponge: *Pseudaxinyssa* sp.		120
360.	$C_{30}H_{52}O_5$	24-Metylcholestane-3β,5α,6β,25-tetrol-25-monoacetate	Soft coral: *Sarcophyton elegans*	+	209

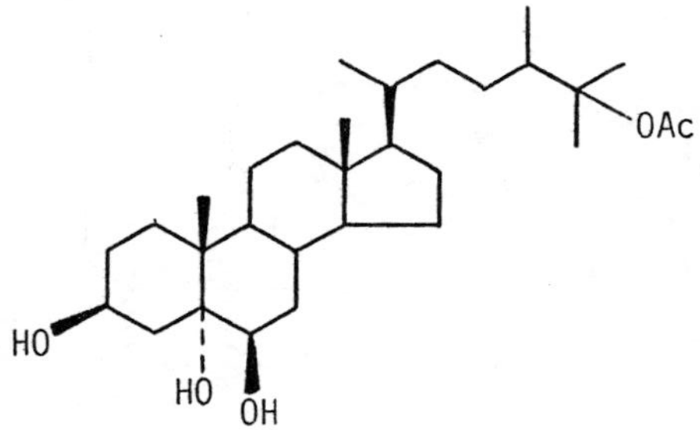

C_{31}

No.	Formula	Name	Source		Ref.
361.	$C_{31}H_{46}O_2$	2-Pentaprenyl-1,4-benzoquinol	Sponge: *Ircinia ramosa*		162

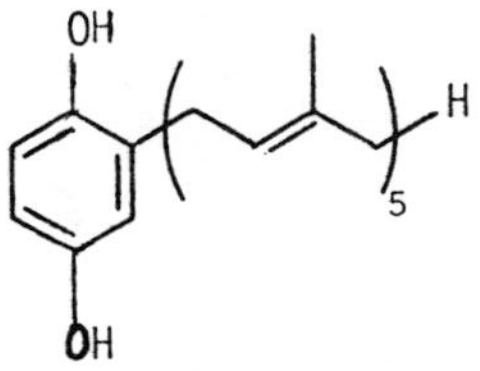

Table 27 (continued)

Molecular formula II/—	Structural formula	Name	Source of compound	Available by synthesis	Available nonmarine sources	Activity	References
362. $C_{31}H_{46}O_2$	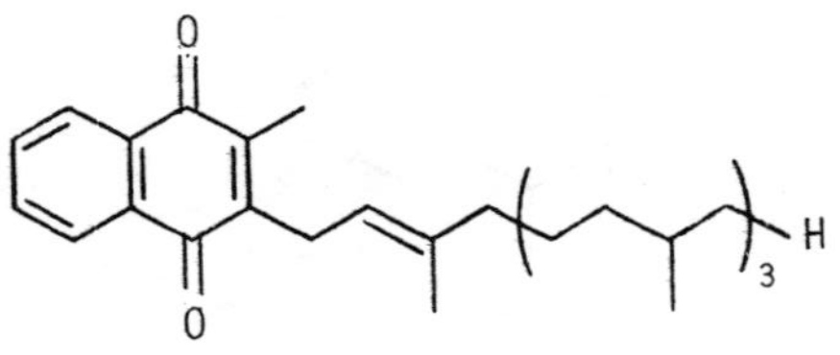 O, O, H, 3	Phylloquinone Vitamin K_1	Algae	+	+	Prophylaxis + treatment of hypoproth-rombinemia	73 202 215
C_{32}							
363. $C_{32}H_{47}BrO_{10}$	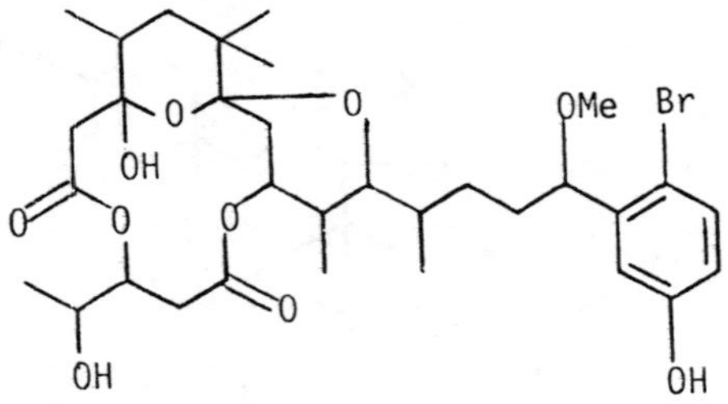 O, OH, O, O, O, O, OH, OMe, Br, OH	Aplysiatoxin	Sea hare: *Stylocheilus longi-cauda*				159 160
364. $C_{32}H_{48}O_{10}$	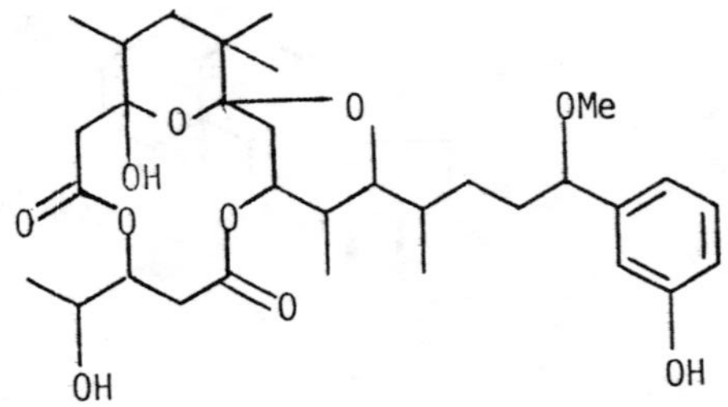 O, OH, O, O, O, O, OH, OMe, OH	Debromoaplysiatoxin	Sea hare: *Stylocheilus longi-cauda*				159 160

C_{33}

No.	Formula	Name	Source	Ref.
365.	$C_{33}H_{32}N_4O_5$	Chlorocruoroporphyrin	Starfish: *Astropecten irregularis*, *Luidia ciliaris*	231
366.	$C_{33}H_{54}O_7$	3β-Hydroxy-5α-cholesta-9(11), 24-dien-6α-yl-β-D-glucoside	Starfish: *Marthasterias glacialis*	265

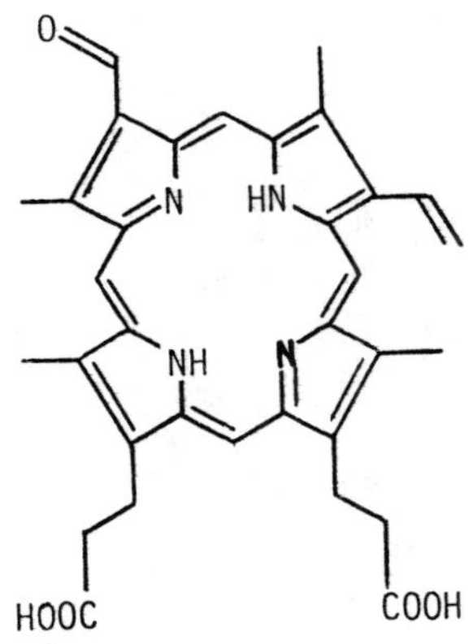

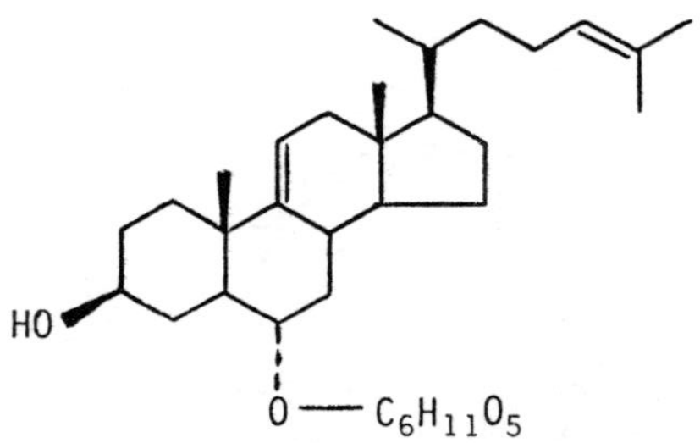

Table 27 (continued)

Molecular formula II/—	Structural formula	Name	Source of compound	Available by synthesis	Available nonmarine sources	Activity	References
C_{34}							
367. $C_{34}H_{33}FeN_4O_5$	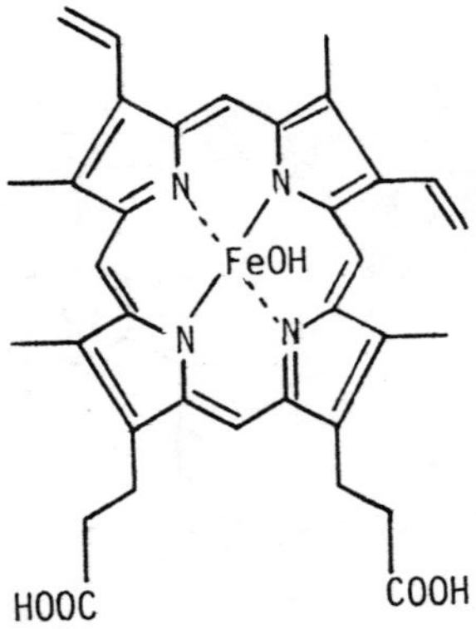 FeOH; HOOC; COOH	Hematin	Sea worms		+		231
368. $C_{34}H_{34}N_4O_4$	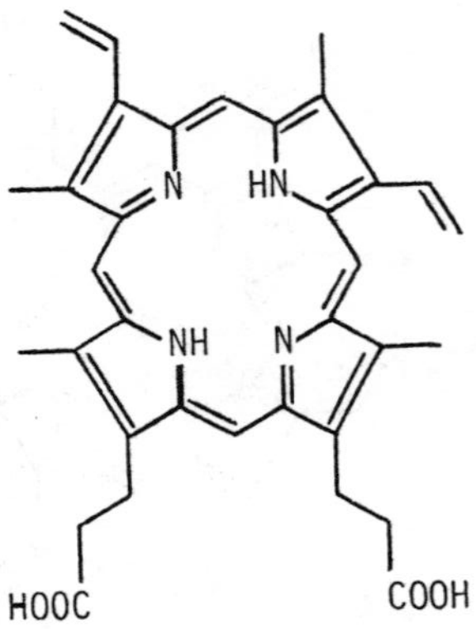 N; HN; NH; N; HOOC; COOH	Protoporphyrin IX	Mollusc: *Bankia setacea* Sea worm: *Lineus longissimus* Starfish		+		103 202 231

369. $C_{34}H_{38}N_4O_6$	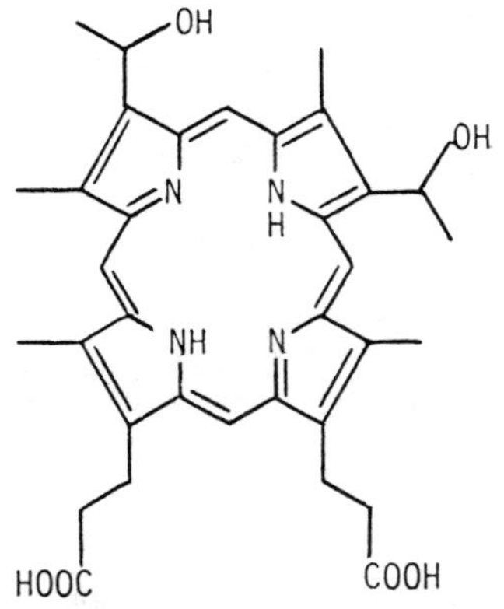	Hematoporphyrin	Corals Jellyfish Sea anemones		Antidepressant	202 231
C_{36} 370. $C_{36}H_{38}N_4O_8$	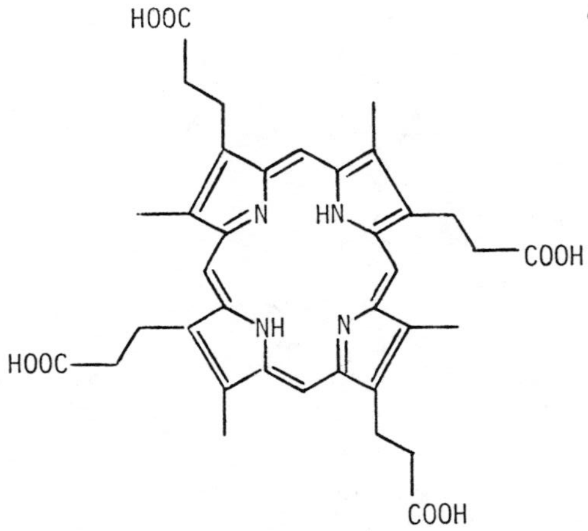	Coproporphyrin I	Sea worms: *Amphitrite johnstoni* *Arenicola marina*	+		231

Table 27 (continued)

Molecular formula II/—	Structural formula	Name	Source of compound	Available by synthesis	Available nonmarine sources	Activity	References
371. $C_{36}H_{38}N_4O_8$	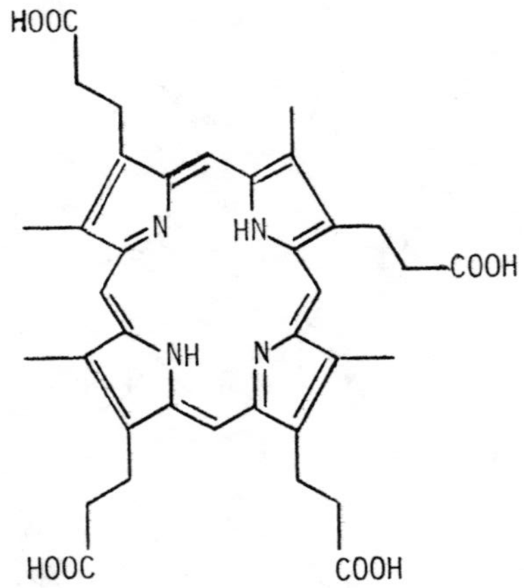	Coproporphyrin III	Sea worms		+		231

C_{40}

No. / Formula	Name	Source		References
372. $C_{40}H_{38}N_4O_{16}$	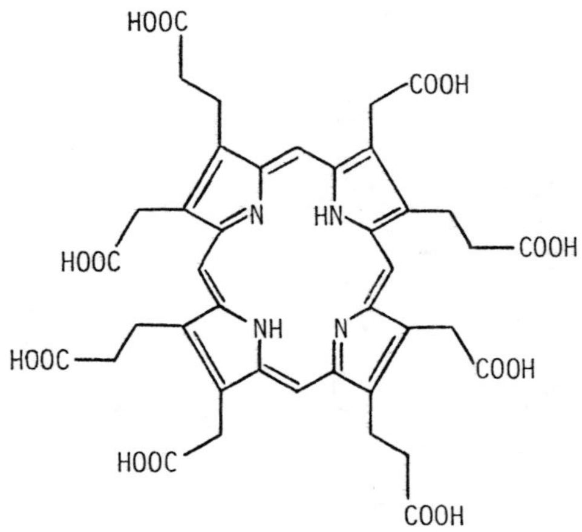 Uroporphyrin I	Molluscs	+	103 231
373. $C_{40}H_{48}O_4$	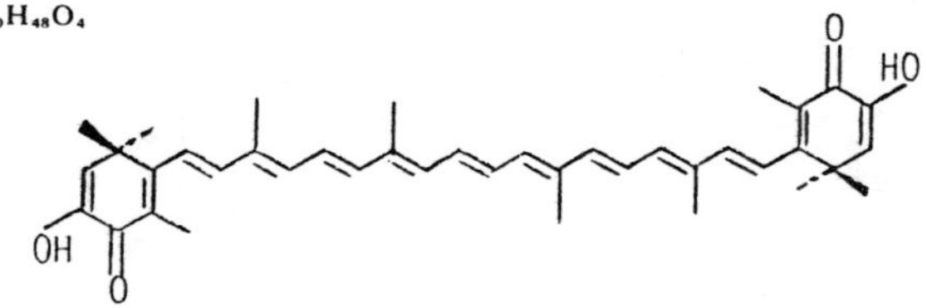 Astacene	Crustaceans Fish Sponge: *Verongia aerophoba*		61, 62, 63, 64 65,133 198,200
374. $C_{40}H_{52}O_3$ 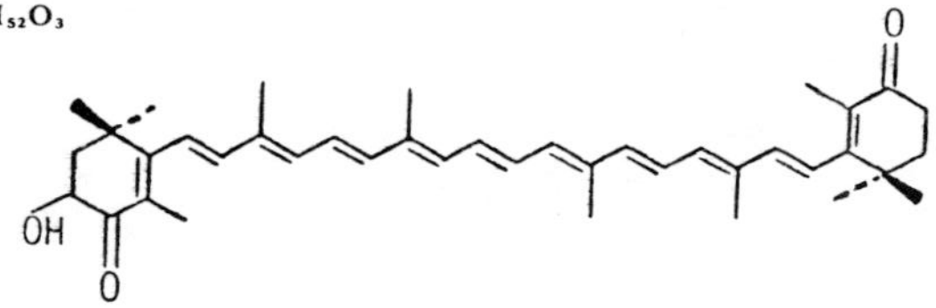	3-Hydroxycanthaxan-thin Phoenicoxanthin	Crustaceans Fish Sea cucumber: *Psolus fabrichii*	+	34,133 153,156 157,158 196,199

Table 27 (continued)

Molecular formula II/—	Structural formula	Name	Source of compound	Available by synthesis	Available nonmarine sources	Activity	References
C_{40}							
375. $C_{40}H_{52}O_3$	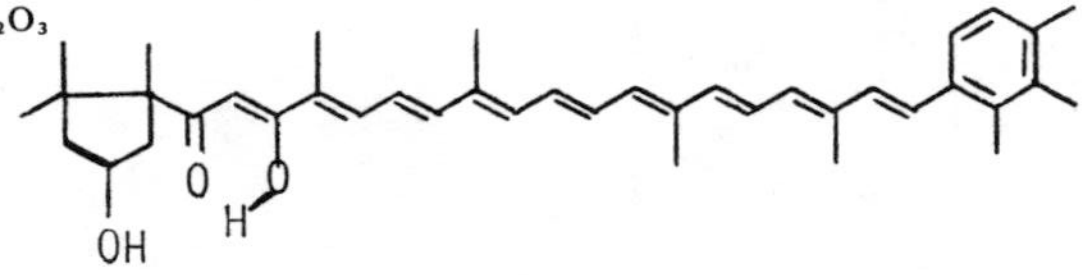	Trikentriorhodin	Sponge: *Trikentrion helium*				5
376. $C_{40}H_{54}O_2$	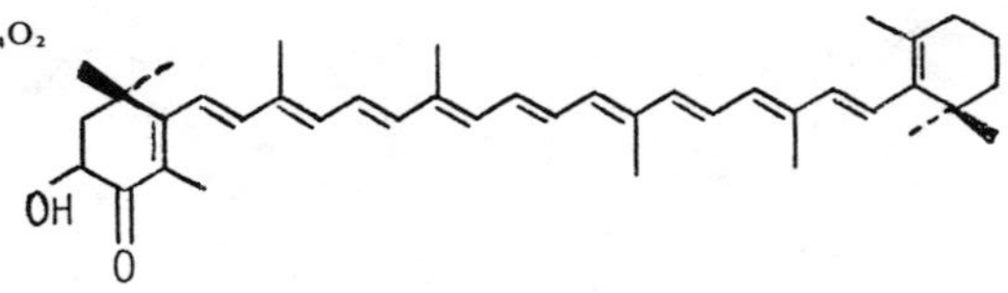	Hydroxyechinenone 3-Hydroxy-4-keto-β-carotene	Crustacean: *Arctodiaptomus salinus* Starfish: *Asterina panceri*		+		133 151
377. $C_{40}H_{54}O_2$	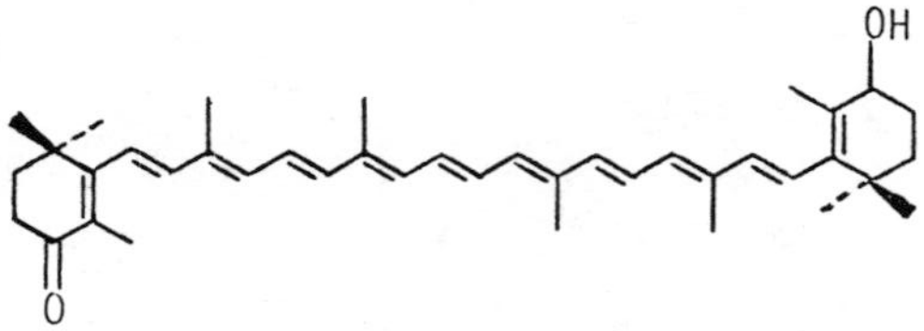	4′-Hydroxyechinenone 4-Hydroxy-4′-keto-β-carotene	Crustaceans Fish		+		64,65 133,151 156
378. $C_{40}H_{54}O_3$	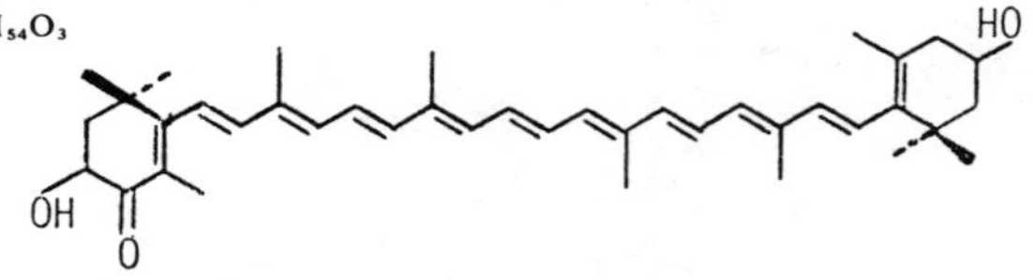	Adonixanthin 3,3′-Dihydroxy-4-keto-β-carotene β-Doradexanthin	Crustaceans Starfish: *Asterina panceri*		+		133 151 196 199

Compound	Name	Source			Activity	References
379. $C_{40}H_{54}O_3$	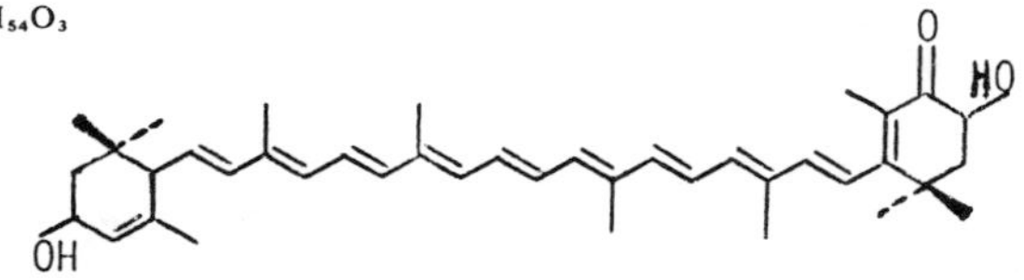 α-Doradexanthin 4-Ketolutein	Sea bream: *Chrysophrys major*				133 157
380. $C_{40}H_{54}O_4$	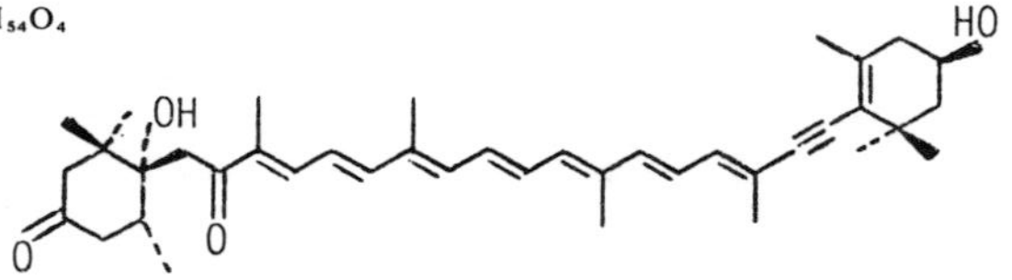 Isomytiloxanthin	Mussel: *Mytilus edulis*				165
381. $C_{40}H_{54}O_4$	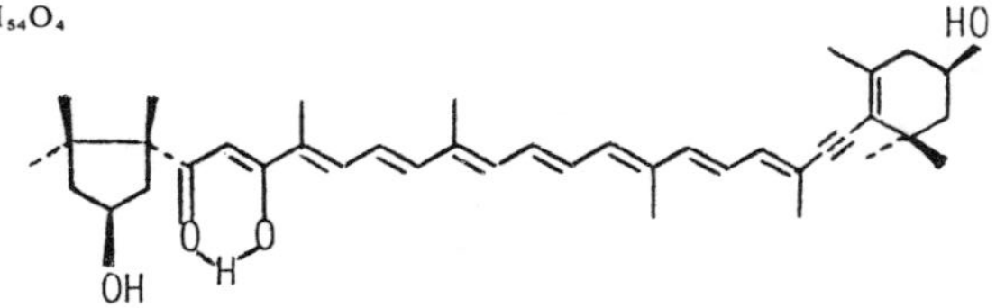 Mytiloxanthin	Mussels: *Mytilus californianus* *M. edulis*				103 165
382. $C_{40}H_{56}$	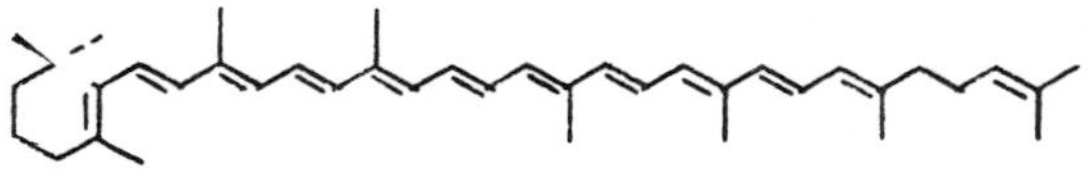 γ-Carotene	Crustaceans Sponges: *Hymeniacidon sanguinea* *Verongia aerophoba*	+	+	Vitamin A activity	61,65 102,133 153,202
383. $C_{40}H_{56}O$	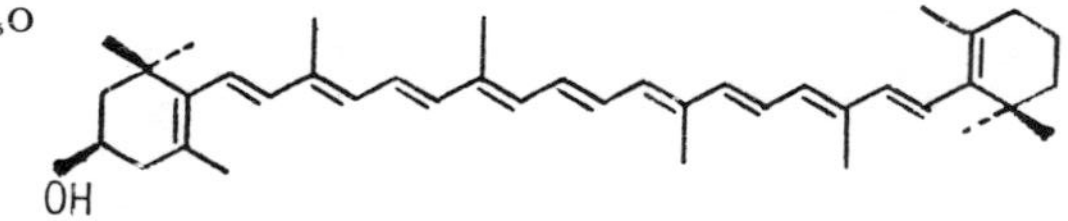 Cryptoxanthin	Crustaceans Fish Molluscs Starfish: *Asterina panceri*	+	+	Vitamin A activity	63 65 103 133 151 198 202
384. $C_{40}H_{56}O$	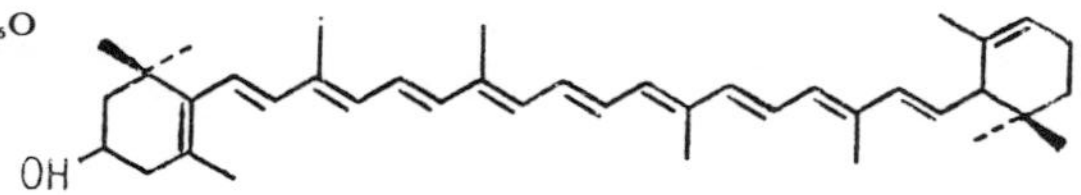 α-Cryptoxanthin	Fish				133 157 158 198

Table 27 (continued)

Molecular formula II/—	Structural formula	Name	Source of compound	Available by synthesis	Available nonmarine sources	Activity	References
C_{40}							
385. $C_{40}H_{56}O$	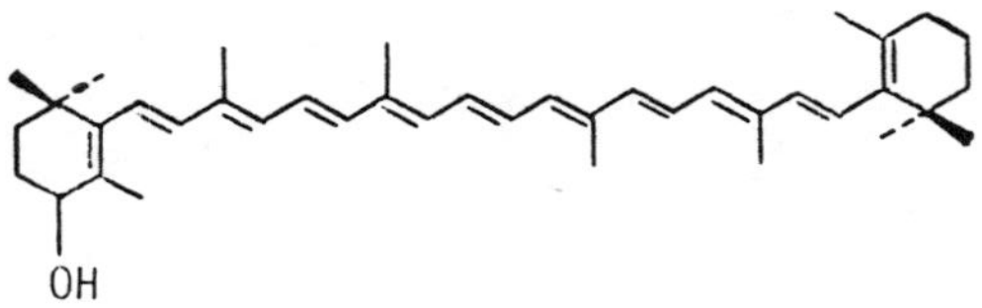	Isocryptoxanthin	Crustaceans Sea worm: *Sabella penicillis*	+	+		63 133 151 156
386. $C_{40}H_{56}O_2$	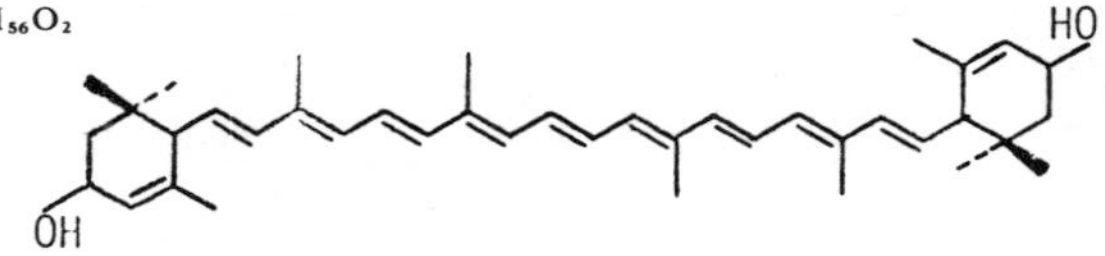	3,3′-Dihydroxy-ε-carotene Tunaxanthin	Fish Prawn: *Penaeus japonicus*				60, 62 64,133 152,154 155,157 158, 198
387. $C_{40}H_{56}O_2$	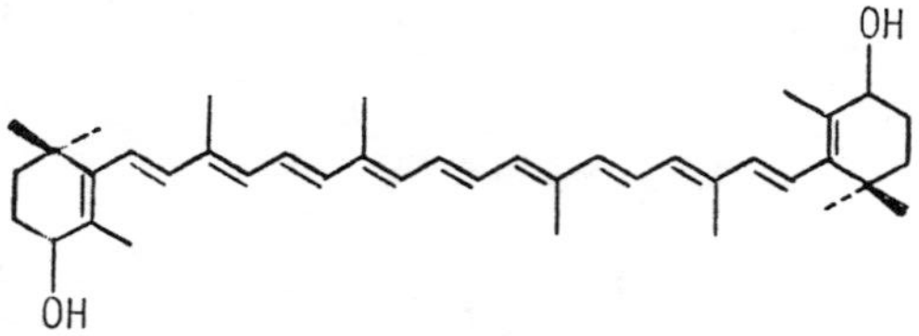	Isozeaxanthin	Crustaceans Fish Sponge: *Verongia aerophoba*	+	+		61,63 64, 65 133,151
388. $C_{40}H_{56}O_3$	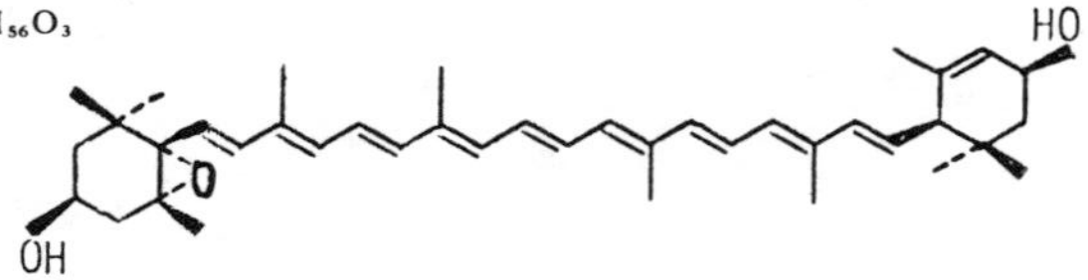	Lutein epoxide Taraxanthin	Crustaceans Fish		+		62 63 64 65 133

389. $C_{40}H_{56}O_3$	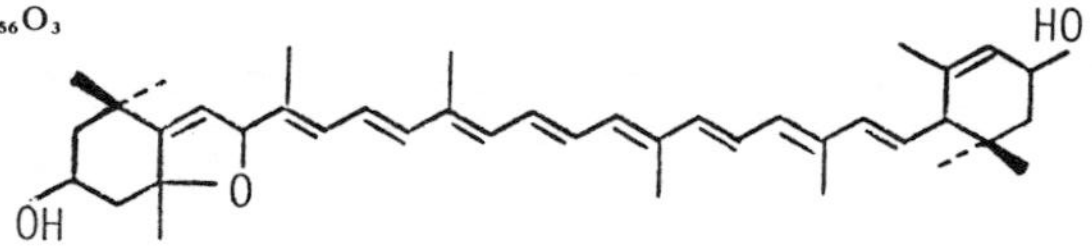 Lutein-5,8-epoxide Flavoxanthin	Crustaceans Mollusc: *Venus japonica*	+	103,133 151,200 202
390. $C_{40}H_{56}O_4$	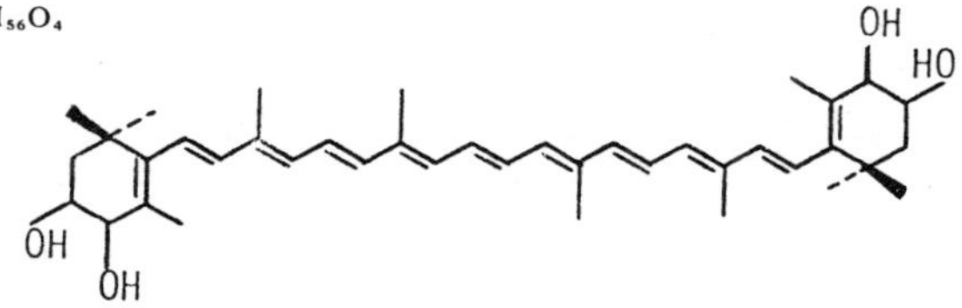 Crustaxanthin	Crustaceans: *Calanoides* sp. *Arctodiaptomus sal-* *inus*		133 151 153
391. $C_{40}H_{56}O_4$	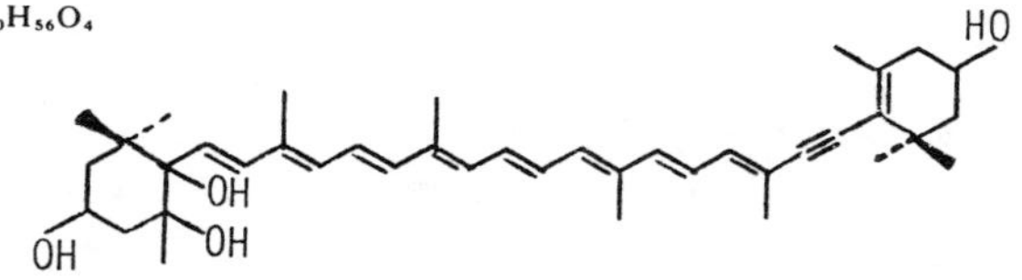 Heteroxanthin	Yellow-brown alga: *Vaucheria sessilis*		133 225
392. $C_{40}H_{56}O_4$	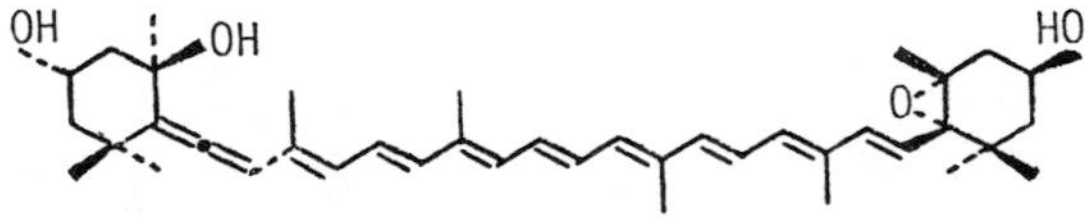 Neoxanthin	Brown alga: *Fucus vesiculosus* Sponge: *Verongia aerophoba*	+	61 133 223
393. $C_{40}H_{56}O_4$	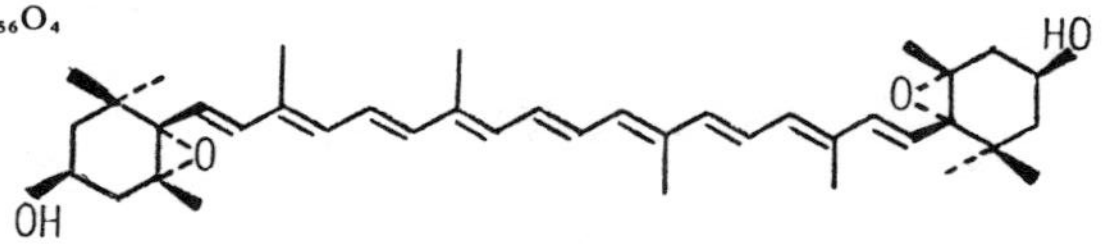 Violaxanthin	Sponge: *Verongia aerophoba*	+	61 133 202

Table 27 (continued)

Molecular formula II/—	Structural formula	Name	Source of compound	Available by synthesis	Available nonmarine sources	Activity	References
C_{42}							
394. $C_{42}H_{58}O_2$	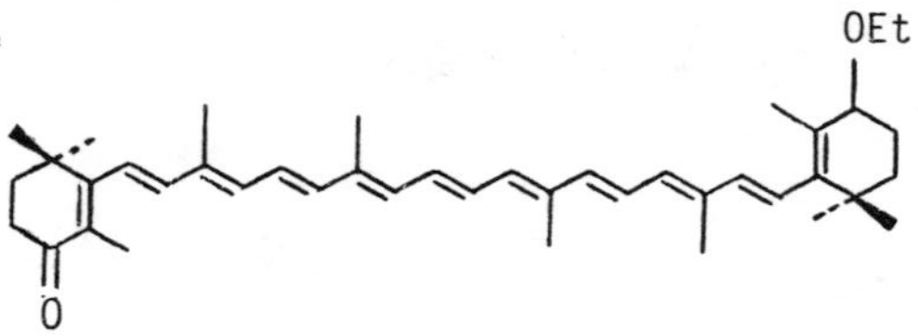	4-Keto-4′-ethoxy-β-carotene	Crustacean: *Eupagurus bernhardus*				65
C_{53}							
395. $C_{53}H_{80}O_2$	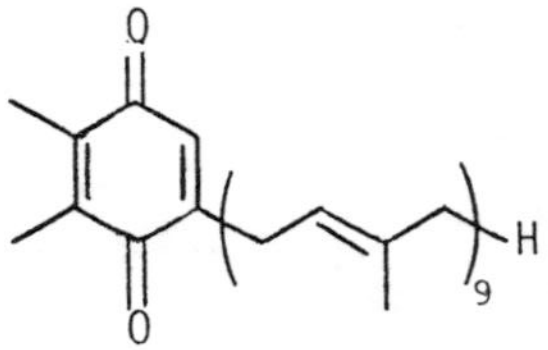	Plastoquinone-9	Algae		+		27 73 215

C_{55}						
396. $C_{55}H_{72}MgN_4O_5$ 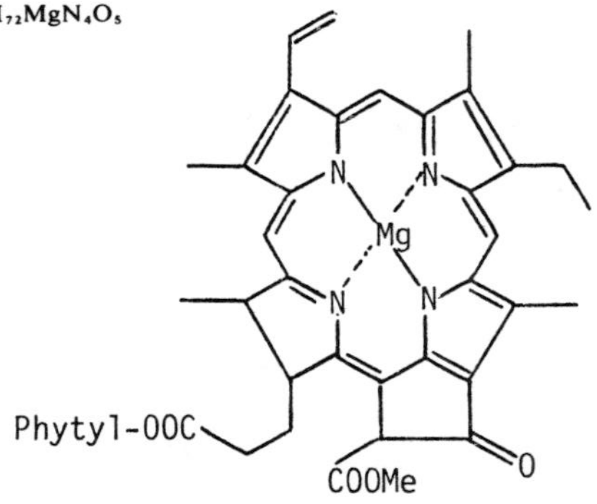	Chlorophyll a	Algae	+	+		38 73 202
C_{59}						
397. $C_{59}H_{94}O_{27}$ 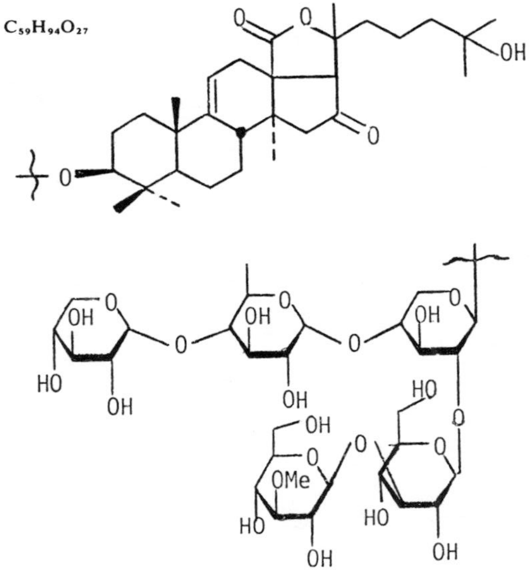	Holotoxin A	Sea cucumber: *Stichopus japonicus*			Antifungal	167 168

COMPOUND NAME INDEX FOR TABLE 27

Name	Compound number (II/—)	Page number
Acanthellin-1	164	26,103
N-Acetoglycolyl-4-methyl-4,9-dideoxyneuraminic acid	128	95
Acetoxyfimbrolide A	102	89
Acetoxyfimbrolide B	103	90
Acetoxyfimbrolide C	101	89
Acetoxyfimbrolide G	100	89
Acetoxyintricatol	174	24,106
Acetylcholine	51	78
N-Acetylneuraminic acid	114	92
Acrylic acid	11	4,5,70
Adenosine diphosphate	96	88
Adenosine triphosphate	99	89
5′-Adenylic acid	91	87
Adonixanthin	378	37,160
AF-350 monosulfate	172	105
Africanol	158	25,26,102
Allo-laurinterol	132	12,96
α-Amino-*n*-butyric acid	22	72
γ-Amino-*n*-butyric acid	23	72
2-Aminoimidazole	14	63,71
Amuresterol	302	43,44,54,135
Anserine	98	66,89
Aplysiatoxin	363	12,154
Astacene	373	34,38,159
Atomaric acid	319	15,29,140
Aucantene	107	1,91
Avarol	229	27,117
Avarone	227	17,27,117
Axisonitrile-2	165	26,103
1,2,4-Benzenetriol	29	16,73
Bifuhalol	116	11,16,93
Bis(3,5-dibromo-4-hydroxyphenyl)methane	119	10,16,93
19, 24-Bisnorcholest-*cis*-22-en-3β-ol	278	46,57,129
Bromoacetone	12	71
4-Bromo-α-chamigren-8,9-epoxide	149	21,100
4-Bromo-α-chamigren-8-one	150	22,100
4-Bromo-β-chamigren-8-one	151	22,101
1-Bromo-3-chloroacetone	9	2,70
6-Bromo-2-chloromyrcene	88	19,87
(E)-9-Bromo-2-chloromyrcene	89	19,87
(Z)-9-Bromo-2-chloromyrcene	90	19,87
3-Bromo-4,5-dihydroxybenzyl alcohol	41	7,8,13,76
Bromoform	1	2,69
3-Bromo-4-hydroxybenzyl alcohol	40	76
2-Bromomyrcene	92	19,87
(E)-9-Bromomyrcene	93	19,88
(Z)-9-Bromomyrcene	94	19,88
(+)-*R*-4-Butylcyclohepta-2,6-dienone	109	1,2,91
(+)-6-Butylcyclohepta-2,4-dienone	110	1,2,91
Candicine chloride	113	62,92
Δ9(12)-Capnellene-3 β,8β,10α-triol	156	25,102
L-Carnitine	49	64,78
Carnosine	76	66,84
γ-Carotene	382	33,36,161
Cartilagineal	79	20,84
Caulerpol	216	28,115
Caulerpol acetate	255	28,124

COMPOUND NAME INDEX FOR TABLE 27 (continued)

Name	Compound number (II/—)	Page number
Cetyl alcohol	171	105
Chimyl alcohol	194	109
Chlorocruoroporphyrin	365	155
1-Chloro-1,3-dibromoacetone	5	2,69
3-Chloro-1,1-dibromoactone	6	2,70
6-Chloro-2,(Z)-9-dibromomyrcene	85	19,86
(Z)-9-Chloro-2,6-dibromomyrcene	86	19,86
2-Chloromyrcene	95	19,88
Chlorophyll a	396	165
3-Chloro-1,1,3-tribromoacetone	3	2,69
5α-Cholesta-7,24-dien- 3β-ol	303	54,135
Cholest-4-en-3-one	304	41,42,57,136
Cholesteryl sulfate	311	43,51,138
Citrulline	35	75
Clerosterol	339	40,52,146
Codisterol	323	40,51,141
Coproporphyrin I	370	157
Coproporphyrin III	371	158
Creatinine	21	72
Crustaxanthin	390	37,163
Cryptoxanthin	383	36,161
α-Cryptoxanthin	384	39,161
(−)-β-Curcumene	153	20,101
Cycloartanol	357	42,152
Cycloeudesmol	159	21,102
Cyclospongiaquinone-1	251	17,122
Cyclospongiaquinone-2	252	18,123
Cypridina luciferin	249	121
5α-Cyprinol	315	47,139
5β-Cyprinol	316	50,139
Cystine	32	74
Dactyloxene-B	155	24,101
Dactylyne	135	2,3,97
Deacetyllaurencin	138	2,98
Debromoaplysiatoxin	364	12,154
1a,2,3,6,7,9,10a,13,14,14a-Decahydro-1a,5,8, 12-tetramethyloxireno[9,10]-cyclotetradeca-[1,2-b]furan	209	32,113
1a,2,3,6,7,9,10a,13,14,14a-Decahydro-1a,5,8,12-tetramethyloxireno[9,10]-cyclotetradeca[1,2-b]furan	210	32,113
Dehydroepoxynephthenol	214	31,114
epi-Dendalone acetate	320	141
Dendalone 3-hydroxybutyrate	353	151
1,3-Dibromoacetone	10	2,70
2,7-Dibromo-8-chloro-octahydro-8,10,10-trimethyl-5-methylene-6H-2,5a-methano-1-benzoxepin-3,4-diol	140	23,98
2,3-Dibromo-4,5- dihydroxybenzyl-*n*-propyl ether	80	13,85
1,3-Dibromo-2-heptanone	48	2,77
3,5-Dibromo-4-hydroxybenzaldehyde	36	9,10,75
3,5-Dibromo-4- hydroxyphenylacetic acid	54	9,79
3,5-Dibromo-4-hydroxyphenylpyruvic acid	66	9,81
1,1-Dibromo-3-iodoacetone	7	2,70
Didehydrocyclospongiaquinone-1	250	17,122
Dihydroflexibilide	215	32,33,115
Dihydrolaurene	142	99
7,8-Dihydroxanthopterin	31	74
3,3′-Dihydroxy- ε-carotene	386	35,39,162

COMPOUND NAME INDEX FOR TABLE 27 (continued)

Name	Compound number (II/—)	Page number
2,4′-Dihydroxy-5-hydroxymethyl-3,3,′5′-tribromodiphenylmethane	124	10,16,95
3,3′-Dihydroxy-4-keto-β-carotene	378	37,160
3,5-Dihydroxyphenylacetic acid	57	13,79
4,4-Dimethyl-5α-cholesta-7, 24-dien-3β-ol	340	146
4,4-Dimethyl-5α-cholesta-8(9),24-dien-3β-ol	341	147
23,24-Dimethylcholesta-5,22-dien-3β-ol	342	53,147
4,4-Dimethyl-5α-cholest-7-en-3β-ol	347	149
[3-(Dimethylsulfonio)propyl]-trimethylammonium dichloride	64	62,81
2-Diprenyl-1,4-benzoquinol	163	14,103
Docosane	258	124
1-*O*-Docosylglycerol	280	129
1-*O*-Dodecylglycerol	162	103
Dopamine	60	13,64,65,80
α-Doradexanthin	379	34,39,161
β-Doradexanthin	378	37,160
Eicosane	222	116
Eicosanol	223	116
1-*O*-Eicosylglycerol	263	125
Elatol	144	22,99
5,8-Epidioxycholesta-6,22-dien-3β-ol	301	45,54,135
Epizoanthoxanthin A	121	63,94
Epizoanthoxanthin B	125	63,95
Epoxynephthenol acetate	256	30,31,124
Ergosterol peroxide	322	45,54,141
Erythropterin	67	81
β-Estradiol	183	12,44,59,107
Estrone	182	12,44,59,107
24-Ethylcholesta-5,22-dien-3β-ol	343	53,147
24-Ethylcholesta-7,22-dien-3β-ol	344	55,148
24-Ethylcholest-5-en-3β-ol	348	53,149
24-Ethyl-3β-hydroxymethyl-A-nor-5α-cholestane	352	45,61,150
24-Ethyl-3β-hydroxymethyl-A-nor-5α-*cis*-cholest-22-ene	349	45,61,149
(E)-24-Ethylidene-5α-cholest-7-en-3β-ol	345	55,148
(Z)-24-Ethylidene-5α-cholest-7-en-3β-ol	346	55,149
24-Ethyl-19-norcholestan-3β-ol	335	46,58,145
24-Ethyl-19-norcholest-*trans*-22-en-3β-ol	328	46,58,143
Ethyl urocanate	58	79
Fasciospongin	277	129
Filiformin	133	96
Filiforminol	134	97
Fimbrolide A	70	82
Fimbrolide B	71	82
Fimbrolide C	68	82
Flavoxanthin	389	34,39,163
Flexibilide	212	32,33,114
Floridoside	77	84
Fucoserratene	62	1,80
L-Fucose-4-sulfate	33	6,7,74
Fumaric acid	18	71
Furospongenol	231	118
Furospongenone	228	117
4-[11-(3-Furyl)-6-hydroxy-4,8-dimethyl-3-undecenyl]-2,6-dioxabicyclo[3.1.0]-hexan-3-one	232	118
4-[11-(3-Furyl)-6-hydroxy-4,8-dimethyl-8-undecenyl]-2,6-dioxabicyclo[3.1.0]-hexan-3-one	233	118

COMPOUND NAME INDEX FOR TABLE 27 (continued)

Name	Compound number (II/—)	Page number
5-[11-(3-Furyl)-6-hydroxy-4,8-dimethyl-3-undecenyl]-2,6-dioxabicyclo[3.1.0]-hexan-3-one	234	118
5-[11-(3-Furyl)6-hydroxy-4,8-dimethyl-8-undecenyl]-2,6-dioxabicyclo[3.1.0]-hexan-3-one	235	119
3-[11-(3-Furyl)-6-hydroxy-4,8-dimethyl-3-undecenyl]-5-hydroxy-2(5H)-furanone	236	119
3-[11-(3-Furyl)-6-hydroxy-4,8-dimethyl-8-undecenyl]-5-hydroxy-2(5H)-furanone	237	119
4[11-(3-Furyl)-6-hydroxy-4,8-dimethyl-3-undecenyl]-5-hydroxy-2(5H)-furanone	238	119
4-[11-(3-Furyl)-6-hydroxy-4,8-dimethyl-8-undecenyl]-5-hydroxy-2(5H)-furanone	239	119
Geranylfarnesol	279	129
Glanduliferol	154	101
Glycerylphosphorylcholine	65	64,81
N-Glycolylneuraminic acid	115	92
Gorgostanol	358	42,43,48,152
Halmiformin-1	276	129
Hematin	367	156
Hematoporphyrin	369	157
Heneicosane	247	121
cis,cis,cis,cis,cis-3,6,9,12,15-Heneicosapentaene	244	1,120
1-*O*-Heneicosylglycerol	270	127
Heptadecane	177	106
Heptadecanol	178	106
1-(Heptadeca-5,8,11,14-tetraenyl)-3,5-dihydroxybenzene	261	13,125
1-*O*-Heptadecylglycerol	224	116
Heteronemin	338	146
Heteroxanthin	391	35,38,163
Hexabromo-2,2′-bipyrrole	53	78
cis,cis-5,9-Hexacosadienoic acid	296	5,134
cis,cis,cis-5,9,19-Hexacosatrienoic acid	295	5,134
Hexadecane	170	105
Hexadecanol	171	105
1-*O*-Hexadecylglycerol	194	109
Hippurin-1	355	48,151
Holotoxin A	397	44,165
Homoserine	24	72
Homoserine betaine	50	66,78
4-Hydroxybenzaldehyde	39	75
3-Hydroxycanthaxanthin	374	37,159
3β-Hydroxy-5α-cholesta-9(11),24-dien-6α-yl-β-D-glucoside	366	56,155
Hydroxyechinenone	376	36,160
4′-Hydroxyechinenone	377	34,37,160
Hydroxyfimbrolide A	72	83
Hydroxyfimbrolide B	73	83
Hydroxyfimbrolide C	69	82
1-*O*-(2-Hydroxy-4-hexadecenyl)glycerol	191	6,109
1-*O*-(2-Hydroxyhexadecyl)glycerol	195	6,109
Hydroxyhydroquinone	29	16,73
24-Hydroxyircinolide	274	128
3-Hydroxy-4-keto-β-carotene	376	36,160
4-Hydroxy-4′-keto-β-carotene	377	34,37,160
3-Hydroxy-L-kynurenine	81	67,85
(−)-*S*-Hydroxymethyl-L-homocysteine	25	65,66,72
3β-Hydroxymethyl-24-methyl-A-nor-5α-cholestane	336	45,61,145

COMPOUND NAME INDEX FOR TABLE 27 (continued)

Name	Compound number (II/—)	Page number
3β-Hydroxymethyl-24-methyl-A-nor-5α-*cis*-cholest-22-ene	329	45,60,143
3β-Hydroxymethyl-A-nor-5α-cholestane	312	45,60,138
3β-Hydroxymethyl-A-nor-5α-*cis*-cholest-22-ene	307	45,60,137
1-*O*-(2-Hydroxy-4-octadecenyl)glycerol	246	6,121
p-Hydroxyphenylacetic acid	56	10,79
1-*O*-(2-Hydroxy-4-tetradecenyl)glycerol	176	6,106
1-*O*-(2-Hydroxytetradecyl)glycerol	180	6,107
Ichthyopterin	74	83
N(4-Imidazolepropionyl)histamine	106	64,90
Inosine	82	85
Inosinic acid	87	86
Iodoacetone	13	71
I-Iodo-3,3-dibromo-2-heptanone	43	2,77
Ircinianin	275	128
Ircinolide	273	128
Isoagatholactone	211	29,113
Isocaespitol	157	24,25,102
Isocryptoxanthin	385	34,36,162
T-Isolaureatin	136	2,97
3-*trans*-Isolaureatin	136	2,97
Isomytiloxanthin	380	35,161
24-Isopropylcholesta-5,22-dien-3β-ol	356	53,152
24-Isopropylcholest-5-en-3β-ol	359	53,153
6-Isopropyl-3,9,13-trimethyl-2,7,9,12-cyclotetradecatetraen-1-ol	213	32,114
Isospongiaquinone	253	18,123
Isoxanthopterin	28	73
Isozeaxanthin	387	34,36,162
24-Ketocholesterol	306	41,51,136
4-Keto-4′-ethoxy-β-carotene	394	35,38,164
4-Ketolutein	379	34,39,161
Δ^4-3-Keto-pregnen-20β-ol	240	57,119
D(−)-Lactic acid	15	5,71
Latimerol	317	47,140
T-Laureatin	137	2,97
3-*trans*-Laureatin	137	2,97
6*R*, 7*R* *cis*-Laurediol	145	2,99
6*S*, 7*S* *cis*-Laurediol	146	2,100
6*R*, 7*R* *trans*-Laurediol	147	2,100
6*S*, 7*S* *trans*-Laurediol	148	2,100
LL-PAA216	120	9,94
Lobophytolide	200	32,111
Lutein epoxide	388	39,162
Lutein-5,8-epoxide	389	34,39,163
Malic acid	20	72
L(−)-Methionine-*l*-sulfoxide	26	65,66,73
1-*O*-(2-Methoxydocosyl)glycerol	297	6,134
1-*O*-(2-Methoxyeicosyl)glycerol	271	6,127
1-*O*-(2-Methoxyheneicosyl)glycerol	281	6,129
1-*O*-(2-Methoxyheptadecyl)glycerol	248	6,121
1-*O*-(2-Methoxy-4-hexadecenyl)glycerol	221	6,116
1-*O*-(2-Methoxyhexadecyl)glycerol	225	6,117
1-*O*-(2-Methoxynonadecyl) glycerol	265	6,126
1-*O*-(2-Methoxy-4-octadecenyl)glycerol	257	6,124
1-*O*-(2-Methoxyoctadecyl)glycerol	260	6,125
1-*O*-(2-Methoxypentadecyl)glycerol	196	6,109
1-*O*-(2-Methoxytetradecyl)glycerol	188	6,108
4α-Methyl-5α-cholesta-7,24-dien-3β-ol	324	55,142

COMPOUND NAME INDEX FOR TABLE 27 (continued)

Name	Compound number (II/—)	Page number
24-Methylcholesta-5,22-dien-3β-ol	325	52,142
24α-Methylcholesta-5,22-dien-3β-ol	326	40,52,142
24-Methylcholesta-7,22-dien-3β-ol	327	55,143
24-Methylcholestane-3β,5α,6β,25-tetrol-25-monoacetate	360	48,153
24-Methylcholesta-7,22,25-trien-3β-ol	321	54,141
4α-Methyl-5α-cholest-7-en-3β-ol	330	55,143
24-Methyl-5α-cholest-5-en-3β-ol	331	52,144
24-Methyl-5α-cholest-7-en-3β-ol	332	55,144
24-Methylene-19-norcholestan-3β-ol	308	46,58,137
1-Methylhistidine	46	65,77
3-Methylhistidine	47	77
N-Methylmethionine sulfoxide	34	65,66,74
Methyl-*trans*-monocyclofarnesate	168	27,28,104
N-Methylmurexine	118	64,93
24-Methyl-19-norcholestan-3β-ol	313	46,58,138
Methylpalmitate	175	19,106
[3-(Methylthio)propyl] trimethylammonium chloride	52	62,78
Methyl-δ-tocotrienol methyl ether	337	14,30,145
Multifidene	108	1,91
Myrcene	97	19,88
Myristyl alcohol	130	96
Mytiloxanthin	381	35,38,161
Neoxanthin	392	39,163
Nephthenol	217	30,31,115
Nidificene	152	22,101
Nidifidiene	143	22,99
Nonadecane	192	109
Nonadecanol	193	109
cis,cis,cis,cis,cis-3,6,9,12,15-Nonadecapentaene	189	1,108
cis,cis,cis,cis-4,7,10,13-Nonadecatetraene	190	1,108
1-*O*-Nonadecylglycerol	259	125
l-Noradrenaline	61	13,64,65,80
22-*cis*-24-Norcholesta-5,22-dien-3β-ol	285	51,131
19-Norcholestan-3β-ol	294	46,58,133
19-Norcholest-*trans*-22-en-3β-ol	293	46,57,133
31-Norcycloartanol	350	42,60,150
19-Nor-5α,10β-ergost-22-en- 3β-ol	309	45,46,59,137
3-Norpseudozoanthoxanthin	104	63,90
Occelasterol	305	46,51,136
Octadecane	185	108
Octadecanol	186	108
cis-9-Octadecen-1-ol	184	107
N-(1,2,3,4,4a,7,8,8a-Octahydro-4-isopropyl-1,6-dimethyl-1-naphthyl)formamide	169	27,104
1,2,3,4,4a,7,8,8a-Octahydro-4-isopropyl-1,6-dimethyl-1-naphthyl isocyanide	166	27,104
1,2,3,4,4a,7,8,8a-Octahydro-4-isopropyl-1,6-dimethyl-1-naphthyl isothiocyanate	167	27,104
2,3,6,7,10a,13,14,14a-Octahydro-1a,5,8,12-tetramethyloxireno[9,10]cyclotetradeca-[1,2,-b] furan-9(1aH)-one	201	32,111
2,3,6,7,10a,13,14,14a-Octahydro-1a,5,8,12-tetramethyloxireno[9,10]cyclotetradeca-[1,2-b]furan-9(1aH)-one	202	32,111
Oleyl alcohol	184	107
Ornithine	27	73
24-Oxocholesterol	306	41,51,136
Pacifenol	139	23,98

COMPOUND NAME INDEX FOR TABLE 27 (continued)

Name	Compound number (II/—)	Page number
Parazoanthoxanthin A	78	84
Parazoanthoxanthin E	126	95
Parazoanthoxanthin F	127	95
Pentadecane	160	103
Pentadecanol	161	103
1-*O*-Pentadecylglycerol	187	108
2-Pentaprenyl-1,4-benzoquinol	361	14,153
Petromyzonol	268	47,127
15*R*-PGA_2 methyl ester	241	120
$PGF_{1\alpha}$	219	4,116
$PGF_{2\alpha}$	218	4,115
Phenylacetic acid	55	10,79
Phloroglucinol	30	11,16,73
Phoenicoxanthin	374	37,159
Phosphoethanolamine	2	69
Phylloquinone	362	18,154
1-*O*-Phytanylglycerol	264	6,126
Phytol	220	28,116
Plastoquinone-9	395	17,164
Pleraplysillin-2	199	27,110
Prepacifenol epoxide	141	22,23,99
Progesterone	230	44,57,118
n-Propyl-4-hydroxybenzoate	83	85
Protoporphyrin IX	368	156
Pseudozoanthoxanthin	117	63,93
Pterin-6-carboxylic acid	38	75
Renilla luciferin	198	110
Retinol	208	28,113
Rhodophytin	131	3,96
Riboflavine	173	105
Sarcophine	203	31,32,111
Scalaradial	300	135
Spongiadiol	204	28,29,112
epi-Spongiadiol	205	28,29,112
Spongiatriol	206	28,29,112
epi-Spongiatriol	207	28,29,113
Spongiadiol diacetate	266	28,29,126
epi-Spongiadiol diacetate	267	28,29,126
Spongiaquinone	254	18,123
Spongiatriol triacetate	282	28,29,130
epi-Spongiatriol triacetate	283	28,29,130
Stearyl alcohol	186	108
Stichopogenin A_4	354	44,151
Succinic acid	19	71
Taraxanthin	388	39,162
Tauroallocholate	289	47,132
Taurochenodeoxycholate	286	49,131
Taurocholate	290	49,132
Taurodeoxycholate	287	49,131
Tauro-3α,12α-dihydroxy-5β-chol-7-en-24-oate	284	56,130
Tauro-3α,7β-dihydroxy-5β-cholan-24-oate	288	49,132
Taurohaemulcholate	291	49,133
Tauro-3α,7β,12α-trihydroxy-5β-cholan-24-oate	292	50,133
1,1,3,3-Tetrabromoacetone	4	2,69
1,1,3,3-Tetrabromo-2-heptanone	42	2,76
Tetrabromopyrrole	17	71

COMPOUND NAME INDEX FOR TABLE 27 (continued)

Name	Compound number (II/—)	Page number
Tetracosane	269	127
Tetradecane	129	95
Tetradecanol	130	96
1-*O*-Tetradecylglycerol	179	106
Tetradehydrofurospongin-1	226	117
3β,6α,15α,24ξ-Tetrahydroxy- 5α-cholestane	314	43,44,47,139
N-(1,5,9,13-Tetramethyl-1-vinyl-4,8,12-tetradecatrienyl)-formamide	245	27,33,120
1,5,9,13-Tetramethyl-1- vinyl-4,8,12-tetradecatrienyl isocyanide	242	27,33,120
1,5,9,13-Tetramethyl-1-vinyl-4,8,12-tetradecatrienyl isothiocyanate	243	27,33,120
Thelepin	123	10,94
α-Tocopherol	351	15,30,150
β-Tocopherol	333	15,30,144
γ-Tocopherol	334	15,30,145
δ-Tocopherol	310	15,30,137
δ-Tocotrienol	298	14,30,134
δ-Tocotrienol epoxide	299	14,30,134
δ-Tocotrienol methyl ether	318	14,30,140
1,1,3-Tribromoacetone	8	2,70
2,3,6-Tribromo-4,5-dihydroxybenzyl alcohol	37	13,75
1,1,3-Tribromo-2-heptanone	44	2,77
1,3,3-Tribromo-2-heptanone	45	2,77
Tricosane	262	125
Tridecane	122	94
Trifuhalol	181	11,16,107
Trikentriorhodin	375	39,160
Trimethylamine oxide	16	62,71
Tunaxanthin	386	35,39,162
Tyramine	59	62,80
cis,trans-1,3,5-Undecatriene	111	1,91
trans,trans,trans-2,4,6-Undecatriene	112	1,91
5′-Uridylic acid	75	83
N-Urocanylhistamine	105	64,90
Uroporphyrin I	372	159
L-Valine betaine	63	66,81
Violacein	197	110
Violacene	84	19,20,86
Violaxanthin	393	37,163
Vitamin A	208	28,113
Vitamin B_T	49	64,78
Vitamin E	351	15,30,150
Vitamin K_1	362	18,154
Watasenia oxyluciferin	272	128

Table 28
COMPOUNDS FROM VOLUME I MENTIONED IN REFERENCES LISTED IN THIS BOOK

Compound No. I/—	Name	Source of compound	Available by synthesis	Activity	References
1.	Methylamine	Fish:			116
		Navodon modestus			
3.	Glycine	Algae			39,117
		Crustaceans			125,175
		Echinoderms			183,186
		Fish			190,208
		Gorgonians			227,254
		Molluscs			
		Sea worm			
4.	Dimethylamine	Fish			116
8.	Taurine	Clam			39,117
		Fish			175,254
		Gorgonians			262
		Red alga			
11.	α-Alanine	Algae			39,117
		Clam			175,183
		Fish			186,190
		Gorgonians			208,254
		Lobster			
12.	β-Alanine	Algae			175
		Fish:			190
		Fugu vermiculare porphyreum			
		Hippoglossoides dubius			
		Paralichthys olivaceus			
15.	Serine	Algae			39,117
		Clam			175,186
		Fish			190,208
		Gorgonians			254
20.	Trimethylamine	Algae			92,116
		Fish			175
21.	*N*-Methyltaurine	Red alga:			262
		Gelidium sesquipedale			
26.	Aspartic acid	Algae			39,117
		Clam			175,183
		Fish			186,190
		Gorgonians			208,254
		Lobster			
27.	Asparagine	Algae			125
		Fish			175
		Octopus:			190
		Hapalochlaena maculosa			
28.	β-Aminoisobutyric acid	Fish:			175
		Fugu vermiculare porphyreum			
		Hippoglossoides dubius			
		Paralichthys olivaceus			
30.	Threonine	Algae			39,117
		Clam			175,186
		Fish			190,254
		Gorgonians			

Table 28 (continued)
COMPOUNDS FROM VOLUME I MENTIONED IN REFERENCES LISTED IN THIS BOOK

Compound No.I/—	Name	Source of compound	Available by synthesis	Activity	References
31.	Creatine	Fish			175, 254
42.	Hypoxanthine	Clam:			117
		Tapes japonica			175
		Fish			179
45.	Guanine	Salmon:			39,117
		Oncorhynchus kisutch			175,183
					186,190
					254
48.	Proline	Algae, clam, fish			39,117
		Clam			175,183
		Fish			186,190
		Gorgonians			254
		Lobster			
49.	4-Hydroxypro-line	Fish:			175
		Fugu vermiculare porphy reum			
		Hippoglossoides dubius			
		Lophius litulon			
50.	Chondrine Artifact?	Algae:			190
		Codium fragile			
		Cystophora moniliformis			
		Ecklonia radiata			
		Martensia elegans			
		Zonaria sinclairii			
		Z. turneriana			
52.	Glutamic acid	Algae			30,39
		Echinoderms			117,175
		Fish			183,186
		Gorgonians			190,208
		Lobster			227,254
		Molluscs			
		Sea worm			
53.	Histamine	Crab			66
		Echinoderms			125
		Molluscs			203
		Sea anemones			
		Sea worms			
		Sponges			
		Tunicate			
54.	Glutamine	Algae			30
		Fish			175
		Sea hare: *Aplysia californica*			190,208
56.	Betaine	Clam: *Tapes japonica*			117
57.	Valine	Algae			39,117
		Clam			175,190
		Fish			208,254
		Gorgonians			
58.	Methionine	Algae			39,117
		Clam			175,186
		Fish			190,254
		Gorgonians			
66.	Choline sulfate	Dinoflagellate:			260
		Amphidinium carteri			262
		Red alga:			
		Gelidium sesquipedale			

Table 28 (continued)
COMPOUNDS FROM VOLUME I MENTIONED IN REFERENCES LISTED IN THIS BOOK

Compound No.I/—	Name	Source of compound	Available by synthesis	Activity	References
71.	Choline	Molluscs:			233
		Concholepas concholepas			234
		Thais chocolata			
		T. haemastoma			
		T. haemastoma floridana			
74.	Urocanic acid	Mollusc:			233
		Concholepas concholepas			
80.	L-Baikiain	Red algae:			190
		Corallina officinalis			
		Gracilaria secundata			
		Pterocladia capillacea			
81.	Histidine	Algae			39,117
		Clam			125,175
		Fish			186,190
		Gorgonians			254
		Octopus			
84.	Isoleucine	Algae			39,117
		Clam			175,186
		Fish			190,208
		Gorgonians			254
85.	Leucine	Algae			39,117
		Clam			175,185
		Fish			190,208
		Gorgonians			254
86.	Lysine	Algae			39,117
		Clam			175,186
		Fish			190,254
		Gorgonians			
87.	Arginine	Algae			117,175
		Clam:			186,190
		Tapes japonica			208,254
		Fish			
90.	2,3-Dibromo-5-hydroxy-4-*O*-sulfate benzyl sulfate, dipotassium salt	Red algae			97
91.	2,3-Dibromo-4,5-dihydroxy-benzaldehyde	Red algae	+	Antibiotic	97
92.	3-Bromo-4,5-dihydroxy-benzaldehyde	Red algae:		Antibiotic	97
		Polysiphonia elongata			
		P. lanosa			
		P. urceolata			
93.	3,5-Dibromo-4-hydroxybenzyl alcohol	Red algae:	+	Antibiotic	97
		Polysiphonia brodiaei			119
		P. lanosa			
		P. nigrescens			
		Sea worm:			
		Thelepus setosus			
94.	Lanosol	Red algae		Antibiotic	97
95.	Homarine	Molluscs			117,164

Table 28 (continued)
COMPOUNDS FROM VOLUME I MENTIONED IN REFERENCES LISTED IN THIS BOOK

Compound No.I/—	Name	Source of compound	Available by synthesis	Activity	References
107.	4-Acetamido-2,6-dibromo-4-hydroxy-cyclohexadienone	Sponge	+		162 274
112.	Octopamine	Lobster: *Homarus americanus* Sea hare: *Aplysia californica*			163 239
113.	4-(β-Carboxy-β-aminoethyl)-1,3-dimethyl-imidazolium chloride	Algae: *Gracilaria secundata* *Zonaria sinclairii*			190
114.	Choline acrylate	Dinoflagellate: *Amphidinium carteri*	+		260
122.	Aeroplysinin-1		+		7,201
127.	Phenylalanine	Algae Clam Fish Octopus			117,125 175, 186 190,208 254
128.	Tyrosine	Algae Clam Fish Octopus			117,125 175,186 190,254
135.	2-(2-Hydroxy-3,5-dibromo-phenyl)-3,4,5-tribromopyrrole	Bacterium: *Chromobacterium* sp.		Antibiotic	6
138.	Spinochrome B	Sea urchins			88 110
148.	Serotonin	Molluscs			58,125 163
158.	Senecioylcholine	Molluscs: *Concholepas concholepas* *Thais chocolata* *T. floridana floridana* *T. haemastoma* *T. haemastoma floridana*			164 233 234
165.	Tryptophan	Brown alga: *Sargassum flavicans* Octopus: *Hapalochlaena maculosa*			125 190
167.	Spongosine		+		16
168.	Dictyopterene B	Brown algae: *Dictyopteris australis* *D. plagiogramma*			214
169.	Dictyopterene D′ Ectocarpene	Brown algae: *Cutleria multifida* *Dictyopteris australis* *D. plagiogramma*			134 214

Table 28 (continued)
COMPOUNDS FROM VOLUME I MENTIONED IN REFERENCES LISTED IN THIS BOOK

Compound No.I/—	Name	Source of compound	Available by synthesis	Activity	References
170.	*trans,cis,cis*-1,3,5,8-Undecatetraene	Brown algae: *Dictyopteris australis* *D. plagiogramma*			214
171.	*trans,trans,cis*-1,3,5,8-Undecatetraene	Brown algae: *Dictyopteris australis* *D. plagiogramma*			214
173.	Dictyopterene A	Brown algae: *Dictyopteris australis* *D. plagiogramma*	+		19 214
174.	Dictyopterene C′	Brown algae: *Dictyopteris australis* *D. plagiogramma*	+		19 214
175.	*trans,cis*-1,3,5-Undecatriene	Brown algae: *Dictyopteris australis* *D. plagiogramma*			214
176.	*trans,trans*-1,3,5,-Undecatriene	Brown algae: *Dictyopteris australis* *D. plagiogramma*			214
177.	Murexine	Molluscs			15,164, 233,234
179.	Dihydromurexine	Molluscs: *Thais haemastoma* *T. haemastoma floridana*			234
189.	3-Acetyl-2,6,7-trihydroxyjuglone Spinochrome S	Sea urchin: *Salmacis sphaeroides*			111 112
190.	6-Acetyl-2,3,7-trihydroxyjuglone Spinochrome G		+		112
191.	Spinochrome A	Sea urchin: *Psammechinus microtuberculatus*			88
211.	Cadalene		+		4
214.	Aplysin	Red alga: *Laurencia nidifica*			270
217.	Laurinterol	Red alga: *Laurencia nidifica*		Antibiotic	270
220.	Laurene	Red alga: *Laurencia filiformis* f. *dendritica*	+		162 188
235.	(−)-α-Curcumene	Gorgonians: *Muricea elongata* *Plexaurella nutans*			138
240.	(+)-β-Bisabolene	Gorgonians: *Muricea elongata* *Plexaurella nutans*			138
254.	Zonarene				129
257.	Oppositol			Now shown to be inactive	79

Table 28 (continued)
COMPOUNDS FROM VOLUME I MENTIONED IN REFERENCES LISTED IN THIS BOOK

Compound No.I/—	Name	Source of compound	Available by synthesis	Activity	References
272.	Isorhodoptilometrin		+		12
275.	AF 350	Jellyfish: *Aequorea aequorea*			55,56, 57,184
287.	Pristane	Fish: *Oncorhynchus gorbuscha* *O. keta* *O. kisutch* *O. nerka* *Priacanthus macracanthus*			131
292.	15*R*-PGA_2	Gorgonian: *Plexaura homomalla*			10 271
293.	15*S*-PGA_2	Gorgonian: *Plexaura homomalla*			10
296.	15*S*-PGE_2	Fish: *Paralichthys olivaceus* *Thunnus thynnus*			224
313.	Furospongin-1	Sponges: *Phyllospongia foliascens* *P. papyracea* *P. radiata* *Spongia* sp.			162 217
315.	1,6,9,12,15,18-Heneicosahexaene	Brown alga: *Cystophora torulosa*			162
317.	5α-Pregn-9(11)-ene-3β-6α-diol-20-one		+		9
318.	15*S*-PGA_2 methyl ester	Gorgonian: *Plexaura homomalla*			10
320.	1,6,9,12,15-Heneicosapentaene	Brown alga: *Cystophora torulosa*			162
323.	Batyl alcohol	Algae Coelenterates Crustaceans Fish Mussel Starfish			114 194 263
335.	*O*-Acetyl(15*R*)-PGA_2 methyl ester	Gorgonian: *Plexaura homomalla*			10
336.	*O*-Acetyl (15*S*)-PGA_2 methyl ester	Gorgonian: *Plexaura homomalla*			10
339.	Aerothionin	Sponge: *Verongia cavernicola*			120
353.	Variabilin	Sponge: *Fasciospongia fovea*			162

Table 28 (continued)
COMPOUNDS FROM VOLUME I MENTIONED IN REFERENCES LISTED IN THIS BOOK

Compound No.I/—	Name	Source of compound	Available by synthesis	Activity	References
354.	Furospinosulin-1	Sponges:			162
		Fasciospongia fovea			
		Ircinia sp.			
		Phyllospongia sp.			
		Spongia sp.			
358.	2-Tetraprenyl-1,4- benzoquinol	Sponge:			162
		Ircinia ramosa			
359.	Asterosterol	Sponge:	+		28
		Halichondria panicea			146
		Starfish:			170
		Asterias rubens			249
		Coscinasterias acutispina			261
		Leiaster leachii			
360.	24-Norcholesta-5,22-dien-3β-ol	Gorgonians			22,68
		Jellyfish			84,127
		Red alga			128,169
		Scallop			172,275
		Sea anemone			276
		Sea urchins			
		Sea worm			
		Sponges			
363.	24-Norcholest-22-en-3β-ol	Tunicate:			269
		Halocynthia roretzi			
367.	Taondiol		+		176
372.	22-*trans*-Cholesta-7,22-dien-3β-ol	Starfish:			146
		Asterias rubens			249
		Coscinasterias acutispina			261
		Leiaster leachii			
374.	22-Dehydrocholesterol	Algae			22,68
		Gorgonians			84, 128
		Jellyfish			145,169
		Scallop			172,240
		Sea anemone			250,275
		Sea urchin			276
		Sea worm			
		Sponges			
375.	Desmosterol	Jellyfish	+		67,84
		Red algae			145,169
		Sea urchins			172,240
		Sea worm			275,276
380.	Cholesterol	Algae			22,68
		Gorgonians			84,128
		Jellyfish			144,145
		Scallop			146,169
		Sea anemone			172,240
		Sea urchins			249,250
		Sea worm			262,275
		Sponges			276
		Starfish			

Table 28 (continued)
COMPOUNDS FROM VOLUME I MENTIONED IN REFERENCES LISTED IN THIS BOOK

Compound No.I/—	Name	Source of compound	Available by synthesis	Activity	References
382.	Lathosterol	Starfish:			146
		Asterias amurensis			249
		A. rubens			261
		Coscinasterias acutispina			263
		Leiaster leachii			
383.	Cholestanol	Jellyfish			84,128
		Red alga			146,249
		Scallop			261,275
		Sea urchin			276
		Starfish			
385.	16-Deoxymyx- inol	Fish bile			258
386.	Myxinol	Fish bile			258
387.	Scymnol	Fish bile			258
393.	Ergosterol	Red algae			145
394.	Brassicasterol	Algae			22
		Gorgonians			68
		Jellyfish			84
		Scallop			128
		Sea anemone			169
		Sea urchins			240
		Sea worm			275
		Sponges			276
395.	Episterol	Starfish:			146
		Asterias rubens			249
		Coscinasterias acutispina			261
		Leiaster leachii			
396.	24-Methylene- cholesterol	Algae			22,68
		Gorgonian			84,128
		Jellyfish			169,172
		Scallop			173,240
		Sea anemone			250,275
		Sea urchins			276
		Sea worm			
		Sponges			
397.	Stellasterol	Starfish:			146
		Asterias rubens			249
		Coscinasterias acutispina			
		Leiaster leachii			
400.	22,23-Dihydro- brassicasterol	Sponges:			68
		Verongia aerophoba			
		V. archeri			
		V. fistularis			
		V. thiona			
407.	24,28- Didehy- droaplysterol	Sponges:			68
		Verongia aerophoba			
		V. archeri			
		V. fistularis			
		V. thiona			

Table 28 (continued)
COMPOUNDS FROM VOLUME I MENTIONED IN REFERENCES LISTED IN THIS BOOK

Compound No.I/—	Name	Source of compound	Available by synthesis	Activity	References
409.	Fucosterol	Algae		Lipase acti-	22,84
		Gorgonian:		vator	128,173
		Plexaura sp.			174,216
		Scallop:			240, 250
		Placopecten magellanicus			
410.	28-Isofucosterol	Jellyfish			128
		Red alga:			240
		Furcellaria fastigiata			275
		Scallop:			
		Placopecten magellanicus			
413.	Stigmasterol	Gorgonians:			22
		Eugorgia ampla			
		Plexaura sp.			
414.	Saringosterol	Brown algae:			240
		Ascophyllum nodosum			
		Laminaria saccharina			
416.	Aplysterol	Sponges:			68
		Verongia aerophoba			
		V. archeri			
		V. fistularis			
		V. thiona			
417.	24ξ-Ethylcholest-7-en-3β-ol	Starfish:			146
		Asterias rubens			249
		Coscinasterias acutispina			261
		Leiaster leachii			
418.	β-Sitosterol	Gorgonians			22
		Red alga			68
		Scallop			128
		Sponges			240
425.	Furospinosulin-2	Sponge			162
435.	Cycloartenol	Starfish:			249
		Asterias rubens			
437.	Gorgosterol	Gorgonian:			22
		Plexaura sp.			
438.	Lanosterol	Starfish:			249
		Asterias rubens			
442.	9,11-Secogorgost-5-en-3,11-diol-9-one Secogorgosterol				74
454.	Biliverdin IXα	Crustaceans:			60
		Cambarus sp.			231
		Peltogaster paguri			
		Fish			
		Sea worm:			
		Nereis diversicolor			
457.	Aplysioviolin	Sea hare:			231
		Aplysia punctata			
462.	Actinioerythrol				8
464.	Ubiquinone Q-6	Fish			215
465.	7,8-Didehydro-isorenieratene		+		130

Table 28 (continued)
COMPOUNDS FROM VOLUME I MENTIONED IN REFERENCES LISTED IN THIS BOOK

Compound No.I/—	Name	Source of compound	Available by synthesis	Activity	References
466.	7,8-Didehydro-renieratene		+		130
472.	Alloxanthin	Alga			93,103
	Cynthiaxanthin	Crustaceans			196,198
		Fish			199,225
		Molluscs			
473.	Canthaxanthin	Crustaceans			34,60
		Fish			62,63
		Sea cucumber:			64,65
		Psolus fabrichii			93,151
					153,154
					156,157
					158,196
					199,200
474.	α-Doradecin	Crustaceans			152,198
		Fish			200
475.	Pectenolone	Scallop:			103
		Pecten maximus			
476.	Astaxanthin	Crustaceans			8,34
		Echinoderms			60,61
		Fish			62,63
		Molluscs			64,65
		Sponge			86,87
					92,102,103
					118,151
					152,153
					154,155
					156,157
					158,196
					197,199
					259,272
477.	Crocoxanthin	Alga:			225
		Chroomonas salina			
478.	Echinenone	Crustaceans			34,63
		Echinoderms			93,102
		Fish			103,151
		Molluscs			153,154
		Sponge			155,156
					157,196
					199,200
479.	Diatoxanthin	Alga			93
		Crab			198
		Fish			225
480.	Monadoxanthin	Alga:			225
		Chroomonas salina			
481.	Diadinoxanthin	Algae:			225
		Pavlova sp.			
		Vaucheria sessilis			
482.	α-Carotene	Algae			60,73,93
		Crab			102,155
		Fish			157,225
		Sea horse			
		Sponge			

Table 28 (continued)
COMPOUNDS FROM VOLUME I MENTIONED IN REFERENCES LISTED IN THIS BOOK

Compound No.I/—	Name	Source of compound	Available by synthesis	Activity	References
483.	β-Carotene	Algae			38,61
		Crustaceans			62,63
		Fish			65,73,86
		Molluscs			93,102
		Sponges			103,118
		Starfish			151,153
					154,155
					156,157
					196,197
					198,199
					200,225
					259,272
484.	δ- Carotene (should be Δ4 not Δ5 as presented in Volume I)	Crab: *Carcinus maenas*			151
485.	ε-Carotene	Sea bream			152
486.	Lutein	Crustaceans			38,61
		Fish			62,63
		Molluscs			65,103
		Red alga			151,152
		Sponge			154,155
					157,158
					196,198
					199,200
487.	Zeaxanthin	Crustaceans			38,60
		Fish			61,62
		Molluscs			63,65
		Red alga			93,103
		Sponge			151,154
		Starfish			155,157
					158,196
					197,198
					199,200
490.	Fucoxanthinol	Brown alga: *Fucus vesiculosus*			223
494.	Fucoxanthin	Alga: *Pavlova* sp.			225
501.	Ubiquinone Q-8				215
503.	Ubiquinone Q-9	Algae			215
		Fish			
504.	Ubiquinone Q-10	Fish			215

COMPOUND NAME INDEX FOR TABLE 28 (continued)

Name	Compound number I/	Page number
4-Acetamido-2,6-dibromo-4-hydroxy-cyclohexadienone	107	178
O-Acetyl(15*R*)-PGA$_2$ methyl ester	335	180
O-Acetyl (15*S*)-PGA$_2$ methyl ester	336	180
3-Acetyl-2,6,7-trihydroxyjuglone	189	179
6-Acetyl-2,3,7-trihydroxyjuglone	190	179
Actinioerythrol	462	183
Aeroplysinin-1	122	178
Aerothionin	339	180
AF 350	275	180
α-Alanine	11	65,175
β-Alanine	12	175
Alloxanthin	472	35,184
β-Aminoisobutyric acid	28	175
Aplysin	214	179
Aplysioviolin	457	183
Aplysterol	416	45,183
Arginine	87	177
Asparagine	27	175
Aspartic acid	26	65,175
Astaxanthin	476	34,184
Asterosterol	359	181
L-Baikiain	80	65,177
Batyl alcohol	323	180
Betaine	56	176
Biliverdin IXα	454	183
(+)-β-Bisabolene	240	20,179
Brassicasterol	394	182
3-Bromo-4,5-dihydroxybenzaldehyde	92	7,8,177
Cadalene	211	179
Canthaxanthin	473	184
4-(β-Carboxy-β-aminoethyl)-1,3-dimethylimidazolium chloride	113	65,178
α-Carotene	482	33,184
β-Carotene	483	33,185
δ-Carotene	484	33,185
ε-Carotene	485	185
22-*trans*-Cholesta-7,22-dien- 3β-ol	372	181
Cholestanol	383	182
Cholesterol	380	41,42,43,46, 181
Choline	71	64,177
Choline acrylate	114	64,178
Choline sulfate	66	176
Chondrine	50	65,176
Creatine	31	176
Crocoxanthin	477	35,184
(−)-α-Curcumene	235	20,179
Cycloartenol	435	183
Cynthiaxanthin	472	35,184
22-Dehydrocholesterol	374	41,181
16-Deoxymyxinol	385	182
Desmosterol	375	181
Diadinoxanthin	481	34,35,184
Diatoxanthin	479	35,184
2,3-Dibromo-4, 5-dihydroxybenzaldehyde	91	7,8,177
3,5-Dibromo-4-hydroxybenzyl alcohol	93	8,9,10,177
2,3-Dibromo-5-hydroxy-4-*O*-sulfate benzyl sulfate, dipotassium salt	90	177
Dictyopterene A	173	1,179

COMPOUND NAME INDEX FOR TABLE 28 (continued)

Name	Compound number I/	Page number
Dictyopterene B	168	178
Dictyopterene C′	174	1,2,179
Dictyopterene D′	169	178
24,28-Didehydroaplysterol	407	45,182
7,8-Didehydroisorenieratene	465	35,183
7,8-Didehydrorenieratene	466	35,184
22,23-Dihydrobrassicasterol	400	182
Dihydromurexine	179	64,179
Dimethylamine	4	175
α-Doradecin	474	34,184
Echinenone	478	184
Ectocarpene	169	178
Episterol	395	182
Ergosterol	393	182
24ξ-Ethylcholest-7-en-3β-ol	417	183
Fucosterol	409	40,41,183
Fucoxanthin	494	34,185
Fucoxanthinol	490	185
Furospinosulin-1	354	181
Furospinosulin-2	425	183
Furospongin-1	313	180
Glutamic acid	52	65,67,176
Glutamine	54	176
Glycine	3	65,67,175
Gorgosterol	437	43,183
Guanine	45	176
1,6,9,12,15,18-Heneicosahexaene	315	180
1,6,9,12, 15-Heneicosapentaene	320	180
Histamine	53	63,176
Histidine	81	66,177
Homarine	95	177
2-(2-Hydroxy-3,5-dibromophenyl)-3,4,5-tribromopyrrole	135	178
4-Hydroxyproline	49	176
Hypoxanthine	42	176
28-Isofucosterol	410	183
Isoleucine	84	177
Isorhodoptilometrin	272	180
Lanosol	94	7,8,177
Lanosterol	438	183
Lathosterol	382	182
Laurene	220	179
Laurinterol	217	22,179
Leucine	85	177
Lutein	486	34,185
Lysine	86	177
Methionine	58	176
Methylamine	1	175
24- Methylenecholesterol	396	41,182
N-Methyltaurine	21	175
Monadoxanthin	480	35,184
Murexine	177	64,179
Myxinol	386	182
24-Norcholesta-5,22-dien-3β-ol	360	46,181
24-Norcholest-22-en-3β-ol	363	181
Octopamine	112	62,178
Oppositol	257	179
Pectenolone	475	34,35,184

COMPOUND NAME INDEX FOR TABLE 28 (continued)

Name	Compound number I/	Page number
15*R*-PGA_2	292	3,4,180
15*S*-PGA_2	293	180
15*S*-PGA_2 methyl ester	318	180
15*S*-PGE_2	296	4,180
Phenylalanine	127	178
5α-Pregn-9(11)-ene-3β-6α-diol-20-one	317	180
Pristane	287	28,180
Proline	48	1,176
Saringosterol	414	183
Scymnol	387	182
9,11-Secogorgost-5-en-3,11-diol-9-one	442	183
Secogorgosterol	442	183
Senecioylcholine	158	64,178
Serine	15	175
Serotonin	148	178
β-Sitosterol	418	183
Spinochrome A	191	179
Spinochrome B	138	178
Spinochrome G	190	179
Spinochrome S	189	179
Spongosine	167	178
Stellasterol	397	182
Stigmasterol	413	183
Taondiol	367	29,181
Taurine	8	66,175
2-Tetraprenyl-1,4-benzoquinol	358	181
Threonine	30	175
Trimethylamine	20	62,175
Tryptophan	165	178
Tyrosine	128	178
Ubiquinone Q-6	464	183
Ubiquinone Q-8	501	185
Ubiquinone Q-9	503	185
Ubiquinone Q-10	504	185
trans,cis-1,3,5-Undecatriene	175	179
trans,trans-1,3, 5-Undecatriene	176	179
trans,cis,cis-1,3,5,8-Undecatetraene	170	179
trans,trans,cis-1,3,5,8-Undecatetraene	171	179
Urocanic acid	74	177
Valine	57	176
Variabilin	353	180
Zeaxanthin	487	185
Zonarene	254	20,21,179

REFERENCES

1. **Abe, H., Uchiyama, M., and Sato, R.**, Isolation of phenylacetic acid and its *p*-hydroxy derivative as auxin-like substances from *Undaria pinnatifida, Agric. Biol. Chem.*, 38, 897, 1974.
2. **Abe, S. and Kaneda, T.**, Occurrence of homoserine betaine in the hydrolyzate of an unknown base isolated from a green alga, *Monostroma nitidum, Bull. Jpn. Soc. Sci. Fish.*, 40, 1199, 1974.
3. **Ackermann, D.**, Über das Vorkommen von Homarin, Taurocyamin, Cholin, Lysin und anderen Aminosäuren sowie Bernsteinsaure in dem Meereswurm-*Arenicola marina, Hoppe-Seyler's Z. Physiol. Chem.*, 302, 80, 1955.
4. **Adachi, K. and Tanaka, J.**, Synthesis of cadalene from bromobenzene, *J. Syn. Org. Chem. (Japan)*, 31, 322, 1973.
5. **Aguilar-Martinez, M. and Liaaen-Jensen, S.**, Animal carotenoids. IX, Trikentriorhodin, *Acta Chem. Scand. Ser. B*, 28, 1247, 1974.
6. **Andersen, R. J., Wolfe, M. S., and Faulkner, D. J.**, Autotoxic antibiotic production by a marine *Chromobacterium, Mar. Biol.*, 27, 281, 1974.
7. **Andersen, R. J., and Faulkner, D.J.**, The synthesis of (±) aeroplysinin-1 and related compounds, *Proc. Food-Drugs from the Sea*, 1974, Marine Technology Society, Washington, D.C., 1976, 263.
8. **Andrewes, A. G., Borch, G., Liaaen-Jensen, S., and Snatzke, G.**, Animal carotenoids. 9. On the absolute configuration of astaxanthin and actinioerythrin, *Acta Chem. Scand. Ser. B*, 28, 730, 1974.
9. **ApSimon, J. W. and Eenkhoorn, J. A.**, Marine organic chemistry. II. Synthesis of 3β,6α-dihydroxy-5α-pregn-9(11)-en-20-one, the major sapogenin of the starfish *Asterias forbesi, Can. J. Chem.*, 52, 4113, 1974.
10. **Baker, J. L.**, U.S. Patent 3,778, 469 (Cl. 260-499), 1972.
11. **Baker, J. T. and Murphy, V.**, *Handbook of Marine Science., Compounds from Marine Organisms*, Vol I, CRC Press, Cleveland, Ohio, 1976.
12. **Banville, J. and Brassard, P.**, The synthesis of some naturally occurring anthraquinones, 9th. I.U.P.A.C. Symp. Natural Products,(Abstr.) Ottawa, 1974, 55A.
13. **Barton, D. H. R., Leclerc, G., Magnus, P. D., and Menzies, I. D.**, An unusual synthesis of ergosterol acetate peroxide, *Chem. Commun.*, p. 447, 1972.
14. **Baxter, J. G.**, Vitamin A, *Comp. Biochem.*, 9, 169, 1963.
15. **Bender, J. A., DeRiemer, K., Roberts, T. E., Rushton, R., Boothe, P., Mosher, H. S., and Fuhrman, F. A.**, Choline esters in the marine gastropods *Nucella emarginata* and *Acanthina spirata;* a new choline ester, tentatively identified as *N*- methylmurexine, *Comp. Gen. Pharmac.*, 5, 191, 1974.
16. **Bergmann, W. and Stempien Jr., M. F.**, Contributions to the study of marine products. XLIII. The nucleosides of sponges. V. The synthesis of spongosine. *J. Org. Chem.*, 22, 1575, 1957.
17. **Bernstein, J., Shmeuli, U., Zadock, E., Kashman, Y., and Neeman, I.**, Sarcophine, a new epoxy cembranolide from marine origin, *Tetrahedron*, 30, 2817, 1974.
18. **Bersis, D.**, Beitrag zur Chemilumineszenz und Biolumineszenz, *Folia Biochim. et Biolog. Graeca*, 11, 30, 1974.
19. **Billups, W. E., Chow, W. Y., and Cross, J. H.**, Synthesis of (±)-dictyopterene A and (±)-dictyopterene C′, *Chem. Commun.*, p. 252, 1974.
20. **Björkman, L. R., Karlsson, K.-A., Pascher, I., and Samuelsson, B. E.**, The identification of large amounts of cerebroside and cholesterol sulfate in the sea star *Asterias rubens, Biochim. Biophys. Acta*, 270, 260, 1972.
21. **Blackman, A. J. and Wells, R. J.**, personal communication.
22. **Block, J. H.**, Marine sterols from some gorgonians, *Steroids*, 23, 421, 1974.
23. **Blumer, M.**, Fossile Kohlenwasserstoffe und Farbstoffe in Kalksteinen. Geochemische Untersuchungen III. *Mikrochemie*, 36/ 37, 1048, 1951.
24. **Boar, R. B. and Widdowson, D. A.**, Biosynthesis, *Annu. Rep. Progr. Chem. Sect. B*, 71, 455, 1974.
25. **Boeryd, B., Hallgren, B., and Ställberg, G.**, Studies on the effect of methoxy-substituted glycerol ethers on tumour growth and metastasis formation, *Brit. J. Exp. Pathol.*, 52, 221, 1971.
26. **Boeryd, B., Hallgen, B., and Ställberg, G.**, Antitumor activity of methoxy-substituted glycerol ethers, *11th Int. Cancer Congr.* p. 139, 1974.
27. **Boger, E. A. and Johansen, H. W.**, Plastoquinones in coralline red algae (Corallinaceae) *Phyton*, 32, 129, 1974.
28. **Boll, P. M.**, Synthesis of asterosterol, a novel C_{26} marine sterol, *Acta Chem. Scand. Ser. B*, 28, 270, 1974.
29. **Bordes, D. B., Morton, G. O., and Wetzel, E. R.**, Structure of a novel bromine compound isolated from a sponge, *Tetrahedron Lett.*, p. 2709, 1974.
30. **Borys, H. K., Weinreich, D., and McCaman, R. E.**, Determination of glutamate and glutamine in individual neurons of *Aplysia californica, J. Neurochem.*, 21, 1349, 1973.

31. **Botticelli, C. R., Hisaw, F. L., Jr., and Wotiz, H. H.,** Estradiol-17β and progesterone in ovaries of starfish *(Pisaster ochraceous), Proc. Soc Exp. Biol. Med.,* 103, 875, 1960.
32. **Botticelli, C. R., Hisaw, F. L., Jr., and Wotiz, H. H.,** Estrogens and progesterone in the sea urchin *Strongylocentrotus franciscanus* and pecten *(Pecten hericius), Proc. Soc. Exp. Biol. Med.,* 106, 887, 1961.
33. **Brooks, D. E., Mann, T., and Martin, A. W.,** The occurrence of carnitine and glycerylphosphorylcholine in the octopus spermatophore, *Proc. R. Soc. London Ser. B.* 186, 79, 1974.
34. **Bullock, E. and Dawson, C. J.,** Carotenoid pigments of the holothurian *Psolus favrichii* Düben and Koren (the scarlet psolus), *Comp. Biochem. Physiol.,* 34, 799, 1970.
35. **Burkholder, P. R.,** The ecology of marine antibiotics and coral reefs, in *Biology and geology of coral reefs: Vol.II, Biology (Part I),* Jones, O. A. and Endean, R., Eds., Academic Press, New York, 1973, 117.
36. **Burreson, B. J. and Scheuer, P. J.,** Isolation of a diterpenoid isonitrile from a marine sponge, *Chem. Commun.,* p. 1035, 1974.
37. **Burreson, B. J., Christophersen, C., and Scheuer, P. J.,** Cooccurrence of a terpenoid isocyanide-formamide pair in the marine sponge *Halichondria* sp., *J. Amer. Chem. Soc.,* 97, 201, 1975.
38. **Calabrese, G. and Felicini, G. P.,** Research on red algal pigments. V. The effect of the intensity of white and green light on the rate of photosynthesis and its relationship to pigment components in *Gracilaria compressa* (C. Ag.) Grev. (Rhodophyceae, Gigartinales), *Phycologia,* 12, 195, 1973.
39. **Cariello, L. and Prota, G.,** Occurrence of 3-hydroxy-L-kynurenine in gorgonians, *Comp. Biochem. Physiol. B,* 41, 195, 1972.
40. **Cariello, L., Crescenzi, S., Prota, G., and Zanetti, L.,** New zoanthoxanthins from the Mediterranean zoanthid *Parazoanthus axinellae, Experientia,* 30, 849, 1974.
41. **Cariello, L., Crescenzi, S., Prota, G., and Zanetti, L.,** Methylation of zoanthoxanthins, *Tetrahedron,* 30, 3611, 1974.
42. **Cariello, L., Crescenzi, S., Prota, G., and Zanetti, L.,** Zoanthoxanthins of a new structural type from *Epizoanthus arenaceus* (Zoantharia), *Tetrahedron,* 30, 4191, 1974.
43. **Chantraine, J.-M., Combaut, G., and Teste, J.,** Phenols bromes d'une algue rouge, *Halopytis incurvus:* acides carboxyliques, *Phytochemistry,* 12, 1793, 1973.
44. **Chapman, D. J. and Fox, D. L.,** Bile pigment metabolism in the sea-hare, *Aplysia, J. Exp. Mar. Biol. Ecol.,* 4, 71, 1969.
45. **Cimino, G., De Stefano, S., and Minale, L.,** Methyl *trans*-monocyclofarnesate from the sponge, *Halichondria panicea, Experientia,* 29, 1063, 1973.
46. **Cimino, G., De Stafano, S., and Minale, L.,** Occurrence of hydroxyhydroquinone and 2-aminoimidazole in sponges, *Comp. Biochem. Physiol. B,* 47, 895, 1974.
47. **Cimino, G., De Stefano, S., and Minale, L.,** Oxidized furanoterpenes from the sponge *Spongia officinalis, Experientia,* 30, 18, 1974.
48. **Cimino, G., De Stefano, S., and Minale, L.,** Pleraplysillin-2, a further furanosesquiterpenoid from the sponge *Pleraplysilla spinifera, Experientia,* 30, 846, 1974.
49. **Cimino, G., De Stefano, S., and Minale, L.,** Scalaradial, a third sesterterpene with the tetracarbocyclic skeleton of scalarin, from the sponge *Cacospongia mollior, Experientia,* 30, 846, 1974.
50. **Cimino, G., De Rosa, D., De Stefano, S., and Minale, L.,** Isoagatholactone, a diterpene of a new structural type from the sponge *Spongia officinalis, Tetrahedron,* 30, 645, 1974.
51. **Coll, J. C., Kazlauskas, R., Murphy, P. T., Wells, R. J., and Hawes, G. B.,** personal communication.
52. **Corey, E. J. and Washburn, W. N.,** The role of the symbiotic algae of *Plexaura homomalla* in prostaglandin biosynthesis, *J. Amer. Chem. Soc.,* 96, 934, 1974.
53. **Cormier, M. J. and Hori, K.,** Studies on the bioluminescence of *Renilla reniformis,* IV. Non-enzymatic activation of renilla luciferin, *Biochim. Biophys. Acta,* 88, 99, 1964.
54. **Cormier, M. J., Hori, K., and Karkhanis, Y.D.,** Studies on the bioluminescence of *Renilla reniformis.* VII. Conversion of luciferin into luciferyl sulfate by luciferin sulfokinase, *Biochemistry,* 9, 1184, 1970.
55. **Cormier, M. J., Wampler, J. E., and Hori, K.,** Bioluminescence: chemical aspects, *Fortschr. Chem. Org. Naturst.,* 30, 1, 1973.
56. **Cormier, M. J., Hori, K., Karkhanis, Y. D., Anderson, J. M., Wampler, J. E., Morin, J. G., and Hastings, J. W.,** Evidence for similar biochemical requirements for bioluminescence among the coelenterates, *J. Cell. Physiol.,* 81, 291, 1973.
57. **Cormier, M. J., Hori, K., and Anderson, J. M.,** Bioluminescence in coelenterates, *Biochim. Biophys. Acta,* 346, 137, 1974.
58. **Cottrell, G. A. and Laverack, M. S.,** Invertebrate pharmacology., *Annu. Rev. Pharmacol.,* 8, 273, 1968.
59. **Crews, P. and Kho, E.,** Cartilagineal. An unusual monoterpene aldehyde from marine alga, *J. Org. Chem.,*39, 3303, 1974.
60. **Crozier, G. F.,** Pigments of fishes, *Chem. Zool.,* 8, 509, 1974.

61. **Czeczuga, B.**, Investigations of carotenoids in some fauna of the Adriatic Sea. I. *Verongia aerophoba* (Porifera : Spongiidae), *Mar. Biol.*, 10, 254, 1971.
62. **Czeczuga, B.**, Carotenoids in fish. II. Carotenoids and vitamin A in some fishes from the coastal region of the Black Sea, *Hydrobiologia*, 41, 113, 1973.
63. **Czeczuga, B.**, Investigations of carotenoids in some fauna of the Adriatic Sea. III. *Leander (Palaemon) serratus* and *Nephrops norvegicus* (Crustacea : Decapoda). *Mar. Biol.*, 21, 139, 1973.
64. **Czeczuga, B.**, Carotenoids in the fish milt, *Bull. Acad. Pol. Sci.*, 22, 211, 1974.
65. **Czeczuga, B.**, Comparative studies of carotenoids in the fauna of the Gullmar Fjord (Bohuslän, Sweden). II. Crustacea: *Eupagurus bernhardus, Hyas coarctatus* and *Upogebia deltaura, Mar. Biol.*, 28, 95, 1974.
66. **Das, N. P., Lim, H. S., and Teh, Y. F.**, Histamine and histamine-like substances in the marine sponge *Suberites inconstans, Comp. Gen. Pharmacol.*, 2, 473, 1971.
67. **Dasgupta, S.K., Crump, D. R., and Gut, M.**, New preparation of desmosterol, *J. Org. Chem.*, 39, 1658, 1974.
68. **De Rosa, M., Minale, L., and Sodano, G.**, Metabolism in porifera II. Distribution of sterols, *Comp. Biochem. Physiol. B*, 46, 823, 1973.
69. *Dictionary of Organic Compounds*, 4th ed., Eyre & Spottiswoode Publishers, London, 1965.
70. **Doig, M. T. III, and Martin, D. F.**, Anticoagulant properties of a red tide toxin, *Toxicon*, 11, 351, 1973.
71. **Doughterty, R. C, Strain, H. H., Svec, W. A, Uphaus, R. A., and Katz, J J.**, Structure of chlorophyll c. *J. Amer. Chem. Soc.*, 88, 5037, 1966.
72. **Dunstan, P. J., Hofheinz, W., and Oberhansli, W. E.**, personal communication.
73. **Egger, K.**, Die Verbreitung von Vitamin K_1 und Plastochinon in Pflanzen, *Planta*, 64, 41, 1965.
74. **Enwall, E. L. and van der Helm, D.**, The crystal structure and absolute configuration of the 3-*p*-iodobenzoate-11-acetate of secogorgosterol, *Rec. Trav. Chim. Pays-Bas*, 93, 53, 1974.
75. **Fagerlumd, U. H. M. and Idler, D. R.**, Marine sterols. VI. Sterol biosynthesis in molluscs and echinoderms, *Can. J. Biochem. Physiol.*, 38, 997, 1960.
76. **Fattorusso, E., Magno, S., Santacroce, C., and Sica, D.**, Sterol peroxides from the sponge *Axinella cannabina, Gazz. Chim. Ital.*, 104, 409, 1974.
77. **Fattorusso, E., Magno, S., Mayol, L., Santacroce, C., and Sica, D.**, Isolation and structure of axisonitrile-2. A new sesquiterpenoid isonitrile from the sponge *Axinella cannabina, Tetrahedron*, 30, 3911, 1974.
78. **Faulkner, D. J., Stallard, M. O., and Ireland, C.**, Prepacifenol epoxide, a halogenated sesquiterpene diepoxide, *Tetrahedron Lett.*, p. 3571, 1974.
79. **Faulkner, D. J.**, personal communication.
80. **Fenical, W.**, Rhodophytin, a halogenated vinyl peroxide of marine origin, *J. Amer. Chem. Soc.*, 96, 5580, 1974.
81. **Fenical, W. and Sims, J. J.**, Cycloeudesmol, an antibiotic cyclopropane containing sesquiterpene from the marine alga *Chondria oppositiclada* Dawson, *Tetrahedron Lett.*, p. 1137, 1974.
82. **Fenical, W.**, Polyhaloketones from the red seaweed *Asparagopsis taxiformis, Tetrahedron Lett.*, p. 4463, 1974.
83. **Fenical, W.**, Geranyl hydroquinone, a cancer-protective agent from the tunicate *Aplidium* sp., Proc. Food-Drugs from the Sea 1974, Marine Technology Society, Washington, D. C., 1976.
84. **Ferezou, J. P., Devys, M., Allais, J. P., and Barbier, M.**, Sur le sterol a 26 atomes de carbone de l'algue rouge *Rhodymenia palmata, Phytochemistry*, 13, 593, 1974.
85. **Firnhaber, H. J. and Wells, R. J.**, personal communication.
86. **Fisher, L. R., Kon, S. K., and Thompson, S. Y.** Vitamin A and carotenoids in certain invertebrates. II. Studies of seasonal variations in some marine crustacea. *J. Mar. Biol. Assoc. U. K.*, 33, 589, 1954.
87. **Fisher, L. R., Kon, S. K., and Thompson, S. Y.**, Vitamin A and carotenoids in certain invertebrates. III. Euphausiacea. *J. Mar. Biol. Assoc. U. K.*, 34, 81, 1955.
88. **Fornasiero, U., Antonello, C., and Guiotto, A.**, Naphthaquinone pigments of *Psammechinus microtuberculatus* Blainville, *Ann. Chim. (Roma)*, 63, 387, 1973.
89. **Fraenkel, G.**, The distribution of vitamin B_T (Carnitine) throughout the animal kingdom, *Arch. Biochem. Biophys.*, 50, 486, 1954.
90. **Fujita, Y.**, Seitai busshitsu no toriatsukaiho, in *Sorui-Jikkenho*, Tamiya, H. and Watanabe, T., Eds., 274, 1965.
91. **Fujita, Y. and Shimura, S.**, Phycoerythrin of the marine blue-green alga *Trichodesmium thiebautii, Plant Cell Physiol.*, 15, 939, 1974.
92. **Fujiwara-Arasaki, T. and Mino, N.**, The distribution of trimethylamine and trimethylamine oxide in marine algae, Proc. 7th Int. Seaweed Symp., University of Tokyo Press, 1971, 506.
93. **Gilchrist, B. M. and Welton, L. L.**, Carotenoid pigments and their possible role in reproduction in the sand crab *Emerita analoga* (Stimpson, 1857), *Comp. Biochem. Physiol, B*, 42, 263, 1972.

94. **Glombitza, K.-W., Rosener, H.-U., Vilter, H., and Rauwald, W.,** Antibiotica aus Algen. 8. Phloroglucin aus Braunalgen, *Planta Med.*, 24, 301, 1973.
95. **Glombitza, K.-W., and Sattler, E.,** Trifuhalol, ein neuer Triphenyldiäther aus *Halidrys siliquosa, Tetrahedron Lett.*, p. 4277, 1973.
96. **Glombitza, K.-W. and Rösener, H.-U.,** Bifuhalol: ein Diphenyläther aus *Bifurcaria bifurcata, Phytochemistry*, 13, 1245, 1974.
97. **Glombitza, K.-W., Stoffelen, H., Murawski, U., Bielaczek, J., and Egge, H.,** Antibiotica aus Algen. 9. Bromphenole aus Rhodomelaceen. *Planta Med.*, 25, 105, 1974.
98. **Goad, L. J., Rubinstein, I., and Smith, A. G.,** The sterols of echinoderms, *Proc. R. Soc. London Ser. B.*, 180, 223, 1972.
99. **González, A. G., Darias, J., Martin, J. D., and Peréz, C.,** Revised structure of caespitol and its correlation with isocaespitol, *Tetrahedron Lett.*, p. 1249, 1974.
100. **González, A. G., Darias, J., Martin, J. D., and Norte, M.,** Atomaric acid, a new component from *Taonia atomaria, Tetrahedron Lett.*, p. 3951, 1974.
101. **Gonzáles, A. G., Darias, J., Martin, J. D., Peréz, C., Sims, J. J., Lin, G. H. Y., and Wing, R. M.,** Isocaespitol, a new halogenated sesquiterpene from *Laurencia caespitosa, Tetrahedron*, 31, 2449, 1975.
102. **Goodwin, T. W.,** Pigments of porifera, *Chem. Zool.*, 2, 37, 1968.
103. **Goodwin, T. W.,** Pigments of mollusca, *Chem. Zool.*, 7, 187, 1972.
104. **Goto, T. and Kishi, Y.,** Luciferins, bioluminescent substances, *Angew. Chem. Int. Ed. Engl.*, 7, 407, 1968.
105. **Goto, T.,** Chemistry of bioluminescence, *Pure Appl. Chem.*, 17, 421, 1968.
106. **Goto, T.,** Cypridina bioluminescence IV. Synthesis and chemiluminescence of 3,7-dihydroimidazo(1,2-a)pyrazin-3-one and its 2-methyl derivative, *Tetrahedron Lett.*, p. 3873, 1968.
107. **Goto, T., Inoue, S., Sugiura, S., Nishikawa, K., Isobe, M., and Abe, Y.,** Cypridina bioluminescence. V. Structure of emitting species in the luminescence of cypridina luciferin and its related compounds, *Tetrahedron Lett.*, p. 4035, 1968.
108. **Goto, T., Isobe, M., Coviello, D. A., Kishi, Y., and Inoue, S.,** Cypridina bioluminescence. VIII. The bioluminescence of cypridina luciferin analogs, *Tetrahedron*, 29, 2035, 1973.
109. **Goto, T., Iio, H., Inoue, S., and Kakoi, H.,** Squid bioluminescence. I. Structure of watasenia oxyluciferin, a possible light-emitter in the bioluminescence of *Watasenia scintillans, Tetrahedron Lett.*, p. 2321, 1974.
110. **Gough, J. and Sutherland, M. D.,** The structure of spinochrome B, *Tetrahedron Lett.*, p. 269, 1964.
111. **Gough, J. H. and Sutherland, M. D.,** Marine pigments. VII. 3-Acetyl-2,5,6,7-tetrahydroxy-1,4-naphthoquinone, a new spinochrome from *Salmacis sphaeroides* (Lovén). *Aust. J. Chem.*, 20, 1693, 1967.
112. **Gough, J. H. and Sutherland, M. D.,** Pigments of marine animals. IX. A synthesis of 6-acetyl-2,3,5,7-tetrahydroxy-1,4-naphthoquinone, its status as an echinoid pigment, *Aust. J. Chem.*, 23, 1839, 1970.
113. **Hallgren, B. and Ställberg, G.,** Methoxy-substituted glycerol ethers isolated from Greenland shark liver oil, *Acta Chem. Scand. Ser. B*, 21, 1519, 1967.
114. **Hallgren B., Niklasson, A., Ställberg, G., and Thorin, H.,** On the occurrence of 1-0-(2-methoxyalkyl)glycerols and 1-0- phytanylglycerol in marine animals, *Acta Chem. Scand.*, 28, 1035, 1974.
115. **Hallgren, B. and Ställberg, G.,** 1-0-(2-Hydroxyalkyl)glycerols isolated from Greenland shark liver oil, *Acta Chem. Scand. Ser. B*, 28, 1074, 1974.
116. **Harada, K. and Yamada, K.,** Distribution of trimethylamine oxide in fishes and other aquatic animals, V. Teleosts and elasmobranchs, *Fish. Inst. Res. Rep. (Japan)*, 22, 77, 1973.
117. **Hashimoto, Y., Konosu, S., Fusetani, N., and Nose, T.,** Attractants for eels in the extracts of short-necked clam. I. Survey of constituents eliciting feeding behaviour by the omission test, *Bull. Jpn. Soc. Sci. Fish.*, 34, 78, 1968.
118. **Herring, P. J.,** Depth distribution of the carotenoid pigments and lipids of some oceanic animals. 2. Decapod crustaceans, *J. Mar. Biol. Assoc. U. K.*, 53, 539, 1973.
119. **Higa, T. and Scheuer, P. J.,** Thelepin, a new metabolite from the marine annelid *Thelepus setosus, J. Amer. Chem. Soc.*, 96, 2246, 1974.
120. **Hofheinz, W.,** personal communication.
121. **Hori, K., Wampler, J. E., Matthews, J. C., and Cormier, M. J.,** Identification of the product excited states during the chemiluminescent and bioluminescent oxidation of *Renilla* (sea pansy) luciferin and certain of its analogs, *Biochemistry*, 12, 4463, 1973.
122. **Hori, K., Wampler, J. E., and Cormier, M. J.,** Chemiluminescence of *Renilla* (sea pansy) luciferin and its analogues, *Chem. Commun.*, p. 492, 1973.
123. **Hori, K, and Cormier, M. J.,** Structure and chemical synthesis of a biologically active form of *Renilla* (sea pansy) luciferin, *Proc. Natl. Acad. Sci. U.S.A.*, 70, 120, 1973.

124. **Hotta, K., Kurokawa, M., and Isaka, S.**, Comparative studies of sialic acids from the jelly coat of sea urchin eggs, *J. Jpn. Biochem. Soc.*, 45, 911, 1973.

125. **Howden, M.E.H. and Williams, P. A.**, Occurrence of amines in the posterior salivary glands of the octopus *Hapalochlaena maculosa* (Cephalopoda), *Toxicon*, 12, 317, 1974.

126. **Ichikawa, N., Naya, Y., and Enomoto, S.**, New halogenated monoterpenes from *Desmia (Chondrococcus) hornemanni*, *Chem. Lett.*, p. 1333, 1974.

127. **Idler, D. R., Wiseman, P. M., and Safe, L. M.**, A new marine sterol, 22-*trans*-24-norcholesta-5,22-dien-3β-ol, *Steroids*, 16, 451, 1970.

128. **Idler, D. R. and Wiseman, P.**, Identification of 22-*cis*-cholesta-5,22-dien-3β-ol and other scallop sterols by gas-liquid chromatography and mass spectrometry, *Comp. Biochem. Physiol. A*, 38, 581, 1971.

129. **Iguchi, M., Niwa, M., and Yamamura, S.**, Stereostructure of zonarene, *Bull. Chem. Soc. Jpn.*, 46, 2920, 1973.

130. **Ike, T., Inanaga, J., Nakano, A., Okukado, N., and Yamaguchi, M.**, Total synthesis of natural acetylenic analogues of isorenieratene and renieratene, *Bull. Chem. Soc. Jpn.*, 47, 350, 1974.

131. **Inoue, N., Hosokawa, Y., and Akiba, M.**, Pristane in salmon muscle lipid, *Hokkaido Univ. Bull. Fac. Fish.*, 23, 209, 1973.

132. **Ishihara, K., Oguri, K., and Taniguchi, H.**, Isolation and characterization of fucose sulfate from jelly coat glycoprotein of sea urchin egg, *Biochim. Biophys. Acta*, 320, 628, 1973.

133. **Isler, O., Ed.**, Carotenoids, *Birkhäuser Verlag, Basel, 1971.*

134. **Jaenicke, L., Müller, D. G. , and Moore, R. E.**, Multifidene and aucantene, C_{11} hydrocarbons in the male-attracting essential oil from the gynogametes of *Cutleria multifida* (Smith) Grev. (Phaeophyta), *J. Amer. Chem. Soc.*, 96, 3324, 1974.

135. **Jaenicke, L. and Seferiadis, K.**, Die Stereochemie von Fucoserraten, dem Gametenlockstoff der Braunalge *Fucus serratus* L., *Chem. Ber.*, 108, 225, 1975.

136. **Jefferts, E., Morales, R. W., and Litchfield, C.**, Occurrence of *cis*-5,*cis*-9-hexacosadienoic and *cis*-5,*cis*-9,*cis*-19 -hexacosatrienoic acids in the marine sponge *Microciona prolifera*, *Lipids*, 9,244, 1974.

137. **Jeffrey, S. W.**, Purification and properties of chlorophyll c from *Sargassum flavicans*, *Biochem. J.*, 86, 313, 1963.

138. **Jeffs, P. W. and Lytle, L. T.**, Isolation of (−)-α-curcumene, (−)-β-curcumene and (+)- β-bisabolene from gorgonian corals. Absolute configuration of (−)-β-curcumene, *Lloydia*, 37, 315, 1974.

139. **Jensen, A.**, Tocopherol content of seaweed and seaweed meal. I. Analytical methods and distribution of tocopherols in benthic algae, *J. Sci. Food Agric.*, 20, 449, 1969.

140. **Jones, O. A. and Endean, R., Eds.**, *Biology and geology of coral reefs*. Vol. II: *Biology*, Part 1., Academic Press, New York, 1973.

141. **Kaisin, M., Sheikh, Y. M., Durham, L. J., Djerassi, C., Tursch, B., Dalo ze, D., Braekman, J. C., Losman, D., and Karlsson, R.**, Capnellane — a new tricyclic sesquiterpene skeleton from the soft coral *Capnella imbricata*, *Tetrahedron Lett.*, p. 2239, 1974.

142. **Kamiya, Y., Ikegami, S., and Tamura, S.**, A novel steroid, 3β, 6α, 15α,24ξ-tetrahydroxy-5α-cholestane from asterosaponins, *Tetrahedron Lett.*, p. 655, 1974.

143. **Kanasawa, A. and Yoshioka, M.** Occurrence of cholest-4-en-3-one in red alga *Meristotheca papulosa*, *Bull. Jpn. Soc. Sci. Fish.*, 37, 397, 1971.

144. **Kanazawa, A. and Yoshioka, M.**, The occurrence of cholest-4-en-3-one in the red alga *Gracilaria textorii*, *Proc. 7th Int. Seaweed Symp.*, University of Tokyo Press, Japan, 1971, 502.

145. **Kanazawa, A., Yoshioka, M., and Teshima, S.**, Sterols in some red algae, *Mem. Fac. Fish. Kagoshima Univ.*, 21, 103, 1972.

146. **Kanazawa, A., Teshima, S., and Ando, T.**, (E)-24-Ethylidene-cholest-7-en-3β-ol and other sterols in asteroids, *Mem. Fac. Fish. Kagoshima Univ.*, 22, 21, 1973.

147. **Kanazawa, A., Teshima, S., Ando, T., and Tomita, S.**, Occurrence of 23,24-dimethylcholesta-5,22-dien-3β-ol in a soft coral, *Sarcophyta elegans*, *Bull. Jpn. Soc. Sci. Fish.*, 40, 729, 1974.

148. **Kanazawa, A., Teshima, S., Tomita, S., and Ando, T.**, Gorgostanol, a novel C_{30} sterol from an asteroid, *Acanthaster planci*, *Bull. Jpn. Soc. Sci. Fish.*, 40, 1077, 1974.

149. **Karpetsky, T. P. and White, E. H.**, The synthesis of *Cypridina* etioluciferamine and the proof of structure of *Cypridina* luciferin, *Tetrahedron*, 29, 3761, 1973.

150. **Kashman, Y., Zadock, E., and Neeman, I.**, Some new cembrane derivatives of marine origin, *Tetrahedron*, 30, 3615, 1974.

151. **Katayama, T., Yokoyama, H., and Chichester, C. O.**, The biosynthesis of astaxanthin. II. The carotenoids in benibuna *Carassius auratus*, especially the existence of new keto carotenoids, α-doradecin and α-doradexanthin, *Bull. Jpn. Soc. Sci. Fish*, 36, 702, 1970.

152. **Katayama, T., Hirata, K., Yokoyama, H., and Chichester, C. O.**, The biosynthesis of astaxanthin. III. The carotenoids in sea breams, *Bull. Jpn. Soc. Sci. Fish.*, 36, 709, 1970.

153. **Katayama, T., Hirata, K., and Chichester, C. O.**, The biosynthesis of astaxanthin. IV. The carotenoids in the prawn *Penaeus japonicus* Bate (Part I), *Bull. Jpn. Soc. Sci. Fish.*, 37, 614, 1971.

154. **Katayama, T., Katama, T., and Chichester, C. O.,** The biosynthesis of astaxanthin. VI. The carotenoids in the prawn *Penaeus japonicus* Bate (Part II), *Int. J. Biochem.*, 3, 363, 1972.
155. **Katayama, T., Shintani, K., and Chichester, C. O.,** The biosynthesis of astaxanthin. VII. The carotenoids in sea bream *Chrysophrys major* Temminck and Schlegel, *Comp. Biochem. Physiol. B*, 44, 253, 1973.
156. **Katayama, T., Kunisaki, Y., Shimaya, M., Sameshima, M., and Chichester, C. O.,** The biosynthesis of astaxanthin. XIII. The carotenoids in the crab *Portunus trituberculatus, Bull. Jpn. Soc. Sci. Fish.*, 39, 283, 1973.
157. **Katayama, T., Miyahara, T., Kunisaki, Y., Tanaka, Y., and Imai, S.,** Carotenoids in the sea bream *Chrysophrys major* Temminck and Schlegel. II. *Mem. Fac. Fish. Kagoshima Univ.*, 22, 63, 1973.
158. **Katayama, T., Miyahara, T., Tanaka, Y., Sameshima, M., Simpson, K. L. , and Chichester, C. O.,** The biosynthesis of astaxanthin. XV. The carotenoids in Chidai, red sea bream, *Evynnis japonica* Tanaka and the incorporation of labelled astaxanthin from the diet of the red sea bream to their body astaxanthin, *Bull. Jpn. Soc. Sci. Fish.*, 40, 97, 1974.
159. **Kato, Y. and Scheuer, P. J.** Aplysiatoxin and debromoaplysiatoxin, constituents of the marine mollusk *Stylocheilus longicauda* (Quoy and Gaimard, 1824), *J. Amer. Chem. Soc.*, 96, 2245, 1974.
160. **Kato, Y. and Scheuer, P. J.,** The aplysiatoxins, *Pure Appl. Chem.*, 41, 1, 1975.
161. **Kazlauskas, R., Murphy, P. T., and Wells, R. J.,** personal communication.
162. **Kazlauskas, R., Murphy, P. T., Quinn, R. J., and Wells, R. J.,** personal communication.
163. **Kerkut, G. A.,** Catecholamines in invertebrates, *Br. Med. Bull.*, 29, 100, 1973.
164. **Keyl, M. J., Michaelson, I. A., and Whittaker, V. P.,** Physiologically active choline esters in certain marine gastropods and other invertebrates, *J. Physiol.*, 139, 434, 1957.
165. **Khare, A., Moss, G. P., and Weedon, B. C. L.,** Mytiloxanthin and isomytiloxanthin, two novel acetylenic carotenoids, *Tetrahedron Lett.*, p. 3921, 1973.
166. **Kishi, Y., Goto, T., Hirata, Y., Shimomura, O., and Johnson, F. H.,** Cypridina bioluminescence. I. Structure of *Cypridina* luciferin, *Tetrahedron Lett.*, p. 3427, 1966.
167. **Kitagawa, I., Sugawara, T., and Yosioka, I.,** Structure of holotoxin A, a major antifungal glycoside of *Stichopus japonicus* Selenka, *Tetrahedron Lett.*, p. 4111, 1974.
168. **Kitagawa, I., Sugawara, T., Yosioka, I., and Kuriyama, K.,** Structure of stichopogenin A4, the genuine aglycone of holotoxin A, isolated from *Stichopus japonicus* Selenka, *Tetrahedron Lett.*, p. 963, 1975.
169. **Kobayashi, M., Nishizawa, M., Todo, K., and Mitsuhashi, H.,** Marine sterols. I. Sterols of annelida, *Pseudopotamilla occelata* Moore, *Chem. Pharm. Bull. Tokyo*, 21, 323, 1973.
170. **Kobayashi, M., Todo, K., and Mitsuhashi, H.,** Marine sterols. III. Synthesis of asterosterol, a novel C_{26} sterol from asteroids, *Chem. Pharm. Bull. Tokyo*, 22, 236, 1974.
171. **Kobayashi, M. and Mitsuhashi, H.,** Marine sterols. IV. Structure and synthesis of amuresterol, a new marine sterol with unprecedented side chain, from *Asterias amurensis* Lütken, *Tetrahedron*, 30, 2147, 1974.
172. **Kobayashi, M. and Mitsuhashi, H.,** Marine sterols. V. Isolation and structure of occelasterol, a new 27-norergostane-type sterol, from an annelida, *Pseudopotamilla occelata, Steroids*, 24, 399, 1974.
173. **Komura, T., Wada, S., and Nagayama, H.,** The identification of fucosterol in the marine brown alga *Hizikia fusiformis, Agr. Biol. Chem.*, 38, 2275, 1974.
174. **Komura, T., Nagayama, H., and Wada, S.,** Studies on the lipase activator in marine algae. IV. Isolation of phytol from the Hiziki unsaponifiable matter and its activating effect on pancreatic liapse activity. *J. Agr. Chem. Soc. Japan*, 48, 459, 1974.
175. **Konosu, S., Watanabe, K., and Shimizu, T.,** Distribution of nitrogenous constituents in the muscle extracts of eight species of fish, *Bull. Jpn. Soc. Sci. Fish.*, 40, 909, 1974.
176. **Kumanireng, A. S., Kato, T., and Kitahara, Y.,** Structure and synthesis of diterpenephenols isolated from marine plants, *17th I.U.P.A.C. Symp. Chem. Natural Products*, Tokyo, p. 94, 1973.
177. **Kurosawa, E., Fukuzawa, A., and Irie, T.,** *trans*-and *cis*- laurediol, unsaturated glycols from *Laurencia nipponica* Yamada, *Tetrahedron Lett.*, p. 2121, 1972.
178. **Lam, J. K.K. and Sargent, M. V.,** Synthesis of methyl tri-0-methylptilometrate (methyl 1,6,8-trimethoxy-3-propylanthraquinone-2-carboxylate), *J. Chem. Soc. Perkin Trans. I*, p. 1417, 1974.
179. **Lee, A. S. K., Vanstone, W. E., Markert, J. R., and Antia, N. J.,** UV-absorbing and UV-fluorescing substances in the belly skin of fry of coho salmon *(Oncorhynchus kisutch), J. Fish. Res. Board, Canada*, 26, 1185, 1969.
180. **Liaaen-Jensen, S.,** Selected examples of structure determination of natural carotenoids, *Pure Appl. Chem.*, 20, 421, 1969.
181. **Lowe, M. E. and Horn, D. H. S.,** Bioassay of the red chromatophore concentrating hormone of the crayfish, *Nature*, 213, 408, 1967.
182. **Lowe, M. E., Horn, D. H. S., and Galbraith, M. N.,** The role of crustecdysone in the moulting crayfish, *Experientia*, 24, 518, 1968.

183. **McBride, W. J., Freeman, A. R., Graham, L. T., Jr., and Aprison, M. H.,** The content of several amino acids in the external cell sheath and four giant axons of a nerve bundle from the CNS of the lobster, *Brain Res.*, 59, 440, 1973.
184. **McCapra, F. and Manning, M. J.,** Bioluminescence of coelenterates: chemiluminescent model compounds, *Chem. Commun.*, p. 467, 1973.
185. **McDonald, F. J., Campbell, D. C., Vanderah, D. J., Schmitz, F. J., Washecheck, D., M., Burks, J. E., and van der Helm, D.,** Marine natural products. Dactylyne, an acetylenic dibromochloro ether from the sea hare *Aplysia dactylomela, J. Org. Chem.*, 40, 665, 1975.
186. **McLachlan, J., Craigie, J. S., Chen, L. C.-M., and Ogata, E.,** *Porphyra linearis* Grev. An edible species of nori from Nova Scotia, *Proc. 7th Int. Seaweed Symp.*, University of Tokyo Press, Japan, 1971, 473.
187. **McMillan, J. A., Paul, I. C., White, R. H., and Hager, L. P.,** Molecular structure of acetoxyintricatol: a new bromo compound from *Laurencia intricata, Tetrahedron Lett.*, p. 2039, 1974.
188. **McMurry, J. E. and von Beroldingen, L. A.,** Ketone methylenation without epimerization: total synthesis of (±) laurene, *Tetrahedron*, 30, 2027, 1974.
189. **Madgwick, J. C., Ralph, B. J., Shannon, J. S., and Simes, J. J.,** Non-protein amino acids in Australian seaweeds, *Arch. Biochem. Biophys.*, 141, 766, 1970.
190. **Madgwick, J. C. and Ralph, B. J.,** Free amino acids in Australian marine algae, *Bot. Mar.*, 15, 205, 1972.
191. **Mann, T., Martin, A. W., Thiersch, J. B., Lutwak-Mann, C., Brooks, D. E., and Jones, R.,** D(−)-Lactic acid and D(−)- lactate dehydrogenase in octopus spermatozoa, *Science*, 185, 453, 1974.
192. **Manning, W. M. and Strain, H. H.,** Chlorophyll d, a green pigment of red algae, *J. Biol. Chem.*, 151, 1, 1943.
193. **Marderosian, A. D.,** Marine pharmaceuticals, *J. Pharm. Sci.* 58, 1, 1969.
194. **Marsh, M. E. and Ciereszko, L. S.,** Alpha-glyceryl ethers in coelenterates, *Proc. Oklahoma Acad. Sci.*, 53, 53, 1973.
195. **Martin, D. F. and Padilla, G. M., Eds.,** *Marine Pharmacognosy*, Academic Press, New York, 1973.
196. **Matsuno, T., Kusumoto, T., Watanabe, T., and Ishihara, Y.,** Carotenoid pigments of spiny lobster, *Bull. Jpn. Soc. Sci. Fish.*, 39, 43, 1973.
197. **Matsuno, T., Akita, T., and Hara, M.,** Carotenoid pigments of Japanese anchovy, *Bull. Jpn. Soc. Sci. Fish.*, 39, 51, 1973.
198. **Matsuno, T., Nagata, S., Sato, Y., and Watanabe, T.,** Comparative biochemical studies of carotenoids in fishes. II. Carotenoids of horse mackerel, swellfishes, porcupine fishes and striped mullet, *Bull. Jpn. Soc. Sci. Fish.*, 40, 579, 1974.
199. **Matsuno, T., Watanabe, T., and Nagata, S.,** Carotenoid pigments of crustacea. II. The carotenoid pigments of *Scyllarides squamosus (= Scyllarus sieboldi)* and *Parribacus antarcticus (= Parribacus ursus-major), Bull. Jpn. Soc. Sci. Fish.*, 40, 619, 1974.
200. **Matsuno, T. and Watanabe, T.** Carotenoid pigments of crustacea. III. The carotenoid pigments of *Sesarma (Holometopus) haematocheir* and *Sesarma (Sesarma) intermedia, Bull. Jpn. Soc. Sci. Fish.*, 40, 767, 1974.
201. **Mazzarella, L. and Puliti, R.,** Crystal structure and absolute configuration of aeroplysinin-1, *Gazz. Chim. Ital.*, 102, 391, 1972.
202. *Merck Index*, 8th Ed., Merck & Co., Rahway, New Jersey, 1968.
203. **Mettrick, D. F. and Telford, J. M.,** The histamine content and histidine decarboxylase activity of some marine and terrestrial animals from the West Indies, *Comp. Biochem. Physiol.*, 16, 547, 1965.
204. **Minale, L. and Sodano, G.,** Marine sterols, 19-nor-stanols from the sponge *Axinella polypoides, J. Chem. Soc. Perkin Trans. I*, p.1888, 1974.
205. **Minale, L. and Sodano, G.,** Marine sterols: unique 3β-hydroxymethyl-A-nor-5α-steranes from the sponge *Axinella verrucosa, J. Chem. Soc. Perkin Trans. I*, 2380, 1974.
206. **Minale, L., Riccio, R., and Sodano, G.,** Acanthellin-1, an unique isonitrile sesquiterpene from the sponge *Acanthella acuta, Tetrahedron*, 30, 1341, 1974.
207. **Minale, L., Riccio, R., and Sodano, G.,** Avarol, a novel sesquiterpenoid hydroquinone with a rearranged drimane skeleton from the sponge *Disidea avara, Tetrahedron Lett.*, p. 3401, 1974.
208. **Miyazawa, K. and Ito, K.,** Isolation of L-methionine-l-sulfoxide and *N*-methylmethionine sulfoxide from a red alga *Grateloupia turuturu, Bull. Jpn. Soc. Sci. Fish.*, 40, 655, 1974.,
209. **Moldowan, J. M., Tursch, B. M., and Djerassi, C.,** 24ξ-Methylcholestane-3β,5α,6β,25-tetrol 25-monoacetate, a novel polyhydroxylated steroid from an alcyonarian, *Steroids*, 24, 387, 1974.
210. **Momzikoff, A.,** Mise en évidence d'une excretion de ptérines par une population naturelle de copépodes planctoniques marins, *Cah. Biol. Mar.*, 14, 323, 1973.
211. **Momzikoff, A. and Gaill, F.,** Mise en évidence d'une émission d'isoxanthoptérine et de riboflavine par différentes espèces d'ascidies, *Experientia*, 29, 1438, 1973.
212. **Moore, R. E. and Scheuer, P. J.,** Palytoxin: a new marine toxin from a coelenterate, *Science*, 172, 495, 1971.

213. **Moore, R. E. and Yost, G.,** Dihydrotropones from *Dictyopteris, Chem. Commun.,* p. 937, 1973.
214. **Moore, R. E., Pettus, J. A., Jr., and Mistysyn, J.,** Odoriferous C_{11} hydrocarbons from Hawaiian *Dictyopteris, J. Org. Chem.,* 39, 2201, 1974.
215. **Morton, R. A.,** Ubiquinones, plastoquinones and vitamins K, *Biol. Rev.,* 46, 47, 1971.
216. **Motzfeldt, A.-M.,** Isolation of 24-oxocholesterol from the marine brown alga *Pelvetia canaliculata* (Phaeophyceae), *Acta Chem. Scand.,* 24, 1846, 1970.
217. **Mulder, J. W. and Wells, R. J.,** personal communication.
218. **Müller, D. G. and Jaenicke, L.,** Fucoserraten, the female sex attractant of Fucus serratus L. (Phaeophyta), *FEBS Lett.,* 30, 137, 1973.
219. **Müller, D. G.,** Sexual reproduction and isolation of a sex attractant in *Cutleria multifida* (Smith) Grev. (Phaeophyta), *Biochem. Physiol. Pflanzen,* 165, 212, 1974.
220. **Mynderse, J. S. and Faulkner, D. J.,** Violacene, a polyhalogenated monocyclic monoterpene from the red alga *Plocamium violaceum, J. Amer. Chem. Soc.,* 96, 6771, 1974.
221. **Nakayama, T. O. M.,** Carotenoids, in *Physiology and Biochemistry of Algae,* Lewin, R. A., Ed., Academic Press, New York, 1962, 409.
222. **Neeman, I., Fishelson, L., and Kashman, Y.,** Sarcophine — a new toxin from the soft coral-*Sarcophyton glaucum* (Alcyonaria), *Toxicon,* 12, 593, 1974.
223. **Nitsche, H.,** Neoxanthin and fucoxanthinol in *Fucus vesiculosus, Biochim. Biophys. Acta,* 338, 572, 1974.
224. **Nomura, T., Ogata, H., and Ito, M.,** Occurrence of prostaglandins in fish testis, *Tohoku J. Agr. Res.,* 24, 138, 1973.
225. **Norgård, S., Svec, W. A., Liaaen-Jensen, S., Jensen, A., and Guillard, R. R. I.,** Algal carotenoids and chemotaxonomy, *Biochem. Syst. Ecol.,* 2, 7, 1974.
226. **Ogura, Y.,** Recent studies on Fugu toxin, *Setai no Kagaku,* 9, 281, 1958.
227. **Osborne, N. N.,** Occurrence of glycine and glutamic acid in the nervous system of two fish species and some invertebrates, *Comp. Biochem. Physiol. B,* 43, 579, 1972.
228. **Patil, V. D., Nayak, U. R., and Dev, S.,** Chemistry of ayurvedic crude drugs. II. Guggulu (resin from *Commiphora mukul*) 2. Diterpenoid constituents, *Tetrahedron,* 29, 341, 1973.
229. **Pedersên, M., Saenger, P., and Fries, L.,** Simple brominated phenols in red algae, *Phytochemistry,* 13, 2273, 1974.
230. **Pettit, G. R., Day, J. F., Hartwell, J. L., and Wood, H. B.,** Antineoplastic components of marine animals, *Nature,* 227, 962, 1970.
231. **Rimington, C. and Kennedy, G. Y.,** Porphyrins: structure, distribution and metabolism, *Comp. Biochem.,* 4, 557, 1962.
232. **Rinehart, K. L., Jr., Johnson, R. D., Paul, I. C., MacMillan, J. A., Siuda, J. F., and Krejcarek, G. A.,** Identification of compounds in selected marine organisms by gas chromatography-mass spectrometry, field desorption mass spectrometry and other physical methods, *Proc. Food- Drugs from the Sea 1974,* Marine Technology Society, Washington, D. C., 1976, 434.
233. **Roseghini, M., Erspamer, V., Ramorino M. L., and Gutierrez, J. E.,** Choline esters, their precursors and metabolites in the hypobranchial gland of prosobranchiate molluscs *Concholepas concholepas* and *Thais chocolata, Eur. J. Biochem.,* 12, 468, 1970.
234. **Roseghini, M. and Fichman, M.,** Choline esters and imidazole acids in extracts of the hypobranchial gland of *Thais haemastoma, Comp. Gen. Pharmacol.,* 4, 251, 1973.
235. **Roseghini, M., Alcala, A. C., and Vitali, T.,** Occurrence of *N*-imidazolepropionyl-histamine in the soft tissues of the Philippine gastropod *Drupa concatenata* Lam., *Experientia,* 29, 940, 1973.
236. **Roseghini, M. and Alcala, A. C.,** Occurrence of *N*-urocanylhistamine in the soft tissues of the gastropod mollusc *Drupa concatenata* Lam., *Biochem. Pharmacol.,* 23, 1431, 1974.
237. **Rubinstein, I. and Goad, L. J.,** Sterols of the siphonous marine alga *Codium fragile, Phytochemistry,* 13, 481, 1974.
238. **Rubinstein, I. and Goad, L. J.,** Occurrence of (24S)-24-methylcholesta-5, 22E-dien-3β-ol in the diatom *Phaeodactylum tricornutum, Phytochemistry,* 13, 485, 1974.
239. **Saavedra, J. M., Brownstein, M. J., Carpenter, D. O., and Axelrod, J.,** Octopamine: presence in single neurons of *Aplysia* suggests neurotransmitter function, *Science,* 185, 364, 1974.
240. **Safe, L. M., Wong, C. J., and Chandler, R. F.,** Sterols of marine algae, *J. Pharm. Sci.,* 63, 464, 1974.
241. **Schantz, E. J.,** Some toxins occurring naturally in marine organisms, in *Microbial Safety of Fishery Products,* Chichester, C. O., Ed., Academic Press, New York, 1973, 151.,
242. **Schmitz, F. J., Vanderah, D. J., and Ciereszko, L. S.,** Marine natural products: nephthenol and epoxynephthenol acetate, cembrene derivatives from a soft coral, *Chem. Commun.,* p. 407, 1974.
243. **Schmitz, F. J. and McDonald, F. J.,** Marine natural products: Dactyloxene-B, a sesquiterpene ether from the sea hare *Aplysia dactylomela, Tetrahedron Lett.,* p. 2541, 1974.
244. **Schmitz, F. J., Campbell, D. C., and McDonald, F. J.,** Dactylyne, a halogenated acetylenic oxetane isolated from the sea hare *Aplysia dactylomela, 9th I.U.P.A.C. Symp. Chem. Natural Products, Abstr.* Ottawa, 1974. 12E.

245. **Sieburth, J. M.,** Acrylic acid, an antibiotic principle in *Phaeocystis* blooms in antarctic waters, *Science*, 132, 676, 1960.

246. **Sims, J. J., Fenical, W., Wing, R. M., and Radlick, P.,** Marine natural products. IV. Prepacifenol, a halogenated epoxy sesquiterpene and precursor to pacifenol from the red alga *Laurencia filiformis*, *J. Amer. Chem. Soc.*, 95, 972, 1973.

247. **Sims, J. J., Lin, G. H. Y., and Wing, R. M.,** Marine natural products. X. Elatol, a halogenated sesquiterpene alcohol from the red alga *Laurencia elata*, *Tetrahedron Lett.*, p. 3487, 1974.

248. **Siuda, J. F., VanBlaricom, G. R., Shaw, P. D., Johnson, R. D., White, R. H., Hager, L. P., and Rinehart Jr., K. L.,** 1-Iodo-3,3-dibromo-2-heptanone, 1,1,3,3-tetrabromo-2-heptanone, and related compounds from the red alga *Bonnemaisonia hamifera*, *J. Amer. Chem. Soc.*, 97, 937, 1975.

249. **Smith, A. G., Rubinstein, I., and Goad, L. J.,** The sterols of the echinoderm *Asterias rubens*, *Biochem. J.*,135, 443, 1973.

250. **Smith. L. L., Dhar, A. K., Gilchrist, J. L., and Lin, Y. Y.,** Sterols of the brown alga *Sargassum fluitans*, *Phytochemistry*, 12, 2727, 1973.

251. **Stallard, M. O. and Faulkner, D. J.,** Chemical constituents of the digestive gland of the sea hare *Aplysia californica*. I. Importance of diet, *Comp. Biochem. Physiol. B*, 49, 25, 1974.

252. **Stoffelen, H., Glombitza, K.-W., Murawski, U., Bielaczek, J., and Egge, H.,** Bromphenole aus-*Polysiphonia lanosa* (L.) Tandy, *Planta Med.*, 22, 396, 1972.

253. **Strain, H. H., Manning, W. M. and Hardin, G.,** Xanthophylls and carotenes of diatoms, brown algae, dinoflagellates and sea anemones, *Biol. Bull.*, 86, 169, 1944.

254. **Suyama, M. and Yoshizawa, Y.,** Free amino acid composition of the skeletal muscle of migratory fish, *Bull. Jpn. Soc. Sci. Fish.*, 39, 1339, 1973.

255. **Suzuki, M., Kurosawa, E., and Irie, T.,** Three new sesquiterpenoids containing bromine minor constituents of *Laurencia glandulifera* Kützing, *Tetrahedron Lett.*, p. 821, 1974.

256. **Suzuki, M., Kurosawa, E., and Irie, T.,** Glanduliferol, a new halogenated sesquiterpenoid from-*Laurencia glandulifera* Kützing, *Tetrahedron Lett.*, p. 1807, 1974.

257. **Takagi, M. and Okumura, A.,** On a new amino acid, S-hydroxymethyl-L-homocysteine isolated from *Chondrus ocellatus*, *Bull. Jpn. Soc. Sci. Fish.*, 30, 837, 1964.

258. **Tammar, A. R.,** Bile salts in fishes, *Chem. Zool.*, 8, 595, 1974.

259. **Tanaka, Y. and Katayama, T.,** Carotenoids in the sea bream, *Chrysophrys major* Temmick and Schlegel. III. The carotenoids in mysis and the internal organs of squid as the food for sea bream, *Mem. Fac. Fish. Kagoshima Univ.*, 23, 117, 1974.

260. **Taylor, F. R., Ikawa, M., Sasner Jr., J. J., Thurberg, F. P., and Andersen, K. K.,** Occurrence of choline esters in the marine dinoflagellate *Amphidinium carteri*, *J. Phycol.*, 10, 279, 1974.

261. **Teshima, S., Kanazawa, A., and Ando, T.,** Isolation of a novel C_{28}-sterol, 24-methycholesta-7,22,25-trien-3β-ol from a starfish, *Leiaster leachii*, *Bull. Jpn. Soc. Sci. Fish.*, 40, 631, 1974.

262. **Torres Pombo, J.,** Contribución al conocimiento quimico del alga "*Gelidium sesquipedale* (Clem.)Thuret" y a la estructura de su agar, *Acta Cient. Compostelana*, 9, 53, 1972.

263. **Toyama, Y. and Takagi, T.,** Sterols and other unsaponifiable substances in the lipids of shell fishes, crustacea and echinoderms. XV. Occurrence of $\Delta 7$: 8-cholestenol as a sterol component of starfish *Asterias amurensis* Lütken, *Bull. Chem. Soc. Japan*,27, 421, 1954.

264. **Tsukuda, N. and Kitahara, T.,** Composition of the esterified fatty acids of astaxanthin diester in the skin of seven red fishes, *Bull. Tokai Reg. Fish. Res. Lab.*, No. 77, 89, 1974.

265. **Turner, A. B.,** Starfish saponins, *9th I.U.P.A.C. Symp. Chem. Natural Products* Abstr., Ottawa, 1974, 8E.

266. **Tursch, B., Braekman, J. C., Daloze, D., Fritz, P., Kelecom, A., Karlsson, R., and Losman. D.,** Chemical studies of marine invertebrates. VIII. Africanol, an unusual sesquiterpene from *Lemnalia africana* (Coelenterata, Octocorallia, Alcyonacea), *Tetrahedron Lett.*, p. 747, 1974.

267. **Tursch, B., Braekman, J. C., Daloze, D., Herin, M., and Karlsson R.,** Chemical studies of marine invertebrates. X. Lobophytolide, a new cembranolide diterpene from the soft coral *Lobophytum cristagalli* (Coelenterata, Octocorallia, Alcyonacea), *Tetrahedron Lett.*, p. 3769, 1974.

268. **Van der Helm, D., Ealick, S. E., and Weinheimer, A. J.,**15(R)-Acetoxy-6(S),10(S)-dibromo-3a(S),4,7,8,11,12,13,14(S),15,15a(R)-decahydro-6,10-14-trimethyl-3-methylene-5(R),9(R)-epoxycyclotetradeca[b]2-furanone,$C_{22}H_{32}Br_2O_5$, *Cryst. Struct. Commun.*, 3, 167, 1974.

269. **Viala, J., Devys, M., and Barbier, M.,**Sur la structure des stérols à 26 atomes de carbone du tunicier *Halocynthia roretzi*, *Bull. Soc. Chim. Fr.*, p. 3626, 1972.

270. **Waraszkiewicz, S. M. and Erickson, K. L.,**Halogenated sesquiterpenoids from the Hawaiian marine alga *Laurencia nidifica*: nidificene and nidifidiene, *Tetrahedron Lett.*, p. 2003, 1974.

271. **Weinheimer, A. J.,** The discovery of 15-*epi* PGA_2 in *Plexaura homomalla*, *Stud. Trop. Oceanogr.*, 12, 17, 1974.

272. **Wilkie, D. W.,** The carotenoid pigmentation of *Pleuroncodes planipes* Stimpson (Crustacea: Decapoda: Galatheidae), *Comp. Biochem. Physiol. B*, 42, 731, 1972.

273. **Wotiz, H. H., Botticelli, C. R., Hisaw, F. L., Jr., and Olsen, A. G.,** Estradiol-17β, estrone and progesterone in the ovaries of dogfish *(Squalus suckleyi)*, *Proc. Natl. Acad. Sci. U.S.A.*, 46, 580, 1960.

274. **Yamada, Y., Kim, J.-S., Iguchi, K., and Suzuki, M.**, An effective synthesis of a bromine-containing antibacterial compound from marine sponges, *Chem. Lett.*, p. 1399, 1974.
275. **Yasuda, S.**, Sterol compositions of jelly fish (Medusae), *Comp. Biochem. Physiol. B.*, 48, 225, 1974.
276. **Yasuda, S.**, Sterol compositions of echinoids (sea urchin, sand dollar and heart urchin), *Comp. Biochem, Physiol. B*, 49, 361, 1974.
277. **Yasumoto, T. and Endo, M.**, Toxicity study on a marine snail, *Turbo argyrostoma* I. Presence of two sulfur- containing amines in the acetone-soluble fraction, *Bull. Jpn. Soc. Sci. Fish.*, 39, 1055, 1973.
278. **Yasumoto, T. and Endo, M.**, Toxicity study on a marine snail, *Turbo argyrostoma* II. Identification of (3-methylthiopropyl)trimethylammonium chloride, *Bull. Jpn. Soc. Sci. Fish.*, 40, 217, 1974.
279. **Yasumoto, T. and Endo, M.**, Toxicity study on a marine snail, *Turbo argyrostoma* III. Occurrence of candicine, *Bull. Jpn. Soc. Sci. Fish.*, 40, 841, 1974.
280. **Yasumoto, T. and Sano, F.**, Occurrence of homoserine betaine and valine betaine in the ovary of shellfish *Callista brevishiphonata, Bull. Jpn. Soc. Sci. Fish.*, 40, 1163, 1974.
281. **Yasumoto, T.**, Toxicity study on a marine snail, *Turbo argyrostoma.* IV. Occurrence of [3-(dimethylsulfonio)propyl]trimethylammonium dichloride, *Bull. Jpn. Soc. Sci. Fish.*, 40, 1169, 1974.
282. **Yoshizawa, T. and Nagai, Y.**, Occurrence of cholesteryl sulfate in eggs of the sea urchin *Anthocidaris crassispina, Jpn. J. Exp. Med.*, 44, 465, 1974.
283. **Youngblood, W. W. and Blumer, M.**, Alkanes and alkenes in marine benthic algae, *Mar. Biol.*, 21, 163, 1973.

Additional General References on Carotenoids

284. **Weedon, B. C. L., Ed.**, *Carotenoids*, Vol. 4, Pergamon, Elmsford, N.Y., 1976; or, *Pure Appl. Chem.*, 47, 97-237, 1976.
285. **Goodwin, T. W., Ed.**, *Carotenoids*, Vol. 5, Pergamon, Elmsford, N.Y., 1979; or, *Pure Appl. Chem.*, 51, 435-871, 1979,
286. **Straub, O.**, *Key to Carotenoids*, Birkhäuser Verlag, Basel, 1971.
287. **Liaaen-Jensen, S.**, Marine Carotenoids, in *Marine Natural Products: Chemical and Biological Perspectives*, Vol. 3, Scheuer, P. J., Ed., Academic Press, New York, in press.
288. **IUPAC-IUB**, *Nomenclature of Carotenoids*, Butterworths, London, 1974.
289. IUPAC-IUB, Nomenclature of Carotenoids, *Pure Appl. Chem.*, 41, 405, 1975.

References From Volume I

290. **Suzuki, M., Kurosawa, E., and Irie, T.** Spirolaurenone, a new sesquiterpenoid containing bromine from *Laurencia glandulifera* Kützing, *Tetrahedron Lett.*, p.4995, 1970.
291. **Faulkner, D. J., and Stallard, M. O.**, 7-Chloro-3,7-dimethyl-1,4,6-tribromo-1-octen-3-ol, a novel monoterpene alcohol from *Aplysia californica, Tetrahedron Lett.*, p. 1171, 1973.
292. **Willcott, M. R., Davis, R. E., Faulkner, D. J., and Stallard, M. O.**, The configuration and conformation of 7-chloro-1, 6-dibromo-3,7-dimethyl-3,4-epoxy-1-octene, *Tetrahedron Lett.*, p. 3967, 1973.
293. **González, A. G., Darias, J., Martin, J. D., Pérez, C.**, Revised structure of caespitol and its correlation with isocaespitol, *Tetrahedron Lett.*, p.1249, 1974.
294. **Cafieri, F., Fattorusso, E., Magno, S., Santacroce, C., and Sica, D.**, Isolation and structure of axisonitrile-1 and axisothiocyanate-1, two unusual sesquiterpenoids from the marine sponge *Axinella cannabina, Tetrahedron*, 29, 4259, 1973.
295. **Cimino, G., De Stefano, S., Minale, L. and Trivellone, E.**, New sesquiterpenes from the marine sponge *Pleraplysilla spinifera, Tetrahedron*, 28, 4761, 1972.
296. **Cimino, G., De Stefano, S., and Minale, L.**, Paniceins, unusual aromatic sesquiterpenoids linked to a quinol or quinone system from the marine sponge *Halichondria panicea, Tetrahedron*, 29, 2565, 1973.
297. **Bergmann, W.**, Comparative biochemical studies on the lipids of marine invertebrates, with special reference to the sterols, *J. Mar. Res.*, 8, 137, 1949.
298. **Bergmann, W.**, Sterols: their Structure and Distribution in *Comparative Biochemistry*, Vol. 3, Florkin, M. and Mason, H. S., Ed., Academic Press, New York, 1962, 103.
299. **De Luca, P., De Rosa, M., Minale, L., and Sodano, G.**, Marine sterols with a new pattern of side-chain alkylation from the sponge *Aplysina (= Verongia) aerophoba, J. Chem. Soc. Perkin I*, 2132, 1972.
300. **Cariello, L. Crescenzi, S., Prota, G., and Zanetti, L.**, New zoanthoxanthins from the Mediterranean zoanthid *Parazoanthus axinellae, Experientia*, 30, 849, 1974.
301. **Prota, G.**, personal communication.
302. **Whittaker, V. P.**, Pharmacologically active choline esters in marine gastropods, *Ann. N. Y. Acad. Sci.*, 90, 695, 1960.

AUTHOR INDEX (VOLUME II)

Name	Ref. no.
Abe, H.,	1
Abe, S.	2
Abe, Y.	107
Ackermann, D.	3
Adachi, K.	4
Aguilar- Martinez, M.	5
Akiba, M.	131
Akita, T.	197
Alcala, A. C.	235, 236
Allais, J. P.	84
Andersen, K. K.	260
Andersen, R. J.	6,7
Anderson, J. M.	56,57
Ando, T.	146—148, 261
Andrewes, A. G.	8
Antia, N. J.	179
Antonello, C.	88
Aprison, M. H.	183
ApSimon, J. W.	9
Axelrod, H.	239
Baker, J. L.	10
Baker, J. T.	11
Banville, J.	12
Barbier, M.	84, 269
Barton, D.H.R.	13
Baxter, J. G.	14
Bender, J. A.	15
Bergmann, W.	16
Bernstein, J.	17
Bersis, D.	18
Bielaczek, J.	97, 252
Billups, W. E.	19
Bjorkman, L. R.	20
Blackman, A. J.	21
Block, J. H.	22
Blumer, M.	23, 283
Boar, R. B.	24
Boeryd, B.	25, 26
Boger, E. A.	27
Boll, P. M.	28
Boothe, P.	15
Borch, G.	8
Borders, D.B.	29
Borys, H. K.	30
Botticelli, C. R.	31, 32, 273
Braekman, J. C.	141, 266, 267
Brassard, P.	12
Brooks, D. E.	33, 191
Brownstein, M. J.	239
Bullock, E.	34
Burkholder, P.R.	35
Burks, J.E.	185
Burreson, B.J.	36, 37
Cafieri, F.	294
Calabrese, G.	38
Campbell, D. C.	185, 244
Cariello, L.	39—42
Carpenter, D. O.	239
Chandler, R. F.	240
Chantraine, J.-M.	43
Chapman, D. J.	44
Chen, L. C.-M.	186
Chichester, C.O.	151—156, 158
Chow, W.Y.	19
Christophersen, C.	37
Ciereszko, L. S.	194, 242
Cimino, G.	45—50
Coll, J.C.	51
Combaut, G.	43
Corey, E. J.	52
Cormier, M. J.	53—57, 121—123
Cottrell, G. A.	58
Coviello, D. A.	108
Craigie, J. S.	186
Crescenzi, S.	40—42
Crews, P.	59
Cross, J. H.	19
Crozier, G. F.	60
Crump, D. R.	67
Czeczuga, B.	61—65
Daloze, D.	141, 266, 267
Darias, J.	99—101
Das, N. P.	66
Dasgupta, S. K.	67
Davis, R. E.	292
Dawson, C. J.	34
Day, J. F.	230
DeLuca, P.	299
DeRiemer, K.	15
De Rosa, D.	50
De Rosa, M.	68
De Stefano, S.	45—50
Dev, S.	228
Devys, M.	84, 269
Dhar, A.K.	250
Djerassi, C.	141, 209
Doig, M. T., III	70
Dougherty, R. C.	71
Dunstan, P. J.	72
Durham, L. J.	141
Ealick, S.E.	268
Eenkhoorn, J. A.	9
Egge, H.	97, 252
Egger, K.	73
Endean, R.	140
Endo, M.	277—279
Enomoto, S.	126
Enwall, E. L.	74
Erickson, K. L.	270
Erspamer, V.	233
Fagerlund, U.H.M.	75
Fattorusso, E.	76, 77
Faulkner, D. J.	6, 7, 78, 79, 220, 251
Felicini, G. P.	38
Fenical, W.	80—83, 246
Ferezou, J. P.	84

AUTHOR INDEX (VOLUME II)
(continued)

Name	Ref. no.
Fichman, M.	234
Firnhaber, H. J.	85
Fishelson, L.	222
Fisher, L. R.	86, 87
Fornasiero, U.	88
Fox, D. L.	44
Fraenkel, G.	89
Freeman, A. R.	183
Fries, L.	229
Fritz, P.	266
Fuhrman, F. A.	15
Fujita, Y.	90, 91
Fujiwara-Arasaki, T.	92
Fukuzawa, A.	177
Fusetani, N.	117
Gaill, F.	211
Galbraith, M. N.	182
Gilchrist, B. M.	93
Gilchrist, J. L.	250
Glombitza, K.-W.	94—97, 252
Goad, L. J.	98, 237, 238, 249
Gonzâlez, A. G.	99—101
Goodwin, T. W.	102, 103, 285
Goto, T.	104—109, 166
Gough, J. H.	110-112
Graham, L. T., Jr.	183
Guillard, R. R. L.	225
Guiotto, A.	88
Gut, M.	67
Gutierrez, J. E.	233
Hager, L.P.	187, 248
Hallgren,B.	25, 26, 113— 115
Hara, M.	197
Harada, K.	116
Hardin, G.	253
Hartwell, J. L.	230
Hashimoto, Y.	117
Hastings, J. W.	56
Hawes, G. B.	51
Herin, M.	267
Herring, P. J.	118
Higa, T.	119
Hirata, K.	152, 153
Hirata, Y.	166
Hisaw Jr., F. L.	31, 32, 273
Hofheinz, W.	72, 120
Hori, K.	53—57, 121—123
Horn, D. H. S.	181, 182
Hosokawa, Y.	131
Hotta, K.	124
Howden, M. E. H.	125
Ichikawa, N.	126
Idler, D. R.	75, 127, 128
Iguchi, K.	274
Iguchi, M.	129
Iio, H.	109
Ikawa, M.	260
Ike, T.	130
Ikegami, S.	142
Imai, S.	157
Inanaga, J.	130
Inoue, N.	131
Inoue, S.	107—109
Ireland, C.	78
Irie, T.	177, 255, 256
Isaka, S.	124
Ishihara, K.	132
Ishihara, Y.	196
Isler, O.	133
Isobe, M.	107, 108
Ito, K.	208
Ito, M.	224
Jaenicke, L.	134, 135, 218
Jefferts, E.	136
Jeffrey, S. W.	137
Jeffs, P. W.	138
Jensen, A.	139, 225
Johansen, H. W.	27
Johnson, F. H.	166
Johnson, R. D.	232, 248
Jones, O. A.	140
Jones, R.	191
Kaisin, M.	141
Kakoi, H.	109
Kamiya, Y.	142
Kanazawa, A.	143—148, 261
Kaneda, T.	2
Karkhanis, Y. D.	54, 56
Karlsson, K.-A.	20
Karlsson, R.	141, 266, 267
Karpetsky, T. P.	149
Kashman, Y.	17, 150, 222
Katama, T.	154
Katayama, T.	151—158, 259
Kato, T.	176
Kato, Y.	159, 160
Katz, J. J.	71
Kazlauskas, R.	51, 161, 162
Kelecom, A.	266
Kennedy, G. Y.	231
Kerkut, G. A.	163
Keyl, M. J.	164
Khare, A.	165
Kho, E.	59
Kim, J.-S.	274
Kishi, Y.	104, 108, 166
Kitagawa, I.	167, 168
Kitahara, T.	264
Kitahara, Y.	176
Kobayashi, M.	169—172
Komura, T.	173, 174
Kon, S. K.	86, 87
Konosu, S.	117, 175

AUTHOR INDEX (VOLUME II) (continued)

Name	Ref. no	Name	Ref. no.
Krejcarek, G. A.	232	Morin, J. G.	56
Kumanireng, A. S.	176	Morton, G. O.	29
Kunisaki, Y.	156, 157	Morton, R. A.	215
Kuriyama, K.	168	Mosher, H. S.	15
Kurokawa, M.	124	Moss, G.P.	165
Kurosawa, E.	177, 255, 256	Motzfeldt, A.-M.	216
Kusumoto, T.	196	Mulder, J. W.	217
Lam, J. K. K.	178	Muller, D. G.	134, 218, 219
Laverack, M. S.	58	Murawski, U.	97, 252
Leclerc, G.	13	Murphy, P. T.	51, 161, 162
Lee, A. S. K.	179	Murphy, V.	11
Liaaen-Jensen, S.	5, 8, 180, 225, 287	Mynderse, J. S.	220
Lim, H. S.	66	Nagai, Y.	282
Lin, G. H. Y.	101, 247	Nagata, S.	198, 199
Lin, Y. Y.	250	Nagayama, H.	173, .74
Litchfield, C.	136	Nakano, A.	130
Losman, D.	141, 266	Nakayama, T. O. M.	221
Lowe, M. E.	181, 182	Naya, Y.	126
Lutwak-Mann, C.	191	Nayak, U. R.	228
Lytle, L. T.	138	Neeman, I.	17, 150, 222
McBride, W. J.	183	Niklasson, A.	114
McCaman, R. E.	30	Nishikawa, K.	107
McCapra, F.	184	Nishizawa, M.	169
McDonald, F. J.	185, 243, 244	Nitsche, H.	223
McLachlan, J.	186	Niwa, M.	129
McMillan, J. A.	187, 232	Nomura, T.	224
McMurry, J. E.	188	Norgård, S.	225
Madgwick, J. C.	189, 190	Norte, M.	100
Magno, S.	76, 77	Nose, T.	117
Magnus, P. D.	13	Oberhänsli, W. E.	72
Mann, T.	33, 191	Ogata, E.	186
Manning, M. J.	184	Ogata, H.	224
Manning, W. M.	192, 253	Ogura, Y.	226
Marderosian, A. D.	193	Oguri, K.	132
Markert, J. R.	179	Okukado, N.	130
Marsh, M. E.	194	Okumura, A.	257
Martin, A. W.	33, 191	Olsen, A. G.,	273
Martin, D. F.	70, 195	Osborne, N. N.	227
Martin, J. D.	99—101	Padilla, G. M.	195
Matsuno, T.	196—200	Pascher, I.	20
Matthews, J. C.	121	Patil, V. D.	228
Mayol, L.	77	Paul, I. C.	187, 232
Mazzarella, L.	201	Pedersen, M.	229
Menzies, I. D.	13	Perez, C.	99, 101
Mettrick, D. F.	203	Pettit, G. R.	230
Michaelson, I. A.	164	Pettus, J. A., Jr.	214
Minale, L.	45—50, 68, 204—207	Prota, G.	39—42
Mino, N.	92	Puliti, R.	201
Mistysyn, J.	214	Quinn, R. J.	162
Mitsuhashi, H.	169—172	Radlick, P.	246
Miyahara, T.	157, 158	Ralph, B. J.	189, 190
Miyazawa, K.	208	Ramorino, M. L.	233
Moldowan, J. M.	209	Rauwald, W.	94
Momzikoff, A.	210, 211	Riccio, R.	206, 207
Moore, R. E.	134, 212—214	Rimington, C.	231
Morales, R. W.	136	Rinehart, K. L., Jr.	232, 248

AUTHOR INDEX (VOLUME II)
(continued)

Name	Ref. no.
Roberts, T. E.	15
Roseghini, M.	233—236
Rosener, H.-U.	94, 96
Rubinstein, I.	98, 237, 238, 249
Rushton, R.	15
Saavedra, J. M.	239
Saenger, P.	229
Safe, L. M.	127, 240
Sameshima, M.	156, 158
Samuelsson, B. E.	20
Sano, F.	280
Santacroce, C.	76, 77, 294
Sargent, M. V.	178
Sasner, J. J., Jr.	260
Sato, R.	1
Sato, Y.	198
Sattler, E.	95
Schantz, E. J.	241
Scheuer, P. J.	36, 37, 119, 159, 160, 212
Schmitz, F. J.	185, 242—244
Seferiadis, K.	135
Shannon, J. S.	189
Shaw, P. D.	248
Sheikh, Y. M.	141
Shimaya, M.	156
Shimizu, T.	175
Shimomura, O.	166
Shimura, S.	91
Shintani, K.	155
Shmeuli, U.	17
Sica, D.	76, 77
Sieburth, J. M.	245
Simes, J. J.	189
Simpson, K. L.	158
Sims, J. J.	81, 101, 246, 247
Siuda, J. F.	232, 248
Smith, A. G.	98, 249
Smith, L. L.	250
Snatzke, G.	8
Sodano, G.	68, 204—207
Stallard, M. O.	78, 251
Ställberg, G.	25, 26, 113—115
Stempien, M. F., Jr.	16
Stoffelen, H.	97, 252
Strain, H. H.	71, 192, 253
Straub, O.	286
Sugawara, T.	167, 168
Sugiura, S.	107
Sutherland, M. D.	110—112
Suyama, M.	254
Suzuki, M.	255, 256, 277
Svec, W. A.	71, 225
Takagi, M.	257
Takagi, T.,	263
Tammar, A. R.	258
Tamura, S.	142
Tanaka, J.	4
Tanaka, Y.	157, 158, 259
Taniguchi, H.	132
Taylor, F. R.	260
Teh, Y. F.	66
Telford, J. M.	203
Teshima, S.	145—148, 261
Teste, J.	43
Thiersch, J. B.	191
Thompson, S. Y.	86, 87
Thorin, H.	114
Thurberg, F. P.	260
Todo, K.	169, 170
Tomita, S.	147, 148
Torres Pombo, J.	262
Toyama, Y.	263
Tsukuda, N.	264
Turner, A. B.	265
Tursch, B. M.	141, 209, 266, 267
Uchiyama, M.	1
Uphaus, R. A.	71
VanBlaricom, G. R.	248
Vanderah, D. J.	185, 242
Van der Helm, D.	74, 185, 268
Vanstone, W. E.	179
Viala, J.	269
Vilter, H.	94
Vitali, T.	235
Von Beroldingen, L.A.	188
Wada, S.	173, 174
Wampler, J. E.	55, 56, 121, 122
Waraszkiewicz, S. M.	270
Washburn, W. N.	52
Washecheck, D. M.	185
Watanabe, K.	175
Watanabe, T.	196, 198—200
Weedon, B. C. L.	165, 284
Weinheimer, A. J.	268, 271
Weinreich, D.	30
Wells, R. J.	21, 51, 85, 161, 162, 217
Welton, L. L.	93
Wetzel, E. R.	29
White, E. H.	149
White, R. H.	187, 248
Whittaker, V. P.	164
Widdowson, D. A.	24
Wilkie, D. W.	272
Williams, P. A.	125
Wing, R. M.	101, 246, 247
Wiseman, P. M.	127, 128
Wolfe, M. S.	6
Wong, C. J.	240
Wood, H. B.	230

AUTHOR INDEX (VOLUME II)
(Continued)

Name	Ref. no.
Wotiz, H. H.	31, 32, 273
Yamada, K.	116
Yamada, Y.	274
Yamaguchi, M.	130
Yamamura, S.	129
Yasuda, S.	275, 276
Yasumoto, T.	277—281
Yokoyama, H.	151, 152
Yoshioka, M.	143—145
Yoshizawa, T.	282
Yoshizawa, Y.	254
Yosioka, I.	167, 168
Yost, G.	213
Youngblood, W. W.	283
Zadock, E.	17, 150
Zanetti, L.	40— 42

TAXONOMIC INDEX

Explanation of Index

I/ = *Compounds from Marine Organisms,* Volume I.
II/ = *Compounds from Marine Organisms,* Volume II.

The numbers following I or II are compound numbers from Volume I (Table 28) or Volume II (Table 27), respectively. The numbers in parentheses are reference numbers in this book (Volume II).

Example: *Aplysia punctata* I/457(231); II/372(103,231).

The compounds isolated from *Aplysia punctata* are I/457 and II/372. Compound I/457 has been listed in Volume I and Compound II/372 is listed in this book (Volume II). Compound I/457 appears in reference No. 231, and Compound II/372 appears in references Nos. 103 and 231 (Volume II).

To verify the spelling of animal genera the following two books have been used:

1. *Nomenclator zoologicus,* Vols. I-VI, Neave, S.A., Ed., Zoological Society of London, 1939-1966.
2. Golvan, Y.J., Catalogue systématique des nomes de generes de poissons actuels, Masson et Cie., Paris, 1965.

Acanthaster planci II/358(148)
Acanthella acuta II/164(206)
Acanthephyra sp. I/476(118)
Acanthina spirata I/177(15)
Acanthocepola limbata I/20, II/16(116)
Acanthogobius flavimanus II/286, 290(258)
Acanthopagrus schlegelii I/20, II/16(116)
Acmaea sp. II/372(231)
Acmoea, see *Acmaea*
Acteon sp. II/372(231)
Aequipecten irradians II/60(58)
Aequorea aequorea I/275(55, 56, 57, 184)
Aiptasia tagetes I/53(203)
Akera bullata II/372(231)
Alaria esculenta II/351(139)
Alosa sapidissima II/16(116)
Amarhicas see *Anarhichas*
Amphidinium carteri I/66, 114(260)
Amphiroa anceps I/3, 11,12, 15, 26, 27, 30, 48, 52, 54, 57, 84, 85, 127(190).
Amphitrite johnstoni II/367, 370, 371(231)
Anarhichas latifrons II/16(116)
Anarhichas lupus II/16(116)
Anguilla japonica I/20,II/16(116) ; 286, 290, 316(258)
Anguilla rostrata II/16(116)
Anguilla vulgaris II/16(116); 173(60)
Anomia sp. II/372(231)
Anomia ephippium I/483(103)
Anoplopoma fimbria II/16(116)
Anthocidaris crassispina I/138 (110); 360, 375, 380, 396(276); II/114, 115(124); 311(282); 348(276)
Anthopleura fuscoviridis I/360, 374, 380, 394, 396,II/331, 348(275)
Aphrodite aculeata I/3, 52(227)
Aplidium sp. II/163(83)
Aplidium crateriferum II/163(162)
Aplysia californica I/52, 54(30); 112(239); II/139(251); 140, 141(78)
Aplysia dactylomela II/135(185, 244); 155(243)
Aplysia depilans I/483(103)
Aplysia punctata I/457(231); II/372(103, 231)
Aplysina aerophoba, see *Verongia aerophoba*
Apogon lineatus I/20,II/16(116)
Arca zebra I/53(203)

TAXONOMIC INDEX (continued)

Arctodiaptomus salinus	I/476, 483,II/376, 390(151)
Arctoscopus japonicus	II/16(116)
Arenicola marina	II/19(3); 367, 370, 371(231)
Argentina silus	II/16(116)
Argyrosomus argentatus	I/20, II/16(116)
Arius dussumieri	II/16(116)
Arius sona	II/16(116)
Artemia salina	I/473, 478(151); 483, II/382(153)
Ascidia nigra	I/53(203)
Ascophyllum nodosum	I/394, 396, 409, 414(240); II/160, 170, 177 (283); 306(240); 310, 333, 334, 351(139)
Aseraggodes kobensis	I/4, 20, II/16(116)
Asparagopsis armata	I/3, 11, 12, 15, 26, 27, 30, 48, 52, 54, 57, 81, 128, II/23, 27(190)
Asparagopsis taxiformis	II/1(85); 3, 4, 5, 6, (82, 85); 7(85); 8(82, 85); 9, 10(82);12, 13, 177(85)
Astea, see *Astraea*	
Asterias amurensis	I/323, 382(263); II/302 (171); 314(142)
Asterias forbesi	I/317(9); II/49(89)
Asterias pectinifera	II/302(171)
Asterias rubens	I/3, 52(227); 359, 372, 380, 382, 383, 395, 397, 417, 435, 438(249); II/51, 60, 61(58); 303(249); 311(20); 324, 330, 332, 340, 341, 344, 345, 346, 347, 350, 357(249); 368(231)
Asterina panceri	I/476, 478, 483, 487, II/376, 378, 383(151)
Astraea undosa	I/483(103)
Astriclypeus manni	I/360, 374, 380, 394, 396, II/331, 348(276)
Astropecten irregularis	II/365, 368(231)
Athennes hians	II/16(116)
Aurelia aurita	I/360, 374, 375, 380, 383, 394, 396, 410, II/331, 348(275).
Auxis tapeinocephalus	I/3, 8, 11, 15, 26, 30, 31, 48, 52, 57, 58, 81, 84, 85, 86, 87, 127, 128(254); II/16(116); 21, 76, 98(254)
Auxis thazard	II/16 (116)
Axinella cannabina	II/165(77); 301, 322(76)
Axinella cristagalli	I/476(102)
Axinella polypoides	II/29(46); 278, 293, 294, 308, 309, 313, 328, 335(204)
Axinella verrucosa	II/307, 312, 329, 336, 349, 352(205)
Bankia setacea	II/368(103)
Belone belone	I/454(60); II/16(116); 290(258)
Beryx splendens	I/476, II/386 (157)
Bifurcaria bifurcata	II/30(94); 116(96)
Blennius sanguinolentus	I/476, 483, II/208, 386(62)
Blennius sphinx	I/476, 483, 486, II/208, 386, 388(62)
Blidingia minima	I/3, 11, 15, 26, 27, 30, 48, 52, 54, 57, 128, II/ 23, 27(190)
Bonnemaisonia hamifera	II/42, 43, 44, 45, 48(232,248)
Box boops	II/16(116)
Branchiostegus japonicus	I/476, 483, 487, II/386(157)
Brasme, see *Brosme*	
Breviraja semirnovi	I/20, II/16(116)
Brosme brosme	II/16(116)
Buccinum undatum	I/3, 52(227); 95(164); 148(58); 483, 486(103); II/ 60(58)
Bulla sp.	II/372(231)
Busycon canaliculatum	II/49(89); 60 (58)
Cacospongia mollior	II/300(49)
Calanoida, see *Calanoides*	
Calanoides sp.	I/476, 483, II /390(153)

TAXONOMIC INDEX (continued)

Calanus helgolandicus II/28, 67, 173(210)
Callionymus beniteguri I/20, II/16(116)
Callionymus japonicus I/20, II/16(116)
Callista brevishiphonata, see *C. brevisiphonata*
Callista brevisiphonata II/50, 63(280)
Callithamnion sp. I/482, 483, II/362, 395, 396(73)
Cambarus sp. I/454 (231)
Capnella imbricata II/156(141)
Capulus hungaricus I/483, 486(103)
Caranx equula I/20, II/16(116)
Carcharhinus gangeticus II/16(116)
Carcharias laticaudas II/16(116)
Carcinus maenas I/3, 52(227); 473, 476, 478, 483, 484, 486, 487(151); II/60(58); 377, 385, 389(151)
Cardium echinatum I/483(103)
Cardium edule I/483(103)
Cassiopea sp. II/369(231)
Cassiopea xaymachana I/53 (203)
Cassiopeia, see *Cassiopea*
Caulerpa brownii II/216, 255(21)
Caulerpa peltata I/3, 11, 12, 15, 26, 30, 48, 52, 54, 57, 127, 128, II/23(190)
Centropages typicus II/28, 67, 173(210)
Centrophorus sp. I/387(258)
Cephalopholis aurantius I/476, II/386 (157)
Cephaloscyllium umbratile II/16(116)
Ceramium rubrum I/482, 483, II/362, 395, 396(73)
Certonardoa semiregularis II/302(171)
Chaetomorpha aerea I/3, 11,12, 15, 26, 27, 48, 52, 54, 57, 128, II/23, 35(190)
Chaetomorpha crassa I/20, II/16(92)
Chaetopterus variopedatus II/49(89); 371(231)
Chaeturichthys hexanema I/20, II/16(116)
Chamaedoris orientalis I/20, II /16(92)
Chelidonichthys kumu II/16(116)
Chiloscyllium griscum II/16(116)
Chirocentrus dorab II/16(116)
Chondria crassicaulis I/20, II/16 (92)
Chondria oppositiclada II/159(81)
Chondrus crispus I/482, 483, II/362, 395, 396(73)
Chondrus ocellatus II/25(257)
Chorda filum II/30(94)
Chromobacterium sp. I/135, II/17, 39, 53, 83, 197(6)
Chroomonas salina I/472, 477, 480, 482(225)
Chrysophrys major I/3, 8, 11, 15, 20, 26, 27, 30, 31, 42, 48, 52, 54, 57, 58, 81, 84, 85, 86, 87, 127, 128(175); 476, 478, 482, 483, 486, 487(155, 157); II/16, 21, 22, 27, 82, 87, 91, 96, 99(175); 379(157); 386(155, 157)
Cirratulus cirratus II/367, 371 (231)
Cladophora sp. I/3, 11, 12, 15, 26, 27, 30, 48, 52, 54, 57, 58, 81, 84, 85, 86, 127, 128, II/27(190)
Cladostephus spongiosus II/30(94)
Cladostephus verticillatus II/30(94)
Cleisthenes pinetorum II/16(116)
Clione limacina I/483(103)
Clupea alosa II/16(116)
Clupea harengus I/323(114); 473, 476(64); II/16(116); 162, 179, 187, 188, 194, 196, 224, 225, 248, 259, 260, 263, 265, 270, 271, 280, 281(114); 290(258); 297(114); 373, 386, 387, 388(64)

TAXONOMIC INDEX (continued)

Clupea lile II/16(116)
Clupea pallasi II/16(116)
Clupea sprattus II/16(116)
Coccolithus huxleyi I/479, 482, 483(225)
Codium dichotomum I/482, 483, II /362, 395, 396(73)
Codium divaricatum I/20, II/16 (92)
Codium fragile I/3, 11, 15(190); 20(92); 26, 27, 48, 50, 52, 54, 57, 84, 85, 127, 128(190); II/16(92);23, 35(190); 323, 339(237)
Coenobita clypeatus I/53(203)
Cololabis saira I/454(60); II/16(116)
Colpomenia sinuosa I/3, 11, 15, 26, 27, 30, 52, 54, 57, 128, II/23(190)
Concholepas concholepas I/71, 74, 158, 177, II/58(233)
Condylactis gigantica I/53(203)
Conger myriaster I/20, II/16(116); 286, 290, 292, 316(258)
Conger vulgaris II/173(60)
Corallina mediterranea I/482, 483, II/ 362, 395, 396(73)
Corallina officinalis I/3, 11, 15, 26, 30, 48, 52, 54, 57, 80, 81, 84, 85, 86, II/27, 35, 46(190); 395(27)
Corax leptolepsis II/16(116)
Coryphaena hippurus I/3, 8, 11, 15, 26, 30, 31, 48, 52, 57, 58, 81, 84, 85, 86, 87, 127, 128, II/21, 76, 98(254)
Coscinasterias acutispina I/ 359, 372, 380, 382, 383, 395, 397, 417, II/332, 344(146)
Crangon allmani I/476, II/208(86)
Crangon vulgaris I/476, 483, II/208(86)
Crenilabrus tinca I/473, 476, 486, II/208, 373, 386(62)
Crepidula fornicata I/ 483(103)
Cryptotethia, see *Cryptotethya*
Cryptotethya crypta I/167(16)
Cutleria multifida I/169(134); II/107(134); 108(134, 219)
Cyclopterus lumpus II/16 (116)
Cymnothorax, see *Gymnothorax*
Cynoglossus bengalensis II/16(116)
Cynoglossus interruptus I/20, II/16(116)
Cynoglossus semifasciatus II/16(116)
Cypraea sp. II/372(231)
Cypraea spadicea I/483(103)
Cypridina hilgendorfii II/249(18, 55, 104, 105, 106, 107, 108, 149, 166, 184)
Cyprina islandica I/483 (103)
Cyproea, see *Cypraea*
Cypselurus sp. II/16(116)
Cystoclonium purpurascens I/482, 483, II/362, 395, 396(73)
Cystophora moniliformis I/3, 11, 12, 15, 26, 27, 30, 48, 50, 52, 54, 57, II/23, 27(190)
Cystophora retroflexa I/11, 12, 15, 26, 27, 48, 52, 54, 57, 81, II/23, 32(190)
Cystophora torulosa I/315, 320, II/261, 318, 337(162)
Cystophyllum trinode I/3, 11, 12, 15, 26, 27, 48, 52, 54, 57, II/23(190)
Cystoseira sp. I/482, 483, II/362, 395, 396(73)
Cystoseira baccata II/30(94)
Cystoseira crinita I/482, 483, II/362, 395, 396(73)
Cystoseira discors II/30(94)
Cystoseira granulata II/30(94)
Cystoseira myriophylloides II/30(94)
Cystoseira tamariscifolia II/30 (94)
Dasyatis akajei I/387(258); II/16(116); 290(258)
Dasyatis kuhlii II/16(116)
Dasyatis zugei II/16(116)
Delesseria sanguinea I/482, 483, II/ 362, 395, 396(73)
Delisea fimbriata II/68, 69, 70, 71, 72, 73, 100, 101, 102, 103(162)

TAXONOMIC INDEX (continued)

Desmarestia aculeata I/482, 483(73); II/351(139); 362, 395, 396(73)
Desmarestia viridis I/20, II/16(92)
Desmia hornemanni II/85, 86, 88, 89, 90, 92, 93, 94, 95, 97, 175(126)
Diadema antillarum I/53(203)
Dictyopteris acrostichoides I/3, 11, 12, 15, 26, 27, 30, 48, 52, 54, 57, II/23, 27, 32(190)
Dictyopteris australis I/168, 169, 170, 171, 173, 174, 175, 176(214); II/109, 110(213); 111, 112(214)
Dictyopteris divaricata I/20, II/16(92)
Dictyopteris plagiogramma I/168, 169, 170, 171, 173, 174, 175, 176(214); II/109, 110(213); 111, 112(214)
Dictyosiphon foeniculaceus II/351 (139)
Dictyota dichotoma I/3, 11, 12, 15, 26, 27, 48, 52, 54, 57(190); 482, 483(73); II/30(94); 362, 395, 396(73)
Dilophus marginatus I/3, 11, 12, 15, 26, 27, 30, 48, 52, 54, 57, II/23, 27, 35(190)
Diodon holacanthus I/474, 483, 486, 487, II/373, 384, 386(198)
Diplodus annularis I/473, 483, 486, 487, II/208, 373, 386(62)
Diplodus sargus I/476, 483, 486, 487, II/208, 386, 388(62)
Discosoma sp. II/369(231)
Disidea, see *Dysidea*
Distolasterias sticantha II/302(171)
Ditrema temmincki II/16(116)
Donax denticulatus I/53(203)
Dosinia exoleta I/486(103)
Drupa concatenata II/105(236); 106(235)
Dumontia incrassata II/351(139)
Durvillea potatorum I/3, 11, 15, 26, 27, 30, 48, 52, 54, 57,81, 127, 128, II/23, 32(190)
Duvaucelia pletia II/372(231)
Dysidea sp. II/229(162)
Dysidea avara II/227, 229(207)
Echinaster echinophorus I/53(203)
Echinocardium cordatum I/360, 374, 380, 383, 394, 396, II/331, 348(276)
Echinus esculentus I/3, 52(227); 138(110); II/60, 61(58)
Ecklonia radiata I/3, 11, 12, 15, 26, 27, 48, 50, 52, 54, 57, 86, 128, II/23(190)
Ectocarpus sp. II/351(139)
Eisenia bicyclis I/20, II/16(92)
Elagatis bipinnulata I/3, 8, 11, 15, 26, 30, 31, 48, 52, 57, 58, 81, 84, 85, 86, 87, 127, 128, II/21, 76, 98(254)
Eledone cirrhosa I/3, 52(227); II/60, 61(58)
Emerita analoga I/472, 473, 476, 478, 479, 482, 483, 487(93)
Endarachne binghamiae I/3, 11, 12, 15, 26, 27, 30, 48, 52, 54, 57, 128, II/23 (190)
Engraulis encrasicholus II/16(116)
Engraulis japonica, see *E. japonicus*
Engraulis japonicus I/ 476, 483, 487(197); II/16(116); 286, 290(258)
Ensis directicus II/49(89); 60(58)
Enteromorpha intestinalis II/351(139)
Enteromorpha linza I/20(92); II/160, 177, 185, 222(283)
Eopsetta grigorjewi II/16(116)
Epiactis japonica II/305(172)
Epizoanthus arenaceus II/104, 117(42); 121(41,42); 125(42)
Eptatretus stoutii I/386(258)
Esox lucius II/290(258)
Eucitharus linguatula II/16(116)
Eugorgia ampla I/360, 374, 380, 394, 396, 413, 418, II/331(22)
Eunicea laciniata I/323, II/171, 184, 186, 194(194)
Eunicea tourneforti I/323, II/171, 184, 186, 194(194)

TAXONOMIC INDEX (continued)

Eunicella cavolini I/3, 8, ll, 15, 26, 30, 48, 52, 57, 58, 81, 84, 85, 86, II/24, 81(39)
Eunicella stricta I/3, 8, 11, 15, 26, 30, 48, 52, 57, 58, 81, 84, 85, 86, II/24, 81(39)
Eunicella verrucosa I/3, 8, 11, 15, 26, 30, 48, 52, 57, 58, 81, 84, 85, 86, II/24, 81(39)
Eupagurus bernhardus I/473, 476, 483, 487, II/373, 388, 394(65)
Euphausia pacifica I/476, II /208(87)
Euphausia superba I/476, II/208(87)
Euthynnus affinis yaito I/3, 8, 11, 15, 26, 30, 31, 48, 52, 57, 58, 81, 84, 85, 86, 87, 127, 128, II/21, 76, 98(254)
Evynnis japonica I/473, 476, 486, 487, II/374, 384, 386(157, 158)
Fasciospongia sp. II/277(162)
Fascispongia fovea I/353, 354, II/279(162)
Fissurella sp. II/372(231)
Flabellum variabile II/369(231)
Fucus distichus II/122, 129, 160, 177, 192, 247, 262, 269(283)
Fucus serratus I/482, 483(73); II/30(94); 62(135, 218); 310, 333, 334, 351(139); 362, 395, 396(73)
Fucus spiralis II/30(94); 310, 333, 334, 351(139)
Fucus vesiculosus I/490(223); II/30(94); 122, 160, 170, 177, 185, 192, 222, 247, 258, 262, 269(283); 310, 333, 334, 351(139); 392(223)
Fugu niphobles I/20, II/16(116)
Fugu rubripes II/287, 289, 290(258)
Fugu vermiculare II/16(116)
Fugu vermiculare porphyreum I/3(175); 4(116); 8, 11, 12, 15(175); 20(116, 175); 26, 27, 28, 30, 31, 42, 48, 49, 52, 54, 57, 58, 81, 84, 85, 86, 87, 127, 128, II/2(175); 16(116, 175); 21, 23, 27, 47, 82, 87, 91, 96(175)
Fungia symmetrica II/369(231)
Furcellaria sp. I/482, 483, II/362, 395, 396(73)
Furcellaria fastigiata I/374, 375, 380, 396, 409, 410, 418(240); II/351(139)
Gadus sp. II/16(116)
Gadus aeglefinus II/16(116)
Gadus callarias I/476(60); II/16(116); 286, 287, 288, 290, 292(258)
Gadus callarius, see *G. callarias*
Gadus macrocephalus II/16(116); 290(258)
Gadus merlangus II/16 (116)
Gadus morhua II/16(116)
Gadus poutassou II/16(116)
Gadus vivens II/16(116)
Gelidium pusillum I/3, 11, 15, 26, 30, 48, 52, 54, 57, 86(190)
Gelidium sesquipedale I/8, 21, 66, 380(262)
Gerres sp. II/16(116)
Gibbula sp. II/372(231)
Gibbula cineraria I/483(103)
Gibbula tumida I/483(103)
Gigartina stellata II/351(139)
Girella melanichthys I/20, II/16(116)
Girella punctata II/16(116)
Glyptocephalus cynoglossus II/16(116)
Gobius melanostomus I/483, II /203, 373(62)
Gorgonia flabellum I/323, II/171, 184, 186, 194(194)
Gorgonia ventalina I/323, II/171, 184, 186, 194(194)
Gracilaria compressa I/483, 486, 487, II/396(38)
Gracilaria edulis I/3, 11, 15, 26, 27, 30, 48, 52, 54, 57, 128, II/27(190)
Gracilaria lichenoides f. *lucasii* I/3, 11, 12, 15, 26, 27, 30, 48, 52, 54, 57, 84, 85, 86, 128, II/23, 27(190)

TAXONOMIC INDEX (continued)

Gracilaria secundata I/3, 11, 12, 15, 26, 27, 30, 48, 52, 54, 57, 80, 81, 86, 113, 127, 128, II/23, 27, 32, 35, 46(190)
Gracilaria textorii I/380(144, 145); II/304(144)
Grateloupia elliptica I/20, II/16(92)
Grateloupia filicina I/20, II/16(92)
Grateloupia turuturu I/3, 11, 15(208); 20(92); 26, 52, 54, 57, 84, 85, 87, 127(208); II/16(92); 26, 27, 34, 35(208)
Gryphaea angulata I/483(103)
Gymnothorax reticularis I/20, II/16(116)
Halichoeres poecilopterus II/286; 290(258)
Halichondria sp. II/166, 167, 169, (36, 37); 242, 243, 245(36)
Halichondria panicea I/359 (28); II/168(45)
Haliclona sp. I/53(203)
Halicondria, see *Halichondria*
Halidrys siliquosa II/30(94); 181(95); 351(139)
Haliotis fulgens I/483 (103)
Halocynthia aurantium I/305(172)
Halocynthia roretzi I/363(269)
Halopithys incurvus II/40, 41(97); 54, 66(43)
Halopitys, see *Halopithys*
Halopytis, see *Halopithys*
Haminea sp. II/372(231)
Hapalochlaena maculosa I/3, 27, 53, 81, 127, 128, 148, 165, II/59(125)
Hapalogenys mucronatus II/286, 290 (258)
Hemicentrotus pulcherrimus II/33(132); 114, 115 (124)
Hemigrapsus sanguineus II/305(172)
Hemilepidotus gilberti II/16(116)
Hemiramphus sajori II/16(116)
Hermodice carunculata I/53(203)
Heterocarpus sp. I/483(118)
Heterodontus francisci I/487(60)
Heterodontus japonicus II/16(116)
Heteromycteris japonicus I/20, II/16(116)
Heteromyoteris, see *Heteromycteris*
Heteronema erecta II/338(162)
Hexagrammas, see *Hexagrammos*
Hexagrammos otakii I/20, II/16(116); 286, 290(258)
Hexagrammos stelleri II/16(116)
Hexagrammus, see *Hexagrammos*
Hilsa kanagurta II/16 (116)
Himanthalia elongata II/30(94)
Himanthalia lorea II/351(139)
Hippocampus erectus I/482(60)
Hippoglossoides dubis, see *H. dubius*
Hippoglossoides dubius I/3, 8, 11, 12, 15, 20, 28, 30, 31, 42, 48, 49, 52, 57, 58, 81, 84, 85, 86, 87, 127, 128(175); II/16(116, 175); 21, 27, 82, 91, 96, 99(175)
Hippoglossoides platessoides II/16(116)
Hippoglossus hippoglossus II/16(116)
Hippoglossus stenolepsis II/16(116)
Hippoglossus vulgaris II/16(116)
Hizikia fusiformis I/20(92); 396(173); 409(173, 174); II/16(92); 220(174)
Holothuria sp. I/53(203)
Homarus americanus I/3, 11, 26, 48, 52(183); 112(163); II/51(164); 60(163)
Homarus vulgaris I/476, II/208(86)
Hormosira banksii I/3, 11, 15, 26, 27, 30, 48, 52, 54, 57, 86, 127, 128, II/32(190)
Hyas coarctatus I/476, 486, II/373, 377, 383, 388(65)

TAXONOMIC INDEX (continued)

Hydatina sp. II/372(231)
Hydrolagus colliei II/16(116)
Hydrolugus, see *Hydrolagus*
Hymeniacidon sanguinea I/478, 482, 483, II/382(102)
Hypodytes rubripinnis I /20, II/16(116)
Idotea granulosa I/473, 478, 483, 486, II/377, 385, 387(151)
Idotea montereyensis I/473, 478, 483, 486, II/377(151)
Idothea, see *Idotea*
Inimicus japonicus II/16(116)
Ircinia sp. I/354(162); II/275(72)
Ircinia halmiformis II/276(162)
Ircinia ramosa I/358, II/361(162)
Isis hippuris II/355(162)
Isurus glaucus II/16(116)
Jania rubens I/3, 11, 15, 26, 27, 48, 52, 54, 57, II/23(190)
Kareius bicoloratus I/3, 8, 11, 15, 20, 26, 27, 30, 31, 42, 48, 52, 54, 57, 58, 81, 84, 85, 86, 87, 127, 128(175); II/16 (116, 175); 21, 27, 82, 87, 91, 96, 99(175)
Katsuwonus pelamis I/3, 8, 11, 15, 26, 30, 31, 48, 52, 57, 58, 81, 84, 85, 86, 87, 127, 128(254); II/16(116); 21, 76, 98(254); 286, 290(258)
Konosirus punctatus I/20, II/16(116)
Kyphosus lembus I/20, II/16(116)
Lagocephalus lunaris spadiceus I/20, II/16(116)
Laiaster, see *Leiaster*
Laminaria digitata II/351(139)
Laminaria hyperborea II/351(139)
Laminaria ochroleuca II/30(94)
Laminaria saccharina I/380, 396, 409, 414(240); II/160, 177, 192, 222, 247, 258, 262, 269(283); 306(240); 351(139)
Lamma, see *Lamna*
Lamna ditropis II/16(116)
Lanice conchilega II/371(231)
Lateolabrax japonicus II/16(116)
Latimeria chalumnae II/315, 317(258)
Laurencia sp. II/131(80)
Laurencia caespitosa II/157(99, 101)
Laurencia elata II/144(247)
Laurencia filiformis f. *dendritica* I/220, II/132, 133, 134, 142(162)
Laurencia glandulifera II/149, 150, 151(255); 154(256)
Laurencia intricata II/174(187)
Laurencia johnstonii II/141(78)
Laurencia nidifica I/214, 217, II/139, 143, 152(270)
Laurencia nipponica II/136, 137, 138, 145, 146, 147, 148(177)
Laurencia obtusa I/3, 11, 12, 15, 26, 27, 30, 48, 52, 54, 57, 84, 85, 86, 127, 128, II/23, 35(190)
Laurencia tasmanica II/139(246)
Leander serratus I/473, 476, 478, 483, 486, 487, II/373, 383, 385, 387(63)
Leiaster leachii I/359, 372(146, 261); 380(146); 382, 383, 395(146, 261), 397(146); 417 (146, 261); II/321, 327(261); 332(146); 344(146, 261); 345 (146)
Leiognathus bindus II/16(116)
Leiognathus nuchalis I/20, II/16(116)
Lemnalia africana II/158(266)
Lepidotrigla microptera I/20, II/16(116)
Lepomis gibbosus II/16(116)
Lethrinus cinereus II/16(116)
Libinia emarginata II/49(89)
Lima doscombei I/483, 486(103)

TAXONOMIC INDEX (continued)

Lima excavata	I/476(103)
Lima loscombi	I/483,486(103)
Limanda ferruginea	II/16(116)
Limanda herzensteini	II/16(116)
Limanda limanda	II/16(116)
Limanda yokohamae	I/4, 20, II/16(116)
Limulus polyphemus	II/49(89)
Lineus longissimus	II/368(231)
Liopsetta obscura	II/16(116)
Lithothamnium sp.	II/395(27)
Littorina brevicula	II/305(172)
Littorina littoralis	I/483(103)
Littorina littorea	I/483, 486(103)
Littorina rudis	I/483(103)
Liza auratus	II/286, 290(258)
Lobophytum sp.	II/214(51)
Lobophytum cristagalli	II/200(267)
Loligo pealei	II/49(89); 51(58)
Loligo pealeii, see *L. pealei*	
Loligo pealii, see *L. pealei*	
Loligo vulgaris	I/148, II/57, 60, 61(163)
Lophiomus setigerus	II/16(116)
Lophius litulon	I/3, 8, 11, 15(175); 20(116, 175); 26, 27, 30, 31, 42, 48, 49, 52, 54, 57, 58, 81, 84, 85, 86, 87, 127, 128(175); II/16(116, 175); 21, 27, 82, 87, 91, 96(175)
Lophius piscatorius	II/16(116)
Lucapina sp.	II/372(231)
Luidia ciliaris	II/365, 368(231)
Lunatia heros	II/60(58)
Lysastrosoma anthostictha	II/302(171)
Lytechinus variegatus	I/53(203)
Macrourus berglax	II/16(116)
Macrozoarces americanus	II/16(116)
Makaira mazara	I/3, 8, 11, 15, 26, 30, 31, 48, 52, 57, 58, 81, 84, 85, 85, 86, 87, 127, 128, II/21, 76, 98(254)
Makaira mitsukurii	I/3, 8, 11, 15, 26, 30, 31, 48, 52, 57, 58, 81, 84, 85, 86, 87, 127, 128, II/21, 76, 98(254)
Marginella sp.	II/372(231)
Martensia elegans	I/3, 11, 15, 26, 27, 30, 48, 50, 52, 54, 57, 86, 127, II/27(190)
Marthasterias glacialis	II/366(265)
Meganyctiphanes norvegica	I/476, II/208(86,87)
Megathura crenulata	I/483(103)
Melanogrammus aeglefinus	I/487(60)
Melongena corona	II/60(58)
Menidia notata	II/16(116)
Mercenaria mercenaria	I/148, II/51, 60(58)
Meretrix eusoria	I/483, 486(103)
Meristotheca papulosa	I/380(145); II/304(143)
Merluccius bilineoris	II/16(116)
Merluccius merluccius	II/16(116)
Metridium dianthus	II/49(89)
Metridium senile	I/323, II/171, 186, 194(194)
Microciona prolifera	II/49 (89); 295, 296(136)
Microdonophis erabo	I/20, II/16(116)
Microstomus achne	II/16(116)
Mnemiopsis leidyi	II/49(89)
Mobula japonica	II/16(116)
Modiolus modiolus	I/483(103); II/60(58)

TAXONOMIC INDEX (continued)

Mola mola	II/16 (116)
Molgula manhattensis	II/49(89)
Monocanthus cirrhifer	II/286, 290(258)
Monostroma sp.	II/244 (283)
Monostroma nitidum	II/50(2)
Mugil sp.	II/16(116)
Mugil aeur	II/16(116)
Mugil auratus	I/473, 483, 486, 487, II/208, 373(62)
Mugil cephalus	I/20(116); 486, 487(198); II/16(116); 286, 290(258); 386 (198)
Mullicides, see *Mulloides*	
Mulloides flavolineatus	II/16(116)
Mullus barbatus	II/16(116)
Muraenesox cinereus	II/286, 290, 316(258)
Murex fulvescens	I/95, 177, II/51(164)
Murex trunculus	II/58(233)
Muricea appressa var. *appressa*	I/360, 374, 380, 394, 418, II/331(22)
Muricea atlantica	I/323, II /171, 184, 186, 194(194)
Muricea elongata	I/235, 240, II/153(138)
Muriceopsis flavida	I/323, II/171, 184, 186, 194(194)
Mustelus canis	II/49(89)
Mustelus griseus	II/16(116)
Mustelus kanekonis	II/16(116)
Mustelus manazo	I/387(258); II/16(116); 290(258)
Mya arenaria	I/483(103); II/60(58)
Mycale laevis	I/53(203)
Myliobatis maculata	II/16(116)
Myoxocephalus groenlandicus	II/16(116)
Myoxocephlus octodecemspinosus	II/16(116)
Myoxocephalus quadricornis	II/284, 286, 287, 290(258)
Mytillus, see *Mytilus*	
Mytilus californianus	II/381(103)
Mytilus californicus, see *M. californianus*	
Mytilus coruscum	II/305(172)
Mytilus edulis	I/323(114); 472, 483(103); II/60(58); 162, 179, 187, 194, 224, 225, 248, 259, 260, 263, 265, 270, 271, 280 (114); 380, 381(165)
Myxicola infundibulum	II/367, 371 (231)
Myxine glutinosa	I/385, 386(258); II/16(116)
Narke japonica	I/20, II/16(116)
Nassa incrassata	I/483(103)
Nassa reticulata	I/483(103)
Natica nitida	I/483, 486(103)
Navodon modestus	I/1, 20, II /16(116)
Nematoscelis difficilis	I/476, II/208(87)
Neobythites fasciatus	I/20, II/16(116)
Neothunnus macropterus	II/16(116)
Nephrops norvegicus	I/3, 52(227); 323(114); 476, 483(63, 86); 486, 487(63); II/179, 187, 188, 194, 196(114); 208(86); 224, 225, 248, 259, 260, 263, 265, 271(114); 373(63)
Nephthea sp.	I/323, II/171, 184, 186, 194(194); 217, 256(242)
Nereis diversicolor	I/454, II/371(231)
Nereis pelagica	II/49(89)
Neritina sp.	II/372(231)
Notacanthus nassus	II/16(116)
Notostomus sp.	I/476(118)
Nucella emarginata	II/118(15)
Nucula sulcata	I/486(103)
Octopus apollyon	II/60, 61(58)
Octopus dofleini	II/51(58)

TAXONOMIC INDEX (continued)

Octopus dofleini martini II/15(191); 49, 65(33)
Octopus vulgaris I/148, II/57, 60(163); 61(58, 163)
Odonthalia dentata II/351(139)
Oncorhynchus gorbuscha I/287(131)
Oncorhynchus keta I/287(131); II/219(224)
Oncorhynchus kisutch I/42, 45(179); 287(131); II/16(116); 31, 38, 74, 87(179); 290(258)
Oncorhynchus masou II/16(116)
Oncorhynchus nerka I/287(131); II/16(116)
Oncorhynchus rhodorus II/290(258)
Oncorhynchus tschawytscha II/16(116)
Onigocia spinosa I/20, II/16(116)
Ophichthus urolophus I/20, II/16(116)
Ophicomina, see *Ophiocomina*
Ophidion barbatum II/16(116)
Ophiocomina nigra II/326(238)
Ophiodon elongatus II/16(116)
Ophisurus macrorhynchus I/20, II/16(116)
Oplegnathus fasciatus I/4, 20, II/16(116)
Oplegnathus punctatus I/20, II/16(116)
Osilinus lineatus I/483(103)
Ostrea edulis I/483 (103)
Pagellus eritrynus II/16(116)
Pagellus mormyrus II/16(116)
Pagrosomus major II/290(258)
Pagurus major I/20, II/16(116)
Pagurus pollicaris II/49(89)
Palinurichthys perciformis II/16(116)
Pampus argenteus II/16(116)
Pandalus borealis I/323, II/179, 187, 194, 224, 225, 248, 259, 260, 263, 265, 271(114)
Panulirus japonicus I/472, 473, 476, 478, 483, 486, 487, II /374, 378(196)
Paracalanus parvus II/28, 67, 173(210)
Paracentrotus lividus I/138(110)
Paralichthys olivaceus I/3, 8, 11, 12, 15(175); 20(116, 175); 26, 27, 28, 30, 31, 42, 48, 52, 54, 57, 58, 81, 84, 85, 86, 87, 127, 128 (175); 296(224); II/16(116, 175); 21, 27, 82, 87, 91, 99(175) ; 286, 290(258)
Paraplagusia japonica I/20, II/16(116)
Parapristipoma trilineatum II/16(116); 286, 290, 291 (258)
Parazoanthus axinellae II/78(40); 126, 127(41)
Parribacus antarcticus I/473, 476, 478, 483, 486, 487, II/374(199)
Patella depressa I/478, 483, 487, II/383 (103)
Patella vulgata I/478, 483, 487, II/383(103)
Pavlova sp. I/481, 483, 494(225)
Pecten hericius II/182, 183, 230, 240(32)
Pecten jacobaeus I/472(103)
Pecten maximus I/472, 475, 483(103)
Pecten opercularis I/483(103)
Pecten strictus I/483(103)
Peltogaster paguri I/454(231)
Pelvetia canaliculata II/306(216); 310, 333, 334, 351(139)
Penaeus japonicus I/473, 476, 478, 483(153, 154); 486, 487(154); II/374 (153); 386(154)
Periophthalamus, see *Periophthalmus*
Periophthalmus cantonensis II/286, 290(258)
Petromyzon marinus II/268(258)
Phaeocystis sp. II/11 (245)
Phaeodactylum tricornutum II/326(238)
Phascolosoma gouldi II/49(89)

TAXONOMIC INDEX (continued)

Philayella, see *Pylaiella*
Philine aperta I/486(103)
Phyllospongia sp. I/354(162)
Phyllospongia dendyi II/320, 353(162)
Phyllospongia foliascens I/313(162)
Phyllospongia papyracea I/313 (217)
Phyllospongia radiata I/313(162)
Phyllospora comosa I/3, 11, 15, 26, 27, 30, 48, 52, 54, 57, 84, 85, 86, 87, 128, II/23, 27, 35(190); 46(189, 190)
Phymatolithon sp. II/395(27)
Pinctada sp. II/372(231)
Pisaster ochraceus II/183, 230(31)
Placopecten magellanicus I/360(127, 128); 374, 380, 383, 394, 396, 409, 410, 418(128)
Platophrys pantherina II/16(116)
Platycephalus indicus II/16(116); 286, 290(258)
Platyrhina sinensis I/20, II/16(116)
Platyrhinoides, see *Platyrhinoidis*
Platyrhinoidis triseriata I/487(60)
Plecoglossus altivelis II/290(258)
Plectorhynchus cinctus II/286, 290, 291(258)
Pleraplysilla spinifera II/199(48)
Plesionika sp. I/476, 483(118)
Pleurobranchus sp. I/476(103)
Pleurogrammus azonus II/16(116)
Pleuroncodes planipes I/476, 483(272)
Pleuronectes microcephalus II/16(116)
Pleuronectes platessa I/3, 52(227); II/290(258)
Pleuronichthys cornutus I/20, II/16(116)
Plexaura sp. I/380, 394, 409, 413, 418, 437, II/331(22)
Plexaura homomalla I/292(10, 271); 293, 318(10); 323(194); 335, 336(10); II/171, 184, 186, 194(194); 241(10)
Plexaurella nutans I/235, 240, II/153(138)
Plocamium cartilagineum II/79(59)
Plocamium coccineum I/482, 483, II/362, 395, 396(73)
Plocamium costatum II/77(162)
Plocamium violaceum II/84(220)
Plotosus anguillaris I/20, II/16(116)
Plurogrammus, see *Pleurogrammus*
Pollachius virens II/16(116)
Polycirrus caliendrum II/371(231)
Polysiphonia brodiaei I/90, 93, 94, II/40(97)
Polysiphonia elongata I/90, 91, 92, 94(97)
Polysiphonia fastigiata II/351(139)
Polysiphonia fruticulosa I/90, 91, 94, II/40(97)
Polysiphonia lanosa I/90, 91, 92, 93, 94, II/37(97); 41(97, 252); 80(97)
Polysiphonia nigra I/90, 91, 94, II/40(97)
Polysiphonia nigrescens I/90, 91, 93, 94, II/41, 80(97)
Polysiphonia thuyoides I/90, 91, 94(97)
Polysiphonia urceolata I/92, 93, II/40, 41(97)
Polysiphonia violacea I/90, 91, 94(97)
Pomolobus chrysochloris II/16(116)
Pomolobus pseudoharengus II/16(116)
Porites porites I/323, II/171, 184, 186, 194(194)
Porphyra columbina I/3, 11, 15, 26, 27, 30, 48, 52, 54, 57, 58, 127, II/35(190)
Porphyra leucostica II/190(283)
Porphyra linearis I/3, 11, 15, 26, 30, 48, 52, 58, 81, 84, 85, 86, 87, 127, 128(186)
Porphyra purpurea I/482, 483, II/362, 395, 396(73)
Porphyra suborbiculata I/20, II/16(92)

TAXONOMIC INDEX (continued)

Porphyra tenera	I/20, II/16(92)
Porphyra umbilicalis	I/3, 11, 15, 26, 30, 48, 52, 81, 86, 87(186)
Porphyridium cruentum	I/374, 375, 380, 393(145)
Portunus trituberculatus	I/473, 476, 478, 483, II/374, 377, 385(156)
Prasiola stipitata	II/160, 177, 185, 192, 222, 247, 258, 262, 269(283)
Priacanthus macracanthus	I/287(131)
Prionace glauca	II/16(116)
Pristiurnas, see *Pristiurus*	
Pristiurus melanostomus	II/16(116)
Prognichthys agoo	II/16(116)
Pronace, see *Prionace*	
Pronotus triacanthus	II/16(116)
Psammechinus microtuberculatus	I/138, 191(88)
Pseudaxinyssa sp.	II/356, 359(120)
Pseudocentrotus depressus	II/33(132); 115, 128(124)
Pseudoceratina crassa	I/122(7)
Pseudopleuronectes americanus	II/16(116)
Pseudopleuronectus, see *Pseudopleuronectes*	
Pseudopotamilla occelata	I/360, 374, 375, 380(169, 172); 394(169); 396(169, 172); II/ 305(172); 331, 343, 348(169, 172)
Pseudopterogorgia acerosa	I/323, II/171, 184, 186, 194(194)
Pseudopterogorgia americana	I/323, II/171, 184, 186, 194(194)
Pseudorhombus pentophthalmus	I/20, II/16(116)
Pseudosciaena manchurica	I/20, II/16(116)
Psolus fabrichii	I/473, 476, 478, II/374(34)
Pteria sp.	II/372 (231)
Pterocladia capillacea	I/3, 11, 12, 15, 26, 27, 30, 48, 52, 54, 57, 58, 80, 127, II/35(190)
Pterogobius elapoides	II/16(116)
Pterogorgia anceps	I/323, II/171, 184, 186, 194(194)
Purpura lapillus	I/483, 486(103)
Pylaiella littoralis	I/482, 483(73); II/189(283); 351(139); 362, 395, 396(73)
Raja sp.	II/16(116)
Raja batis	II/16(116)
Raja clavata	II/16(116)
Raja erinacea	II/16(116)
Raja hollandi	II/16(116)
Raja laevis	II/16(116)
Raja scabrata	II/16(116)
Raja senta	II/16(116)
Rastrelliger kanagurta	II/16(116)
Reniera cratera	II/14(46)
Renilla reniformis	II/198(53, 54, 55, 56, 57, 104, 121, 122, 123, 184)
Rhinoptera sp.	II/16(116)
Rhizostoma octopus	II/369(231)
Rhodactis sanctithomae	I/ 53(203)
Rhodomela confervoides	II/160, 170, 177, 185, 192, 222, 247, 258, 262, 269(283)
Rhodomela subfusca	I/ 90, 91, 94(97); 482, 483(73); II/37(97); 351(139); 362, 395, 396(73)
Rhodymenia ligulata	I/482, 483, II/362, 396 (73)
Rhodymenia palmata	I/360, 374, 375, 380, 383, 394, 396, 409, II/350(84); 351(139); 357(84)
Rhombus maximus	II/16(116)
Rhynchobatus djeddensis	II/16(116)
Rhynoptera, see *Rhinoptera*	
Rissoa sp.	I/483, 486(103)
Roccus saxatilis	II/16(116)
Rumphella antipathes	I/323, II/171, 184, 186, 194(194)

TAXONOMIC INDEX (continued)

Sabella penicillis	II/385(133)
Saccorhiza polyschides	II/30 (94)
Salmacis sphaeroides	I/138(110); 189(111)
Salmo gairdneri	I/473(60)
Salmo solar	II/16(116)
Sarcophyta, see *Sarcophyton*	
Sarcophyton elegans	II/342(147); 360(209)
Sarcophyton glaucum	II/201, 202 (150); 203(17, 222); 209, 210, 213(150)
Sarcophyton trocheliophorum	II/203(162)
Sarcophytum, see *Sarcophyton*	
Sardinella albella	II/16(116)
Sardinella fimbriata	II/16(116)
Sardinella longiceps	II/16(116)
Sardinops melanosticta	I/20, II/16(116)
Sargassum flavicans	I/11, 15, 26, 48, 52, 54, 57, 86, 128, 165, II/23, 35(190)
Sargassum fluitans	I/374, 380, 396, 409, II/ 325, 331, 343, 348(250)
Sargassum thunbergii	I/20, II/16(92)
Sargassum tortile	I/367, II/298, 299(176)
Saurida tumbil	II/16(116)
Saurida undosquamis	I/4, 20, II/16(116)
Scaphechinus mirabilis	I/360, 374, 375, 380, 394, 396, II/331, 348(276)
Scatophagus argus	II/16(116)
Sciaena coitor	II/16(116)
Sciaena mitsukuri	II/286, 290(258)
Scomber japonicus	I/3, 8, 11, 15(175); 20(116, 175); 27, 30, 31, 42, 48, 52, 54, 57, 58, 81, 84, 85, 86, 87, 127, 128 (175); II/16(116, 175); 21, 27, 82, 87, 91, 96, 99(175)
Scomber scomber, see *S. scombrus*	
Scomber scombrus	I/323(114); II/16(116); 179, 187, 194, 196, 224, 225, 248, 259, 260, 263, 265(114); 286, 290(258)
Scomberesox saurus	II/16(116)
Scomberomorus commersonii	II/16(116)
Scomberomorus niphonius	II/16(116); 290(258)
Scophthalamus rhombus	II/290(258)
Scorpaenichthys marmoratus	II/16(116)
Scrobicularia plana	I/483(103)
Scyliorhinus canicula	II/290(258)
Scyllarides squamosus	I/472, 473, 476, 478, 483, 486, 487, II/374,378(199)
Scylliorhinus, see *Scyliorhinus*	
Scyllium canicula	I/3, 52(227); II/16(116)
Scytosiphon lomentaria	I/20, II/16(92); 160, 177, 185, 189, 190(283)
Scytosiphon lomentarius, see *S. lomentaria*	
Sebastes sp.	II/16(116)
Sebastes baramenuke	II/16(116)
Sebastes inermis	II/16(116)
Sebastes marinus	I/476(60); II/16(116)
Sebastes vulpes	II/16(116)
Sebastiscus marmoratus	I/20, II/16(116)
Sebastodes sp.	II/386(60)
Sebastodes inermis	II/286, 290(258)
Sebastodes matsubarae	II/286, 290(258)
Sebastodes melanops	II/16(116)
Sebastodes ruberrimus	II/16(116)
Sebastolobus macrochir	II/16(116)
Sepia officinalis	I/148, II/60, 61(163)
Sergestes sp.	I/476(118)
Seriola aureovittata	I/3, 8, 11, 15, 26, 30, 31, 48, 52, 57, 58, 81, 84, 85, 86, 87, 127, 128, II/21, 76, 98 (254)

TAXONOMIC INDEX (continued)

Seriola dumerili II/16(116)
Seriola purpurascens I/3, 8, 11, 15, 26, 30, 31, 48, 52, 57, 58, 81, 84, 85, 86, 87, 127, 128, II/21, 76, 98(254)
Seriola quinqueradiata I/3, 8, 11, 15, 26, 30, 31, 48, 52, 57, 58, 81, 84, 85, 86, 87, 127, 128(254); II/16(116); 21(254); 286, 290(258)
Serranus diacanthus II/16(116)
Sesarma haematocheir I/474, 478, 483, 486, 487, II/373, 389(200)
Sesarma intermedia I/474, 483, 486, 487, II/373, 389(200)
Sillago sihama I/20, II/16(116); 286, 290(258)
Sinularia flexibilis II/212, 215(161)
Solaster paxillatas II/302(171)
Solea solea II/16(116)
Solen ensis I/483, 486(103)
Somniosus microcephalus I/387(258) ; II/176(113, 115); 180(115); 191(113, 115); 195(115); 221, 225(113); 246(113, 115); 257(113)
Spheroides niphobles I /472, 474, 479, 486, 487, II/373, 383, 384, 386(198)
Sphyrna malleus II/16(116)
Sphyrna zygaena II/16(116)
Spirocodon saltatrix I/360, 374, 375, 380, 394, 396, 410, II/331, 348(275)
Spisula solida II/60, 61(58)
Spisula solidissima II/60(58)
Spisula subtruncata I/483(103)
Splanchnidium rugosum I/3, 11, 15, 26, 27, 30, 48, 52, 54, 57, 84, 85, 127, 128, II/23, 27, 35(190)
Spongia sp. I/313, 354, II/204, 205, 206, 207, 226, 228, 231, 266, 267, 282, 283(162)
Spongia officinalis II/211(50; 232, 233, 234, 235, 236, 237, 238, 239(47)
Spongomorpha arcta II/160, 177, 185, 244(283)
Sprattus sprattus I/473, 476, II/373, 377, 387(64)
Squalus acanthias II/16 (116); 290(258)
Squalus acanthus, see *S. acanthias*
Squalus suckleyi II/16(116); 182, 183, 230(273)
Stelephorus japonicus II/16(116)
Stenotomus chrysops II/49(89)
Stephanolepis cirrhifer II/16(116)
Stichopus japonicus II/305(172); 354(168); 397(167, 168)
Stomolophus sp. I/360, 374, 380, 394, 396, 410, II/331, 348(275)
Stromateus cinereus II/16(116)
Stromateus niger II/16(116)
Strongylocentrotus franciscanus II/183, 230, 240(32)
Strongylocentrotus intermedius II/305(172)
Strongylocentrotus pulcherrimus I/138(110)
Stylocheilus longicauda II/363, 364(159, 160)
Stylocheiron elongatum I/476, II/208(87)
Stylocheiron maximum I/476, II/208(87)
Stylochus megalops I/53(203)
Suberites inconstans I/53(66)
Suggrundus meerdervoorti I/20, II/16(116)
Synodus hoshinonis I/4, 20, II/16(116)
Taeniotoca lateralis II/16(116)
Taius tumifrons II/16(116)
Taonia atomaria II/319(100)
Tapes japonica I/3, 8, 11, 15, 26, 30, 42, 48, 52, 56, 57, 58, 81, 84, 85, 86, 87, 95, 127, 128, II/18, 19, 20, 32, 75, 82, 87, 91, 96(117)
Tapes pullastra I/483, 486(103)
Tarpon atlanticus II/16(116)

TAXONOMIC INDEX (continued)

Tautoga onitis II/49(89)
Tellina crassa I/483(103)
Temnopleurus toreumaticus I/375, 380, 396(276)
Terapon oxyrhynchus II/16(116); 286, 290(258)
Terebella lapidaria II/367, 371(231)
Tetranarce occidentalis II/16(116)
Tetraodon prophyreus II/289, 290(258)
Tetrodon, see *Tetraodon*
Thais chocolata I/71, 158(233)
Thais ermarginata, see *Nucella emarginata*
Thais floridana floridana I/95, 158(164)
Thais haemastoma I/71, 158, 177, 179(234)
Thais haemastoma floridana I/71, 158, 177, 179(234)
Thais lapillus I/95, 177, II/51(164)
Thalassoma cupido II/286, 290(258)
Thelepus setosus I/93, II/36, 119, 123, 124(119)
Theodoxis sp. II/372(231)
Theodoxus, see *Theodoxis*
Theragra chalcogramma II/16(116); 286, 290(258)
Therapon, see *Terapon*
Thione briareus II/49(89)
Thorecta marginalis II/273, 274(162)
Thunnus alalunge II/16(116)
Thunnus albacares I/3, 8, 11, 15, 26, 30, 31, 48, 52, 57, 58, 81, 84, 85, 86, 87, 127, 128(254); II/16(116); 21, 76, 98(254)
Thunnus maccoyii I/3, 8, 11, 15, 26, 30, 31, 48, 52, 57, 58, 81, 84, 85, 86, 87, 127, 128, II/21, 76, 98(254)
Thunnus obesus I/3, 8, 11, 15, 26, 30, 31, 48, 52, 57, 58, 81, 84, 85, 86, 87, 127, 128(254); II/16(116); 21, 76, 98(254)
Thunnus thynnus I/296(224); II/16(116); 218(224); 290(258); 386(60)
Thysanoessa gregaria II/208(87)
Thysanoessa inermis I/476, II/208(87)
Thysanoessa raschii I/476(86); II/208(86, 87)
Thysanoessa spinifera I/476, II /208(87)
Thysanopoda acutifrons I/476(87)
Trachinocephalus myops I/4, 20, II/16(116)
Trachinus draco II/16(116)
Trachurus japonicus I/3, 8, 11, 15, 20, 26, 30, 31, 42, 48, 52, 57, 58, 81, 84, 85, 86, 87, 127, 128(175); 486, 487(198); II/16(116, 175); 21, 22, 27, 82, 87, 91, 96, 99(175); 384, 386(198)
Triakis scyllia II/16(116)
Trichiurus lepturus I/20, II/16(116)
Trichiurus savala II/16(116)
Tricolia sp. II/372(231)
Trigla corax II/16(116)
Trikentrion helium II/375(5)
Trivia europaea I/483(103)
Trochus sp. II/372(231)
Trochus zizyphinus I/483, 486(103)
Trygon imbricata II/16(116)
Trygon microps II/16(116)
Trygon urnak II/16(116)
Turbo argyrostoma II/50(280); 52(277, 278); 64(277, 281); 113(279)
Ulva lactuca I/11, 15, 26, 48, 52, 54, 57, 128(190); 482, 483(73); II/23(190); 160, 177, 185, 192, 222(283); 351(139); 362, 395, 396(73)
Ulva pertusa I/20, II/16(92)
Ulva reticulata I/20(92)

TAXONOMIC INDEX (continued)

Umbraculum sp. II/372(231)
Undaria pinnatifida I/20, II/16(92); II/55, 56(1)
Upeneus bensasi I/4, 20, II/16(116)
Upogebia deltaura I/473, 476, 487, II/373, 377, 382, 383, 387(65)
Uranoscopus scaber II/16(116)
Urolophus aurantiacus II/16(116)
Uronoscopus, see *Uranoscopus*
Urophycis sp. II/16(116)
Urophycis chuss II/16(116)
Uroplycis see *Urophycis*
Urosalpinx cinereus I/95, 177, II/51(164)
Vaucheria sessilis I/481, 483, II/391(225)
Velutina sp. II/372(231)
Venus sp. II/372(231)
Venus fasciata I/483(103)
Venus gallina I/483(103)
Venus japonica I/483, 487, II/389(103)
Venus ovata I/483(103)
Verongia aerophoba I/122(7, 201); 360, 374, 380, 394, 396, 400, 407, 416, 418(68); 476, 483, 486, 487(61); II/285(68); 373, 382, 387, 392, 393(61)
Verongia archeri I/360, 374, 380, 394, 396, 400, 407, 416, 418, II/285(68)
Verongia cavernicola I/339(120)
Verongia fistularis I/360, 374, 380, 394, 396, 400, 407, 416, 418, II/285(68)
Verongia lacunosa II/120(29)
Verongia thiona I/360, 374, 380, 394, 396, 400, 407, 416, 418, II/285(68)
Volsella barbata I/483(103)
Watasenia scintillans II/172, 272(109)
Xiphias gladius II/16(116); 290(258)
Zeus japonicus I/20, II/16(116)
Zonaria sinclairii I/3, 11, 15, 26, 27, 48, 50, 52, 54, 57, 81, 86, 113, 127, 128, II/23, 27(190)
Zonaria turneriana I/3, 11, 12, 26, 27, 48, 50, 52, 54, 57, 84, 85, 86, 128, II/23, 35, 46(190)
Zygaena blochii II/16(116)